Paradox Role of Oxidative Stress in Cancer: State of the Art

Paradox Role of Oxidative Stress in Cancer: State of the Art

Editors

Cinzia Domenicotti
Barbara Marengo

MDPI • Basel • Beijing • Wuhan • Barcelona • Belgrade • Manchester • Tokyo • Cluj • Tianjin

Editors

Cinzia Domenicotti
University of Genoa
Italy

Barbara Marengo
University of Genoa
Italy

Editorial Office
MDPI
St. Alban-Anlage 66
4052 Basel, Switzerland

This is a reprint of articles from the Special Issue published online in the open access journal *Antioxidants* (ISSN 2076-3921) (available at: https://www.mdpi.com/journal/antioxidants/special_issues/Paradox_Oxidative_Stress_Cancer).

For citation purposes, cite each article independently as indicated on the article page online and as indicated below:

LastName, A.A.; LastName, B.B.; LastName, C.C. Article Title. *Journal Name* **Year**, *Volume Number*, Page Range.

ISBN 978-3-0365-4421-2 (Hbk)
ISBN 978-3-0365-4422-9 (PDF)

Contents

 *antioxidants*

Editorial

Paradox Role of Oxidative Stress in Cancer: State of the Art

Cinzia Domenicotti * and Barbara Marengo

Department of Experimental Medicine, General Pathology Section, University of Genoa, 16132 Genoa, Italy; barbara.marengo@unige.it

* Correspondence: cinzia.domenicotti@unige.it

Citation: Domenicotti, C.; Marengo, B. Paradox Role of Oxidative Stress in Cancer: State of the Art. *Antioxidants* **2022**, *11*, 1027. https://doi.org/10.3390/antiox11051027

Received: 26 April 2022
Accepted: 18 May 2022
Published: 23 May 2022

Publisher's Note: MDPI stays neutral with regard to jurisdictional claims in published maps and institutional affiliations.

The modulation of oxidative stress is essential for the maintenance of redox homeostasis in healthy and cancer cells. In fact, although cancer cells are characterized by the overproduction of reactive oxygen species (ROS) with respect to healthy cells, they are able to survive and proliferate under pro-oxidant conditions by activating redox-sensitive signaling pathways, and by inducing the expression of many antioxidant genes.

Furthermore, oxidative stress can play a fundamental role during all phases of carcinogenesis as well as under therapy-induced stress conditions. In fact, on one hand, ROS over-production contributes to the carcinogenic process by impairing cellular macromolecules and on the other, chemo- and radio-therapeutic agents kill cancer cells via pro-oxidant action. Unfortunately, long-term anticancer therapy has been demonstrated to stimulate antioxidant adaptive responses, contributing to therapy refractoriness.

In addition, antioxidants derived from natural products can exert a chemopreventive action by counteracting cancer development or a chemosensitive effect by potentiating the cytoxicity of anti-cancer therapies. Therefore, focusing the attention on the paradox role of oxidative stress, this Special Issue has collected five research articles, four review articles, and one perspective article that investigates the double-edge role of oxidative stress, which can determine beneficial or detrimental outcomes in cancer development and treatment.

In this context, Nitti et al. have focused their attention on the clinical significance of Heme-oxygenase 1 (HO-1) in cancer progression, and reported contrasting evidence on its role in tumor biology [1]. The authors supported the notion that the pro- or anti-tumor activity of HO-1 is related to its subcellular localization and catalytic activity, and showed a direct correlation between HO-1 over-expression and cancer therapy resistance. Based on the collected findings, they suggest that HO-1 can be a promising biomarker of cancer progression, and in some cases, an interesting target to inhibit in order to increase the cytotoxic effect of standard anti-cancer drugs.

The resistance to anti-cancer therapy is often observed in the clinic, and considering that both iron metabolism and accumulation have been found dysregulated in cancer, the selective induction of ferroptosis could be an alternative anti-cancer strategy. Ferroptosis is an iron-dependent, non-apoptotic regulated cell death characterized by lipid membrane peroxidation and triggered by the depletion of glutathione (GSH), the most important intracellular antioxidant, and by glutathione peroxidase 4 (GPX4) inhibition. In their perspective article, Fujihara and co-authors highlighted that although several studies have described the molecular mechanisms underlying ferroptosis; the clinical use of ferroptosis inducers is limited [2]. However, in this perspective, the authors discussed the possibility to utilize ferroptosis in the clinic (i) by inducing directly ferroptosis in tumor cells and (ii) by lowering the threshold for ferroptosis induction in cancer cells, in order to enhance the efficacy of conventional therapies including chemo, radio, and immunotherapy.

A suggestion in this direction has been given by Monteleone et al., who demonstrated that the combination of etoposide, a known chemotherapeutic drug, with a protein kinase C (PKC)-α inhibitor is able to induce the ferroptosis of cancer stem cells (CSCs) resistant to etoposide [3]. CSCs are characterized by low levels of ROS, high amounts of GSH and over-expression of xCT, a transporter of cystine which is crucial for GSH biosynthesis.

Since xCT was found to be stabilized on cell membrane through interaction with CD44, a stemness marker whose expression is modulated by PKC-α, the authors utilized a combined approach of etoposide with sulfasalazine, a direct xCT inhibitor, or with an inhibitor of PKC-α (C2-4). The results obtained demonstrated that the co-treatment with the PKC-α inhibitor was able to induce the ferroptosis of resistant CSCs by inducing aerobic glycolysis, decreasing intracellular GSH levels, down-regulating GPX-4 activity, and stimulating lipid peroxidation.

According to the evidence that the modulation of ROS production might be a valid strategy to counteract tumor development, Menegazzi et al. reported that natural compounds such as *Hypericum perforatum* L. (St. John's wort, SJW) and its active compound hyperforin (HPF) could exert both a prophylactic and a therapeutic effect [4]. These compounds, by decreasing mitochondrial ROS production and restoring pH imbalance, were able to inhibit cancer cell proliferation, induce apoptosis, down-regulate inflammatory mediators, and inhibit angiogenic factors, limiting tumor growth and spread. The large bioavailability, together with the absence of adverse effects, confer to SJW and HPF a biological relevance for both tumor prevention and treatment, suggesting their potential use in association with current chemotherapeutic drugs.

However, considering other anti-cancer natural compounds, Peng et al. demonstrated that Withaferin (WFA), a triterpenoid isolated from *Whithania somnifera*, combined with a low dose of non-ionizing radiation such as ultraviolet–C (UVC), can act as a therapeutic sensitizer [5]. In fact, low-dose UVC/WFA co-treatment was able to induce ROS overproduction, oxidative DNA damage, the inhibition of cell proliferation, and the apoptosis of oral cancer cells without effects on healthy oral cells. Interestingly, this selective therapy deserves in vivo investigations in order to validate the promising in vitro effects of WFA as a UVC chemosensitizer.

Several studies support the notion that the increase in ROS generation is exploited as a valuable anti-cancer strategy. In this regard, Van Loenhout and co-authors focused their attention on therapies based on oxidative stress induced by exogenous pro-oxidant compounds (chemotherapeutic drugs), as well as by targeting endogenous antioxidant molecules [6]. In particular, in this review, the authors provided evidence that ROS-inducing anti-cancer treatments could have a direct effect on tumor microenvironment, exerting both immunostimulatory as well as immunosuppressive effects that must be taken into account during anticancer treatment.

Among ROS-centered anticancer therapies, ionizing radiation (IR) plays a critical role in the management of hematological cancers, even if it is well known that IR is able to impair bone marrow and diverse other organs leading to post-irradiation mortality and morbidity. In this context, Allegra and co-authors reported that an increasing number of compounds able to act as radioprotectors or radiosensitizers has been identified [7]. Among natural and synthetic radioprotective agents, the phytochemical compounds and in particular phenolics (simple phenols, benzoic acid derivatives, flavonoids, stilbenes and tannins) revealed a promising protective effect on healthy cells due to their ability to scavenge ROS production while preserving the sensitivity of cancer cells to IR. On the other side, other phytochemicals such as curcumin, quercetin and genistein have been identified as enhancers of IR treatment or radiosensitizers via oxidative stress induction on cancer cells. In the future perspective to increase the radiosensitivity of cancer cells or to enhance the radioprotection of healthy cells, the authors conclude that patients with lymphomas treated with ultra-high dose rate radiation positively responded to therapy with reduced toxic effects probably due to acute oxygen decrease within the irradiated tissue. Interestingly, it has been demonstrated that miR-139-5p controls the IR therapy response in cancer, and a miR-139-5p mimetic can synergize with IR by increasing ROS production, impairing DNA repair mechanisms and triggering the apoptosis of cancer cells.

Therefore, based on the efficacy of chemotherapy and radiotherapy relying on ROS release, the design of novel ROS modulators could offer new promises for the development of selective and efficient anti-cancer therapies. In this context, Sardella et al. demonstrated that the use of plasma-treated water solutions (PTWS) could be a valid tool able to generate balanced amounts of ROS and reactive nitrogen species (RNS) in liquids [8]. In this study, the authors showed that the synergic action of H_2O_2 and NO_2^- can induce the death of osteosarcoma cells, but does not affect endothelial cells of the tumor microenvironment.

Furthermore, Yang et al. described the innovative use of sonodynamic therapy as a sensitizer of ultrasound treatment [9]. They demonstrated that carbon-doped titanium dioxide nanoparticles, made of biocompatible material, are able to inhibit the proliferation of low-intensity ultrasound-treated breast cancer cells through ROS over-production, which sensitizes them to the anti-cancer effect of sonodynamic therapy.

Another approach based on ROS release and used as a chemosensitizer of anticancer therapy is described by Cordani et al. [10]. The authors synthesized triazole-based coordination trimers made with Fe(II) in aqueous media and tested them as adjuvants for the treatment of pancreatic cancer. These coordination complexes were able to stimulate ROS generation in pancreatic cancer cells and enhanced the cytotoxic effect of gemcitabine, an approved drug for pancreatic cancer treatment, through the apoptosis induction and down-regulation of the mTOR pathway.

Altogether, these studies, although in vitro, report different ROS-centered approaches in order to chemosensitize cancer cells to conventional anti-cancer therapies without affecting the viability of healthy cells.

Although many above-described therapeutic strategies need to be tested in animal models, we hope that the articles collected in this Special Issue can clarify the role of ROS modulation in cancer prevention and treatment. Moreover, we believe that the multidisciplinary approach used to address this issue could help to dissect the molecular mechanisms underlying the paradoxical role of oxidative stress in order to counteract carcinogenesis or enhance the sensitivity to anticancer therapy.

We thank all the authors who have contributed to the Special Issue and shared their recent findings, having a potential favorable impact on human health. Finally, we would like to thank the reviewers for their review processing and the Editorial staff for their kind support.

Author Contributions: Writing—review and editing, B.M. and C.D. All authors have read and agreed to the published version of the manuscript.

Funding: This research received no external funding.

Conflicts of Interest: The authors declare no conflict of interest.

References

1. Nitti, M.; Ivaldo, C.; Traverso, N.; Furfaro, A.L. Clinical Significance of Heme Oxygenase 1 in Tumor Progression. *Antioxidants* **2021**, *10*, 789. [CrossRef] [PubMed]
2. Fujihara, K.M.; Zhang, B.Z.; Clemons, N.J. Opportunities for Ferroptosis in Cancer Therapy. *Antioxidants* **2021**, *10*, 986. [CrossRef]
3. Monteleone, L.; Speciale, A.; Valenti, G.E.; Traverso, N.; Ravera, S.; Garbarino, O.; Leardi, R.; Farinini, E.; Roveri, A.; Ursini, F.; et al. PKCα Inhibition as a Strategy to Sensitize Neuroblastoma Stem Cells to Etoposide by Stimulating Ferroptosis. *Antioxidants* **2021**, *10*, 691. [CrossRef]
4. Menegazzi, M.; Masiello, P.; Novelli, M. Anti-Tumor Activity of *Hypericum perforatum* L. and Hyperforin through Modulation of Inflammatory Signaling, ROS Generation and Proton Dynamics. *Antioxidants* **2020**, *10*, 18. [CrossRef] [PubMed]
5. Peng, S.Y.; Wang, Y.Y.; Lan, T.H.; Lin, L.C.; Yuan, S.F.; Tang, J.Y.; Chang, H.W. Low Dose Combined Treatment with Ultraviolet-C and Withaferin a Enhances Selective Killing of Oral Cancer Cells. *Antioxidants* **2020**, *9*, 1120. [CrossRef]
6. Van Loenhout, J.; Peeters, M.; Bogaerts, A.; Smits, E.; Deben, C. Oxidative Stress-Inducing Anticancer Therapies: Taking a Closer Look at Their Immunomodulating Effects. *Antioxidants* **2020**, *9*, 1188. [CrossRef] [PubMed]
7. Allegra, A.G.; Mannino, F.; Innao, V.; Musolino, C.; Allegra, A. Radioprotective Agents and Enhancers Factors. Preventive and Therapeutic Strategies for Oxidative Induced Radiotherapy Damages in Hematological Malignancies. *Antioxidants* **2020**, *9*, 1116. [CrossRef] [PubMed]
8. Sardella, E.; Veronico, V.; Gristina, R.; Grossi, L.; Cosmai, S.; Striccoli, M.; Buttiglione, M.; Fracassi, F.; Favia, P. Plasma Treated Water Solutions in Cancer Treatments: The Contrasting Role of RNS. *Antioxidants* **2021**, *10*, 605. [CrossRef] [PubMed]

9.   Yang, C.C.; Wang, C.X.; Kuan, C.Y.; Chi, C.Y.L.; Chen, C.Y.; Lin, Y.Y.; Chen, G.S.; Hou, C.H.; Lin, F.H. Using C-doped TiO$_2$ Nanoparticles as a Novel Sonosensitizer for Cancer Treatment. *Antioxidants* **2020**, *9*, 880. [CrossRef] [PubMed]
10.   Cordani, M.; Resines-Urien, E.; Gamonal, A.; Milán-Rois, P.; Salmon, L.; Bousseksou, A.; Costa, J.S.; Somoza, Á. Water Soluble Iron-Based Coordination Trimers as Synergistic Adjuvants for Pancreatic Cancer. *Antioxidants* **2021**, *10*, 66. [CrossRef] [PubMed]

 *antioxidants*

Review

Clinical Significance of Heme Oxygenase 1 in Tumor Progression

Mariapaola Nitti, Caterina Ivaldo, Nicola Traverso and Anna Lisa Furfaro *

Department of Experimental Medicine, University of Genoa, Via L.B. Alberti 2, 16132 Genova, Italy; mariapaola.nitti@unige.it (M.N.); caterina.ivaldo@edu.unige.it (C.I.); nicola.traverso@unige.it (N.T.)
* Correspondence: annalisa.furfaro@unige.it

Abstract: Heme oxygenase 1 (HO-1) plays a key role in cell adaptation to stressors through the antioxidant, antiapoptotic, and anti-inflammatory properties of its metabolic products. For these reasons, in cancer cells, HO-1 can favor aggressiveness and resistance to therapies, leading to poor prognosis/outcome. Genetic polymorphisms of HO-1 promoter have been associated with an increased risk of cancer progression and a high degree of therapy failure. Moreover, evidence from cancer biopsies highlights the possible correlation between HO-1 expression, pathological features, and clinical outcome. Indeed, high levels of HO-1 in tumor specimens often correlate with reduced survival rates. Furthermore, HO-1 modulation has been proposed in order to improve the efficacy of antitumor therapies. However, contrasting evidence on the role of HO-1 in tumor biology has been reported. This review focuses on the role of HO-1 as a promising biomarker of cancer progression; understanding the correlation between HO-1 and clinical data might guide the therapeutic choice and improve the outcome of patients in terms of prognosis and life quality.

Keywords: HO-1; Nrf2; cancer progression; patients; therapy; prognosis; biomarker

Citation: Nitti, M.; Ivaldo, C.; Traverso, N.; Furfaro, A.L. Clinical Significance of Heme Oxygenase 1 in Tumor Progression. *Antioxidants* **2021**, *10*, 789. https://doi.org/10.3390/antiox10050789

Academic Editor: Stefan W. Ryter

Received: 30 March 2021
Accepted: 10 May 2021
Published: 17 May 2021

Publisher's Note: MDPI stays neutral with regard to jurisdictional claims in published maps and institutional affiliations.

1. Introduction

Heme oxygenase (HO) is an evolutionarily conserved enzyme that, in the presence of molecular oxygen (O_2) and reduced nicotinamide adenine dinucleotide phosphate (NADPH), catalyzes the degradation of heme into equimolar amounts of biliverdin, carbon monoxide (CO), and free iron (Fe^{2+}), releasing $NADP^+$ and H_2O [1].

Two different isoforms of HO have been described in mammalian cells (HO-1 and HO-2) and, heme oxygenase 1 (HO-1) represents the inducible form [2]. The *HMOX-1* gene maps on the human chromosome 22q12.3 [3], on a region of approximately 13,148 bp, containing five exons and four introns [4], and codifies for a 32 kDa stress protein present at low levels in physiological conditions in most mammalian tissues [2]. HO-1 induction generally occurs in response to different endogenous and exogenous stimuli, mainly related to oxidative stress and inflammation, as well as to iron metabolism imbalance [5–8]. In tissues responsible for heme metabolism, such as spleen, liver, and bone marrow, HO-1 is highly expressed [9].

The induction of HO-1 exerts pleiotropic effects. It is well known that HO-1 is involved in the adaptive response to cellular stress and in attenuating inflammation, and, in healthy cells, HO-1 maintains redox homeostasis and prevents carcinogenesis. Importantly, in cancer cells, its expression correlates with tumor growth, aggressiveness, metastatic and angiogenetic potential. Recently, a crucial role of HO-1 in tumor immune escape has also been highlighted [10–13].

All the above-mentioned functions are ascribed mainly to the activity of HO-1 metabolic products [14–16]. Bilirubin (BR), derived by biliverdin reduction catalyzed by biliverdin reductase (BVRA), is a powerful antioxidant [17–20], able to scavenge reactive oxygen species (ROS) [21], therefore preventing protein and lipid peroxidation [17,22–24]. BR plays a key role in the regulation of inflammation and adaptive immunity, exerting immuno-suppressive effects and promoting immune tolerance [25–27]. It is important to remark that BR is an important modulator of endothelial cell activity also in the microvasculature. Indeed, BR is able to reduce leukocyte transmigration and to prevent leucocyte rolling by decreasing the expression of P- and E-selectin, VCAM, and ICAM [28–31]. CO is well known as an antiapoptotic, anti-inflammatory, antiproliferative, and anticoagulant factor [32–35] and modulates the mitogen-activated protein kinase pathway (MAPK), soluble guanylyl cyclase (sGC) and the level of intracellular cGMP [36–38]. HO-1-derived CO is involved in blood vessel development [39] and VEGF synthesis [37], and enhances the proliferation of endothelial cells [38]. In addition, CO is able to attenuate inflammation [40,41], acting on both T cells [42] and antigen-presenting cells [11,12,33,43]. Finally, HO-1-derived free iron induces the synthesis of the heavy chain of the iron-chelating protein ferritin and activates the membrane transporter Fe-ATPase, which is crucial for decreasing the intracellular concentration of free Fe^{2+} and for preventing ROS production through the Fenton reaction [44,45]. Notably, HO-1 overactivation, if not balanced by the induction of ferritin and iron transporters or quenching systems, can trigger ferroptosis. Indeed, in this condition, iron accumulation leads to cell death through excessive ROS production and consequent lipid peroxidation [46,47].

Among HO-1 metabolic products, only CO has been recognized to be directly involved in tumor progression, promoting cancer cell proliferation, migration, angiogenesis, and immune escape [11]. The role of HO-1-derived bilirubin in cancer biology has been hypothesized considering its pro-surviving, pro-angiogenetic, and anti-inflammatory activity [31,48]. Instead, the generation of free iron due to HO-1 activation has been proved to favor non-canonical ferroptosis and is considered a therapeutic approach.

This review touches on the relevance of HO-1 expression in cancer progression, with a particular interest in the correlation with clinical features of tumors, taking into account data from histopathological analysis of tumor specimens.

2. HO-1 Gene Transcription and Protein Localization

2.1. HO-1 Transcriptional and Post-Transcriptional Regulation

The regulation of HO-1 expression occurs mainly at the transcriptional level (Figure 1). The promoter region of HO-1 contains several binding sites for different transcription factors activated in oxidative stress conditions, such as AP-1, HIF-1, NF-kB, and Nrf2 [49,50]. Thus, HO-1 is under the control of different signaling pathways. Moreover, two kinds of polymorphisms are present in its promoter region: the length of (GT)n repeats and the single nucleotide polymorphism (SNP) at the codon −413. Further, HO-1 protein levels can be regulated post-transcriptionally. Here, the main aspects of HO-1 synthesis regulation will be in brief as they are already reviewed elsewhere [51,52]; in particular, we will focus on the roles of HO-1 in cancer biology.

Among the HO-1 promoter polymorphisms, the (GT)n microsatellite repeats are crucial in modulating HO-1 expression. In particular, (GT)n polymorphisms are usually classified as short and long according to the number of the GT repeats: individuals with long (GT)n repeats show lower HO-1 inducibility due to a decreased promoter activity compared to individuals with short (GT)n repeats who have higher transcriptional activity, higher HO-1 inducibility and thus higher HO-1 levels [53]. The presence of this polymorphism correlates with the development of various pathologies, such as cardiovascular diseases, pulmonary disease [53–55], and cancer. However, contrasting results have been reported in different types of cancers [56].

Moreover, the SNP rs2071746 (−413A > T) polymorphism can also modulate HO-1 inducibility, being the higher HO-1 expression associated with the 413-A variant [57]. This polymorphism correlates with a reduced incidence of ischemic heart disease [58] and with graft survival after liver transplantation when present in the donor [59]. To our knowledge, only recently, the role of SNPs −413A > T in cancer risk has been analyzed by Bukowska [60]. The role of HO-1 polymorphisms in cancer will be discussed later.

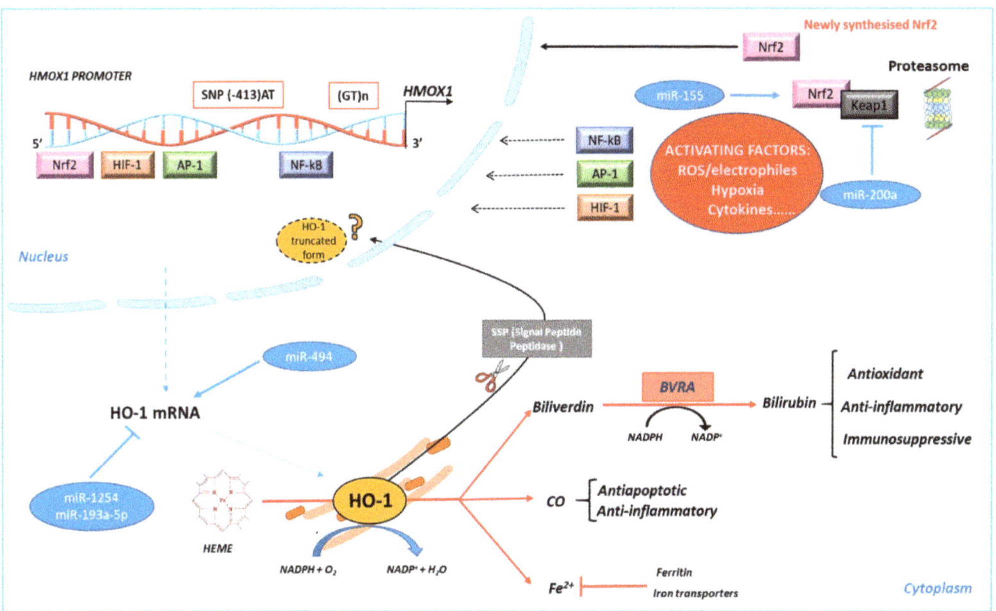

Figure 1. Schematic representation of heme oxygenase 1 (HO-1) activity and regulation. HO-1 induction can be regulated at the transcriptional level by several stress-related transcription factors (Nrf2, AP-1, NF-kB, and HIF-1). Two polymorphisms that modify HO-1 inducibility have been indicated. Post-transcriptional regulation can involve miRNA. HO-1 regulates intracellular heme level catalyzing its degradation into biliverdin, carbon monoxide (CO), and ferrous iron (Fe^{2+}). Biliverdin is converted into bilirubin by biliverdin reductase A (BVRA). Free iron activates iron transporters and induces the expression of ferritin. HO-1 metabolic products exert pro-survival activities, as indicated. A truncated form of HO-1, formed by signal peptide peptidase (SSP) cleavage, with nuclear localization and no enzymatic activity, has been described.

The transcription factor nuclear erythroid 2-related factor-2 (Nrf2) is recognized to be the master regulator of HO-1 activation. Under nonstressed conditions, Nrf2 is bound to Kelch-like ECH-associated protein 1 (Keap1), which continuously targets Nrf2 for proteasome degradation. When cells are exposed to electrophiles and/or oxidants, Keap1 is inactivated and the newly synthetized Nrf2 is free to move into the nucleus, where it dimerizes with small Maf proteins and binds to the antioxidant/electrophile responsive elements (ARE/EpRE), leading to HO-1 gene transcription [5,61].

Of note, in cancer cells, genetic and epigenetic modifications of Nrf2/Keap1 have been described [5,62,63]. Indeed, gain-of-function mutations in Nrf2 or loss-of-function mutations in Keap1 lead to constitutive activation of Nrf2 and of its downstream target genes [5,62]. In particular, Nrf2 gain-of-function mutations have been identified in lung, head and neck, and bladder cancer, while Keap1 loss-of-function mutations have been identified in esophageal, head and neck, liver, gastric, and colorectal cancer [64]. In addition, epigenetic, especially TET-dependent demethylation of the Nrf2 promoter or Keap1 and CUL3 hypermethylation, favors Nrf2 activation, as demonstrated in lung, colorectal, and ovarian cancer [65–68].

Furthermore, HO-1 transcriptional regulation specifically involves the BTB domain and CNC homolog 1 (Bach1), a heme-binding protein that represents a major transcriptional repressor of HO-1. Indeed, Bach1 competes with Nrf2 for the binding to ARE sequences and impairs Nrf2-DNA binding activity. In response to oxidative stress, and in particular, to high levels of intracellular heme, Bach1 detaches from ARE sequences and is degraded by proteasome; in this condition, HO-1 transcription is allowed [10,69,70]. Of note, it has been demonstrated that in lung cancer metastasis, Bach1 can be stabilized in terms of protein expression and correlates with poor overall survival [71,72]. In the same works, high levels of HO-1 have been observed, meaning that the activity of HO-1 can halt Bach1 proteasomal degradation by reducing heme content. Thus, Bach1 stabilization can be observed both in the absence of Nrf2 activity or in the presence of Nrf2 activity, being dependent on the content of intracellular free heme.

Different kinase pathways (i.e., MAPKs and PI3K/AKT) are involved in HO-1 induction in cancer cells, not only by acting on Nrf2 but also by favoring Nrf2 independent HO-1 activation. p38 MAPK is responsible for Nrf2-dependent HO-1 activation in human MCF-7 breast cancer cells exposed to cadmium chloride [73] and cooperates with ERK for Nrf2-independent HO-1 activation in MKN-45 and in MKN-28 human gastric cancer cells [74]. Moreover, PI3K/AKT has been proved to be involved in HO-1 induction in SH-SY5Y neuroblastoma cancer cells in response to guanosine [75] and in cholangiocarcinoma cells treated with piperlongumine [76].

The regulation of HO-1 expression also occurs at the post-transcriptional level and microRNAs (miRNAs) play a key role. miRNAs can directly regulate HO-1 or indirectly modulate Nrf2, as already reviewed by Cheng and coworkers [77]. More recently, the involvement of miRNAs in regulating HO-1 in cancer cells has been proved. In particular, miR-155 favors lung cancer resistance to arsenic trioxide through Nrf2/HO-1 activation [78]. miR200a, in breast cancer, regulates HO-1 via Nrf2 activation by targeting Keap1 mRNA [79]. miR-1254 or miR-193a-5p, in non-small cell lung cancer (NSCLC) and prostate cancer, respectively, act on HO-1, reducing its expression and contributing to decreasing cancer cell growth [80,81]. We also demonstrated the involvement of miR494 in favoring neuroblastoma cell adaptation to oxidative stress through HO-1 up-regulation [82].

2.2. HO-1 Sub-Cellular and Extra-Cellular Localization

As far as HO-1 localization is concerned, HO-1 is mainly present at the endoplasmic reticulum (ER), where co-localizes with cytochrome P-450 reductase [83,84]. In addition, HO-1 can co-localize with caveolin 1/2 on plasma membrane caveolae [85] and a mito-chondrial localization has been also demonstrated [86]. Of note, HO-1 can move into the nucleus, and nuclear translocation is favored by the signal peptide peptidase (SSP)-mediated intra-membrane cleavage, which leads to a C-terminal truncated form of HO-1 without catalytic activity but with transcriptional function [87–89]. Indeed, the truncated form of HO-1 interacts with Nrf2, increasing its stabilization [90]. Moreover, it has been demonstrated that the acetylation of the truncated form of HO-1 significantly enhances JunD-mediated AP-1 transcriptional activity leading to cancer cell proliferation, invasion, and resistance to therapy [91], indicating that post-translational modification of nuclear HO-1 plays an important role in cell proliferation, migration, and metastasis [92]. HO-1 nuclear compartmentalization is associated with cancer progression and chemoresistance, as demonstrated in chronic myeloid leukemia (CML) [93,94]; however, some opposite observations are reported in the literature [95–98]. A deeper review of the significance of HO-1 nuclear-truncated form has been recently published [92].

Furthermore, an extracellular localization of HO-1 in body fluids, including plasma, serum, milk, and cerebrospinal fluid, has been described [99–101]. In this context, a potential role of HO-1 as a disease biomarker has been suggested [94]. To date, the mechanisms of HO-1 release in biological fluids have not been understood. It has been hypothesized that plasma levels of HO-1 are the result of an active secretion and not the consequence of cell necrosis since it has been demonstrated, in patients with acute myocardial infarction, that HO-1 plasma levels are independent of necrosis biomarkers [102]. Interestingly, in acute kidney injury (AKI), HO-1 plasma and urinary levels parallel the level of HO-1 expression in renal tissue in response to damage [103]. Moreover, in both serum and urine, a truncated form of HO-1 was detected, suggesting that proteolytic cleavage occurs, even though the causes and consequences of this cleavage remain unknown [103]. More recently, the involvement of extracellular vesicles (EVs), such as exosomes and micro-vesicles, as potential sources of extracellular biomarkers has been considered [104,105]. In this context,

HO-1 mRNA and protein have been detected in exosomes isolated from peripheral blood mononuclear cells (PMBC) of psoriasis patients [106]. Schipper and coworkers detected HO-1 protein in EVs from various human bio fluids [107]. With regard to cancer, HO-1 protein is found in EVs from the culture medium of several types of cancer cells, such as breast, lung, melanoma, and kidney [108]. However, this aspect needs further investigation.

3. Role of HO-1 in Cancer Progression

HO-1 overexpression has been described in several types of cancers and is associated with cancer cell proliferation, angiogenesis, invasiveness, immune escape, and resistance to therapy. However, opposite evidence has been reported as well, correlating HO-1 expression with inhibition of cancer cell proliferation, induction of apoptosis, and reduction of invasiveness; this suggests that the role of HO-1 in tumors could be tissue- and cell-specific [10].

3.1. HO-1 in Cancer Cell Growth, Metastasis, and Angiogenesis

The overexpression of HO-1 correlates with an increase in proliferation of cell viability in many types of cancer, such as human renal adenocarcinoma and in murine melanoma [109,110]. It favors the proliferation of malignant prostate tissues [111], pancreatic cancer, hepatoma, and lymphosarcoma [112], as well as brain and hematological cancers, as widely reviewed [11,113,114].

The acquisition of a metastatic phenotype, characterized by more aggressive features, is a key step in cancer growth and progression. In this context, HO-1 overexpression has been shown to favor metastasis development in melanoma [110], pancreatic cancer [115], oral squamous cell carcinoma [116], and prostate cancer [117]. In non-small cell lung cancer (NSCLC), the invasive and migratory abilities of cancer cells significantly increase after HO-1 overexpression, decrease after HO-1 silencing and correlate with the expression of metastasis-associated protein EGFR, CD147, and MMP9 [118]. In gastric cancer, the Nrf2-dependent HO-1 activation is involved in metastatic potential both in vitro and in vivo models [119]. Furthermore, HO-1 is involved in the epithelial-to-mesenchymal transition, a critical step in the metastasis process. Indeed, in ovarian cancer cells, HO-1 inhibition by zinc II protoporphyrin IX (ZnPPIX) down-regulates the expression of the mesenchymal markers vimentin, N-cadherin, and Zeb1, while up-regulates the expression of epithelial markers [120]. Consistently, it has been demonstrated that the down regulation of GRP78 increases the migration and invasiveness of colon cancer cells by the activation of Nrf2/HO-1, the induction of vimentin, and the reduction of E-cadherin expression [121].

Moreover, tumor invasiveness and metastasis development are strictly related to the stimulation of angiogenesis. In this regard, the role played by HO-1 in pathological angiogenesis of cancer is well documented both in vitro and in vivo. The up-regulation of VEGF expression in response to prostaglandin in human microvascular endothelial cells (HMEC-1) is mediated by the activation of HO-1 [122], and CO seems to be the main mediator in stimulating blood vessel formation [39]. It has been shown that HO-1 overexpression promotes angiogenesis in urothelial carcinoma cells [123] as well as in human pancreatic cancer [115]; in bladder cancer, HO-1 overexpression correlates with HIF-1α and VEGF expression [124]. Moreover, HO-1 inhibition by ZnPPIX suppresses VEGF production in GC9811-P gastric cancer cells, a cellular line characterized by high peritoneal metastatic potential [125], and in HCT-15-induced xenografts model of colorectal cancer reduces VEGF release and tumor angiogenesis [126]. In addition, inhibition of the Nrf2/HO-1 pathway by oxysophocarpine treatment suppresses the migration, the invasion potential, and the angiogenesis of oral squamous cells carcinoma [127].

3.2. HO-1 in Cancer Immune Escape

Recently, an important role of HO-1 in cancer immune escape has been highlighted. Indeed, HO-1 expression in infiltrating immune cells, including macrophages, dendritic cells (DC), neutrophils, natural killer cells (NK), and T and B lymphocytes, leads to their polarization toward a tumor-promoting and immunosuppressive phenotype. Moreover, HO-1 expression in cancer cells can be associated with the recruitment of specific subsets of infiltrating leucocytes and to the generation of specific cytokines that favor tumor progression.

Indeed, HO-1 expression is involved in macrophages polarization towards a pro-tolerogenic, pro-angiogenic, IL-10 producing, M2 phenotype [128], and HO-1-derived CO keeps DCs immature and modulates their cytokines secretion towards a tolerogenic phenotype [129].

In particular, it has been demonstrated that HO-1 is highly expressed in monocytes within the tumor microenvironment once they differentiate to TAMs, which indicates that HO-1 promotes their immunosuppressive function [130]. Furthermore, HO-1 detection in TAMs of prostate and breast cancers correlates with accelerated tumor growth [131,132].

Interestingly, in aggressive and metastatic prostate cancer, both in vivo and in ex vivo models, HO-1 positive macrophages were mainly detected outside the tumor tissue at the invasive zone of prostate tumors. These data suggest that extra tumor HO-1 positive macrophages could be involved in cancer aggressiveness, probably by playing a prominent role in stimulating tumor growth and metastasis [117].

Furthermore, in HO-1 overexpressing solid tumors, as well as in hematological malignancies, a high number of T regulatory cells (T_{reg}) are present and act to suppress the immune response against the tumor mass [133–135]. For instance, in 4T1 breast cancer and in breast and melanoma bearing mice, it has been demonstrated that T_{reg} recruitment is increased in an HO-1 dependent manner [136], and HO-1 expressing T_{reg} accumulates during glioma progression [137].

Regarding the role played by HO-1 in regulating NK lymphocytes, crucially involved in the early immune response to tumor cells [138], little data are available in the literature. In a co-culture of an HO-1 positive cervical cancer cell (CCC) line and NK cells, pretreatment with various HO-1 inhibitors, tin II protoporphyrin IX (SnPPIX) and ZnPPIX, restores the expression of NKG2D, NKp30, and NKp46, markers of NK activation, and increases the production of IFN-γ and TNF-α, enhancing NK killing activity towards cancer cells [139]. Furthermore, we have recently demonstrated in BRAFv600 melanoma cells that HO-1 inhibition with tin mesoporphyrin IX (SnMPIX) and HO-1 siRNAdown-regulation favors cell death induced by vemurafenib, and increases NK cancer cell recognition by up-regulating B7H6 and ULBP3 ligands of NK cells [140]. To the best of our knowledge, no studies have been reported so far on the expression of HO-1 in NK cells.

3.3. HO-1 in the Resistance to Therapy

An important aspect of HO-1 expression in cancer cells is the gain of a resistant phenotype. It is well known that conventional anticancer treatments such as chemo- and radiotherapies can act to induce oxidative stress by increasing intracellular ROS levels [141] in order to favor apoptosis, as recently reviewed by Aggarwal and co-workers [142]. However, cancer cells, by up-regulating their antioxidant defenses, including HO-1, can counteract oxidative stress. Thus, the increase in HO-1 expression attenuates the efficacy of anticancer therapy as shown in different types of tumor where high levels of HO-1 are associated with a lower sensitivity to anticancer treatment. For instance, HO-1 overexpression is involved in resistance to chemo- and radio-therapy in central nervous system malignancies [113] and in resistance to cisplatin in hepatoma cells and ovarian cancer cells [143,144]. This aspect will be discussed later in Section 5, in the context of the possible modulation of HO-1 to favor antitumor therapies [145–154].

4. HO-1 Promoter Polymorphisms and Cancer Risk

As reported above, two major polymorphisms in the HO-1 promoter have been identified and linked to the modulation of HO-1 transcription: the ($-413A > T$) SNP and the presence of long/short (GT)n repeats. So far, no association between SNP-413 and cancers has been demonstrated, as indicated by Wang et al., who analyzed studies conducted on digestive neoplasms [155]. Moreover, recently, no prognostic significance has been shown for ($-413A > T$) SNP in children with acute lymphoblastic leukemia (ALL) [60].

Considering the length of GT repeats, an association has been found considering only East-Asian carriers of long (GT)n repeats, who show a high incidence of cancers in the digestive tract compared to carriers of short repeats. In fact, in Caucasian, American, and West-Asian populations, this association has not been demonstrated. Notwithstanding the small number of samples and the lack of uniformity of the studies analyzed, it seems evident that for the East-Asian populations, the presence of long (GT)n repeats is a risk factor for digestive tract cancers, probably in association with environmental factors. Indeed, in some studies, an association with alcohol consumption has been shown for the development of laryngeal squamous cell carcinoma (LSCC) for L-allele carriers in male Chinese [156]. Exposure to carcinogenic chemical compounds is a determinant to be considered; for instance, the role of smoking in male Japanese carriers of long repeats (GT)n who developed lung adenocarcinoma has been proven [157]; moreover, in asbestos-exposed Japanese subjects, the frequency of L-genotype correlates with an increased risk of developing mesothelioma [158].

An interesting study from Wu and collaborators, conducted in a cohort of patients in the area of Taiwan in which arsenic poisoning is endemic, demonstrated that (GT)n polymorphisms modify the risk of cancer due to arsenic exposure. Indeed, the risk of developing the different subtypes of arsenic-dependent tumors (skin cancer and urothelial carcinomas) is differently affected by (GT)n length. In particular, the S/S genotype carriers show a high risk of skin cancer, while no association is found for the risk of developing urothelial carcinoma among the three genotypes (S/S, L/S, and L/L) [159].

Based on this evidence, the analysis of (GT)n polymorphisms may represent a tool for evaluating an individual risk profile for a specific type of cancer, also considering the specific patient ethnicity.

5. HO-1 Expression, Tumor Aggressiveness, and Disease Outcome. Evidence from Immunohistochemistry

To date, the most consistent data regarding the correlation among HO-1 expression, cancer progression, patient prognosis, and outcome derive from immunohistochemistry studies on specimens from surgical patients. The data available in the literature are synthesized in Table 1 at the end of this paragraph. It is important to underline that, since Nrf2 is crucially involved in the regulation of HO-1 transcription, its expression has been considered as well. Both solid and hematopoietic malignancies have been taken into consideration, and the possible existence of negative association has also been analyzed.

5.1. HO-1 Expression and Disease Outcome

HO-1 expression in tumor mass is associated with poor prognosis/outcome and with high grade/stage in several types of tumors. In serous ovarian cancer, the association of HO-1 expression with FIGO stage III-IV and with poor overall survival has been proven [160]. In non-muscle-invasive bladder cancer (NMIBC), HO-1 expression is associated with grade 3, and poor prognosis or low recurrence/progression-free survival [161,162]. Similarly, in astrocytoma, high levels of HO-1 have been associated with tumor grade II and III and poor overall survival [163], and NSCLC at stage III-IV, high levels of HO-1 have been associated with high mortality risk and short overall survival [118]. In gallbladder cancer, the positivity for Nrf2, together with high expression of HO-1, has been shown to correlate with high grade/stage and poor prognosis [164], highlighting the role played by Nrf2 in the induction of HO-1 during tumor progression. Similar observations have been provided for clear cell renal cell carcinoma (ccRCC) [165], even though without correlation with tumor grade or stage. Indeed, patients with ccRCC showing high levels of HO-1 and Nrf2 have lower median survival time and shorter post-operative overall survival, with no proven correlation with tumor grade/stage.

In some studies, the expression level of HO-1 in tumors has been associated with clinical outcomes but without reference to the histopathological analysis. Thus, cholangiocarcinoma [166], acute myeloid leukemia (AML) [167], and neuroblastoma [168] show a correlation between high HO-1 expression and poor disease outcomes. Furthermore, HO-1 positivity in chronic myeloid leukemia [169], acute myeloid leukemia [170], and myelodysplastic syndrome [171] correlate with disease progression, resistance to therapy, and relapse.

5.2. HO-1 Expression and Tumor Grade/Stage

Vice versa, in other reports, HO-1 expression has been correlated with grade and stage and with invasion potential, but the clinical outcomes have not been analyzed. For instance, HO-1 overexpression in papillary thyroid cancer positively correlates with the TNM stage and cancer progression [172].

The intensity of HO-1 positivity has also been analyzed in order to find a possible correlation with the progression of a disease or with clinical outcomes. Interestingly, in NSCLC, the levels of HO-1 correlates with advanced stage (III-IV), T3, and T4 status and with lymph node metastasis; however, no association with overall survival has been demonstrated when patients were divided into two different subgroups related to HO-1 intensity of expression. Thus, no differences in patient survival were observed with regard to HO-1 intensity, highlighting that HO-1 positivity also at a low degree correlates with disease severity [98].

5.3. Correlation between HO-1 Expression and Tumor Markers

In many studies, HO-1 positivity has been correlated with other tumor markers. In localized prostatic cancer, HO-1 positivity associates with relapse frequency and PTEN deletion [173]. In NMIBC bladder cancers, HO-1 expression in tumor mass correlates with HIF-1α expression and microvessel density [123], and in particular, Nrf2 and HO-1 positivity correlates with HIF-1α, HIF-2α, and VEGF expression in the tumor, and with VEGF and interleukin levels in the plasma [124]. Similarly, in gastric cancer [174] and hepatocellular carcinoma [175], HO-1 positivity is associated with VEGF expression, poor differentiation, and microvascular density.

It is worth noting, in melanoma [176], thyroid cancer [172], and acute myeloid leukemia [167], HO-1 positivity correlates with the gain of function mutations of specific oncogenes B-Raf and RET. Moreover, in high-risk and very high-risk myelodysplastic syndrome, HO-1 expression correlates with overexpression of the enhancer of the zeste homologue 2 (EZH2) gene [171].

It is remarkable to note that HO-1 expression can be detected not only in tumor cells but also in cancer-associated cells, where it can contribute to the generation of a tumor-permissive environment. The number of HO-1 positive cancer-associated cells correlates with the tumor grade, metastatic competence, and neoangiogenesis. Indeed, in NMIBC bladder cancer HO-1 positivity has been detected not only in tumor cells but also in infiltrating fibroblasts and endothelial cells, in association with an increased risk of metastasis but without association to recurrence [177]. Further, high levels of HO-1 in infiltrating macrophages show a positive correlation with vascular density and high tumor grade in glioblastoma [178], with stage II, lymph node metastasis, and poor prognosis in colorectal cancer [179], and with a high Gleason score and bone metastasis in prostate cancer [117]. High HO-1 expression in lymphocyte T_{reg} shows a correlation with a high tumor grade in glioma [137].

5.4. Contrasting Evidence

Although a great deal of literature highlights the correlation between HO-1 overexpression and cancer progression and often with the poor clinical outcomes, it seems important to consider that opposite evidence has also been provided. Indeed, it has been demonstrated that high HO-1 expression level correlates with a better prognosis and better overall survival in colorectal cancer [180,181], in gastric cancer [182], in small intestinal adenocarcinoma [183], and in oral squamous carcinoma [184].

An important observation concerning HO-1 subcellular localization comes from three different studies on head and neck cancer [185], breast cancer [186], and colorectal cancer [187] that analyzed the correlation of histological features with HO-1 positivity in cytosol or nuclei. In these studies, high expression of HO-1 in cytosol correlated with low grade and differentiation without correlation with invasiveness. However, nuclear localization of HO-1 was associated with a high grade and poor differentiation. Moreover, in breast cancer, Gandini showed that cytosolic HO-1 is enzymatically active, while the nuclear form is truncated and with no catalytic activity [186]. These observations appear to be interesting and helpful in understanding the contrasting observation of the role of HO-1 in tumor progression and lead to speculation that HO-1 pro- or antitumor activity may depend on its subcellular localization and catalytic activity.

Table 1. Correlation among HO-1 expression, aggressiveness, and outcomes in histological specimens.

Tumor	HO-1	Nrf2	Grade and Stage	Additional Markers	Metastasis, Lymph Node, Angiogenesis	Clinical and Pathological Features	Disease Outcome/Prognosis	Ref.
Positive correlation among HO-1 expression and tumor aggressiveness/poor prognosis								
-Solid tumors								
Astrocytoma	High level	n.e.	Grade II and III	n.e.	n.e.	n.e.	Poor OS	[163]
Clear cell renal cell carcinoma	High level	High level	No correlation with ISUP grade and T stage	n.e.	No correlation with lymph node metastasis	No significant correlation with age, gender	Poor prognosis Low MST Low post operative OS	[165]
Colangiocarcinoma	High level	n.e.	n.e.	n.e.	No association with metastasis	No significant association with age, gender, histological type	Poor OS	[166]
Gastric cancer	High level	High level	Poor differentiated tumors	Positive correlation with VEGF	Positive correlation with MVD	n.e.	n.e.	[174]
Gallbladder cancer	High level	High level	Moderately differentiated and poorly differentiated tumors (G2-G3) Correlation with Nevin classification (III-IV-V)	Positive correlation with MRP3	Metastasis	No significant correlation with gender, age, and histology type (SCC and AD)	Poor OS	[164]
Hepatocellular carcinoma	High level	n.e.	Poor differentiated tumors Edmondson-Steiner grade 2–4	n.e.	Microvascular and capsular invasion	High levels of preoperative AFP	No significant correlation with OS and recurrence	[175]
Hormone refractory prostate cancer	High level	n.e.	n.e.	n.e.	n.e.	Cancer progression	n.e.	[188]
Laryngeal cancer	High level	High level	No correlation with tumor stage (clinical stage III and IV), size tumor	High level Keap1 and NQO1	No correlation with lymph node metastasis	No correlation with age	n.e.	[189]
Melanoma	High level	n.e.	n.e.	Positive correlation with B-Raf and ERK	n.e.	n.e.	n.e.	[176]

Table 1. *Cont.*

Tumor	HO-1	Nrf2	Grade and Stage	Additional Markers	Metastasis, Lymph Node, Angiogenesis	Clinical and Pathological Features	Disease Outcome/Prognosis	Ref.
Neuroblastoma	High level	n.e.	n.e.	n.e.	n.e.	n.e.	Poor OS	[168]
Non-muscle-invasive bladder cancer	High level	n.e.	Tumor grade G3 tumor stage pT1	Ki-67 and p53	n.e.	No significant correlation with age and gender	Poor prognosis No correlation with RFS and PFS	[161]
	High level	n.e.	Tumor grade G3 Tumor stage T1	Positive correlation with S100A4	Lymph vascular invasion	n.e.	Low RFS Low PFS	[162]
	High level	n.e.	n.e.	Positive correlation with HIF-1α	High MVD	n.e.	n.e.	[123]
	High level	High level	n.e.	Correlation with HIF-1α, HIF-2α, VEGF	n.e.	Increased serum/plasma level of IL-6, IL-8, VEGF	n.e.	[124]
Non-small cell lung cancer	High level	n.e.	Stage III-IV	Positive correlation with MMP-9	High metastatic rate	No correlation with age and gender	Poor prognosis Low OS High mortality risk	[118]
	High level	n.e.	Stage III-IV T status (T3-T4)	n.e.	Lymph node metastasis	No correlation with gender	No significant difference in patient survival between high and low staining group	[98]
Ovarian cancer	High level	n.e.	Serous undifferentiated tumors Correlation with FIGO stage (III-IV)	n.e.	Lymph node metastasis	Non optimal-debulking	Poor OS	[160]
Prostate cancer	High level	n.e.	Localized tumor	PTEN deletion	n.e.	n.e.	Relapse after radical prostatectomy	[173]
Thyroid cancer	High level	n.e.	Positive correlation with TNM (1,2,3,4) and with MACIS score	BRAFV600E mutation	No significant association with lymph node metastasis	Correlation with age and tumor aggressiveness	n.e.	[172]

Table 1. *Cont.*

Tumor	HO-1	Nrf2	Grade and Stage	Additional Markers	Metastasis, Lymph Node, Angiogenesis	Clinical and Pathological Features	Disease Outcome/Prognosis	Ref.
-Hematopoietic tumors								
Acute myeloid leukemia	High level	n.e.	n.e.	Positive correlation with HIF-1α and GLUT-1	n.e.	n.e.	Correlation with relapse and refractory	[170]
	High level	n.e.	Correleation with M5 patients	Correlation with RET gene	n.e.	Correlation with leukocytosis at diagnosis	n.e.	[167]
Chronic myeloid leukemia	Higher level in peripheral blood cells	n.e.	n.e.	n.e.	n.e.	Tumor progression	Correlation with relapse	[169]
Myelodysplastic Syndrome	High level	n.e.	Correlation with high-risk and very high-risk patients	Positive correlation with EZH2	n.e.	Progression to AML and decreased response to decitabine	n.e.	[171]
Positive correlation among HO-1 expression in tumor-associated cells and tumor aggressiveness/poor prognosis								
Colorectal cancer	High level in cancer cells and in macrophages	n.e.	Stage III	n.e.	Lymph node metastasis	No significant difference between the HO-1-positive and negative with gender, age, tumor size, histological type, and depth of tumor invasion	Poor prognosis Short DSF	[179]
Glioblastoma	High level in infiltrating macrophages	n.e.	Grade IV	n.e.	Positive correlation with vascular density	n.e.	n.e.	[178]
Glioma	HO-1 positive Treg	n.e.	Correlation with grade glioma (II-III-IV)	n.e.	n.e.	n.e.	n.e.	[137]

Table 1. *Cont.*

Tumor	HO-1	Nrf2	Grade and Stage	Additional Markers	Metastasis, Lymph Node, Angiogenesis	Clinical and Pathological Features	Disease Outcome/Prognosis	Ref.
Non-muscle-invasive bladder cancer	High level in cancer cells and fibroblast-like, tumor-infiltrating, and endothelial cells	n.e.	Correlation with high grade tumors and with stage (T1)	COX-1	MVD, LVD, PI, increased risk of metastasis	No association with age and gender	No association with recurrence	[177]
Prostate cancer	HO-1 positive macrophages infiltrate and in bone metastasis	n.e.	High-grade tumors Gleason score 7–10	n.e.	Bone metastasis	n.e.	n.e.	[117]
Negative correlation among HO-1 expression and tumor aggressiveness/poor prognosis								
Colorectal cancer	High level	n.e.	Invasive CRC	Significant correlation with K-ras	n.e.	Significant correlation with normal CEA level	Better prognosis, increased MTS	[181]
	High level	n.e.	n.e.	n.e.	Low vascular invasion and lymph node metastasis	n.e.	Better survival rate	[180]
Gastric cancer	High level	n.e.	Well and moderate differentiated	n.e.	Negative lymph node metastasis	n.e.	Better prognosis	[182]
Oral squamous cell carcinoma	High level	n.e.	Well-differentiated Grade G1 No association with T stage	n.e.	Low lymph node metastasis	No association with age and sex No association with clinical stage	n.e.	[184]
Small intestinal adenocarcinoma	High level	n.e.	Low T stage (T1, T2, T3)	n.e.	Low pancreatic invasion	n.e.	Tend to have longer OS (difference not significative)	[183]

Table 1. *Cont.*

Tumor	HO-1	Nrf2	Grade and Stage	Additional Markers	Metastasis, Lymph Node, Angiogenesis	Clinical and Pathological Features	Disease Outcome/Prognosis	Ref.
Different correlation among HO-1 expression and tumor aggressiveness/poor prognosis depending on HO-1 subcellular localization								
Breast cancer	High level in malignant epithelial cells	n.e.	Grade I-II (>80%)	Positive correlation with E-cadherin	Negative correlation with lymph node metastasis	Reduced tumor size	Longer OS with increased MST	[186]
Colorectal cancer	High level in cancer cells and in stromal cells (fibroblasts, neutrophils, and macrophages)	n.e.	Well-differentiated adenocarcinoma Nuclear HO-1 localization in moderate and poor differentiated No association with TNM	n.e.	No correlation with lymph node and liver metastasis	n.e.	n.e.	[187]
Head and neck squamous cell carcinoma	High level	n.e.	High rate of HO-1 positivity in well-differentiated and moderately differentiated (<90%) Poor-differentiated high rate of nuclear HO-1	n.e.	n.e.	No association with age, gender, tumor location	n.e.	[185]

Tumors are listed alphabetically. List of table abbreviations. n.e., not evaluated; AD, adenocarcinoma; AFP, alpha feto protein; CEA, carcinoempryonic antigen; ISUP, International Society of Urologic Pathologists; LVD, lymph vascular density; MTS, median survival time; MVD, microvascular density; OS, overall survival; PI, proliferation index: PFS, progression free survival; RFS, recurrence free survival; SCC, squamous cell carcinoma.

6. HO-1 and Tumor Therapies

It has been widely reported that the induction of HO-1 in response to anticancer treatments can attenuate the efficacy of therapy, increasing cancer cell survival. Indeed, HO-1 expression is increased in response to different chemotherapeutic agents that act through the imbalance of intracellular oxidative state. For instance, in neuroblastoma cells, HO-1 expression is induced by exposure to etoposide through the activation of Nrf2 [145], and by the exposure to proteasome inhibitors bortezomib or carfilzomib [148–150], and mediates cell survival. To note, doxorubicin or pharmorubicin promote HO-1 expression increasing cell survival in breast cancers through the activation of Src/STAT3 or PI3K/AKT, respectively [146,147].

Remarkably, HO-1 induction mediates cancer cell resistance not only to chemotherapeutic agents but also to radio-, photodynamic-, and non-thermal-plasma (NTP) therapies, as demonstrated in non-small cell lung carcinoma [152–154].

As far as hematological malignancies are concerned, HO-1 expression significantly increases in myeloid neoplasms both in chronic and acute myeloid leukemia. Its overexpression occurs mainly after therapeutic intervention and induces chemoresistance. Recently, it has been demonstrated that PI3K/AKT-dependent HO-1 induction drives drug resistance to imatinib in CML [190] as well as to panobinostat in AML [191] by modulating the expression of HDACs. HO-1 overexpression enhances the viability and decreases the apoptotic rate in AML cell lines treated with cytarabine. Accordingly, the derived xenograft mouse model shows a significantly shorter survival and a great extent of organ invasion, while HO-1 down regulation significantly increases the survival rate [192]. Moreover, HO-1 up-regulation in myelodysplastic syndromes is closely related to resistance to decitabine-induced apoptosis [193], and in multiple myeloma, HO-1 up-regulation is involved in bortezomib chemoresistance [194].

In this context, pharmacological and genetic tools to reduce HO-1 activity have been proposed, and their use has been hypothesized in therapy, as described later and summarized in Table 2.

6.1. Inhibition of HO-1 by Pharmacological Compounds

Among the pharmacological tools, metalloporphyrins and imidazole-based compounds are the most well-known and have been recently reviewed [195].

Briefly, metalloporphyrins represent the first generation of HO-1 inhibitors and include deuteroporphyrin, mesoporphyrin, and protoporphyrin [196]. Structurally similar to heme, metalloporphyrins strongly inhibit HO-1 by a competitive mechanism [197]. The most used metalloporphyrins are ZnPPIX, SnPPIX, and SnMPIX, and their efficacy in favoring conventional tumor therapies has been widely demonstrated in vitro and in vivo. For instance, ZnPPIX favors the sensitivity of nasopharyngeal carcinoma cells to radiotherapy [198] and of neuroblastoma to glutathione depletion and etoposide [145]. Moreover, ZnPPIX sensitizes C-26 colon and MDAH2774 ovarian carcinoma cells to photodynamic therapy-mediated cytotoxicity [199] and increases the effects of cisplatin in liver cancers [143]. It has also been demonstrated that treatment with ZnPPIX reduces cell growth in hepatoma, sarcoma, lung cancer, and B cell lymphoma [52,125]. Furthermore, in melanoma cells, SnPPIX enhances the efficacy of photodynamic therapy [200] and in BRAF^{V600}-mutated melanoma cells SnMPIX increases cell death induced by vemurafenib/PLX4032 [140].

Unfortunately, metalloporphyrins are able to act on other heme-dependent enzymes, such as nitric oxide synthase (NOS), sGC, and cytochrome P450 [201,202]. Moreover, even though they efficiently inhibit HO-1 activity, they can often favor HO-1 protein synthesis, as demonstrated in liver cells and fibroblasts, and more recently, in prostate cancer PC-3 cells by a compensatory mechanism [203–205]. Of note, another important disadvantage of using metalloporphyrins is related to their photo reactivity, which is responsible for side effects and even tissue and organ damage [196]. Another strong drawback for the potential clinical use of some metalloporphyrins (e.g., ZnPPIX) is represented by their poor solubility in aqueous solutions, which limits translational applicability. However, this inconvenience has been overcome by conjugation with specific molecules, e.g., polyethylene-glycol or amphiphilic styrene-maleic acid copolymer, generating water-soluble molecules [206–210].

Imidazole-based compounds represent the second generation of HO-1 inhibitors. These molecules are non porphyrin-based and non competitive water-soluble inhibitors of HO-1 and exhibit low or even no inhibitory action on NOS, sGC, and CYP [211,212]. The first reported was Azalanstat [213], but other molecules and novel azole-based compounds derived from the structural modification of Azalanstat have been recently discovered [214,215]. Imidazole-based compounds have shown potent antitumor activity in prostate and breast cancer cell lines [216]; in a preclinical model of hormone-refractory prostate cancer, the small molecule imidazole-derived OB-24 acts in synergism with the conventional chemotherapy drug Taxol, preventing tumor growth and formation of lymph node and lung metastasis [188]. However, imidazole-based compounds have not been tested in clinical studies so far.

6.2. Inhibition of HO-1 by RNA Interference and CRIPR/Cas9 Technology

With regard to genetic tools to modulate HO-1 activity, the most consistent data derive from studies on RNA interference, including small interfering RNA and short hairpin RNA, able to inhibit HO-1 activity by targeting HO-1 transcription and consequently protein synthesis. Thus, HO-1 silencing increases the effect of chemotherapeutic drugs in pancreatic cancer [217], neuroblastoma [148,149], and melanoma cancer cells [140], as well as in myeloid leukemia [170]. Moreover, HO-1 silencing sensitizes cancer cells to apoptosis, as demonstrated in lung, colon, and leukemic cancer cells [195]. Similar results have been obtained in an in vivo experimental mouse model of hepatocellular carcinoma, where injection of siRNA-HO-1 results in the diminished growth of the tumor [218]. Furthermore, HO-1 is considered a survival factor in ALL, regardless of Philadelphia chromosome positivity; indeed, the down-regulation of HO-1 expression by siRNA increases apoptosis and arrests cell growth [219]. Consistently, in chronic lymphocytic leukemia (CLL), it has been demonstrated that HO-1 silencing directly leads to apoptosis of MEC-1 cells and enhances the effects of the combined therapy fludarabine plus entinostat [220].

A new approach in the inhibition of HO-1 activity is represented by genetic ablation of HO-1 with the CRISPR/Cas9 editing system. It has been recently demonstrated that homozygous HO-1 knock-out in BRAF-WT melanoma cells is able to decrease clone formation and to lower tumor cell growth [176]; further, in pancreatic ductal adenocarcinoma cells, HO-1 CRISPR/Cas9 is able to suppress cell proliferation and improve the efficacy of gemcitabine treatment [151]. Importantly, in in vivo experiments on C57/BL6 mice, HO-1 CRISPR/Cas9 editing blocks lymphocyte B development [221].

Table 2. HO-1 inhibitory tools.

Pharmacological Inhibitors		Benefits	Drawbacks	Ref.
	Porphyrin-Based Compounds			
	Metalloporphyrins - Zinc II protoporphyrin IX (ZnPPIX) - Tin protoporphyrin IX (SnPPIX) - Tin mesoporphyrin IX (SnMPIX)	- Competitive inhibitors - Well proved activity in vitro and in vivo	- Non selective on HO-1 isoform - Active on other heme-dependent enzymes (NOS, sGC, and CYP) - HO-1 inducers - Photo reactive - Poor soluble	[196,201,202]
	Modified protoporphyrins - Polyethylene-glycol (PEG-ZnPPIX) - Amphiphilic styrene-maleic acid copolymer (SMA-ZnPPIX)	- Water-soluble		[206–210]
	Imidazole-based compounds			
	- Azalanstat - Other imidazole-derived compounds - (OB-24)	- Non competitive inhibitors - Selective on HO-1 isoforms - Limited inhibitory activity on NOS, sGC, and CYP - Water-soluble	- Not well studied and not tested in clinical trials	[211,212]
Genetic tools				
	Small interfering RNA and short hairpin RNA	- Specific targeting HO-1 mRNA	- Limited therapeutic application (delivery methods)	[195]
	CRISPR/Cas9	- Genetic ablation of HO-1 gene - Stable knock-down - High efficiency of HO-1 inhibition	- Limited therapeutic application (delivery methods)	[195]

6.3. Induction of HO-1 as a Therapeutic Strategy

Thus, a great deal of literature shows a direct correlation between the overexpression of HO-1 and the gain of resistance of cancer cells and tumor progression. However, it must be taken into account that in some tumors, the over expression of HO-1 exerts opposite effects by inhibiting tumor growth and cancer progression. In particular, it has been shown in some types of prostate cancer that HO-1 expression and carbon monoxide generation are associated with significant inhibition of cell proliferation and invasiveness [96]. Moreover, in non-small-cell lung carcinoma NCI-H292 cells, the stable HO-1 overexpression is able to up-regulate tumor-suppressive miRNAs and to down-regulate the expression of oncomirs and angiomirs, leading to the inhibition of cell proliferation, invasiveness, and angiogenesis [222]. It has been highlighted that this tumor-suppressive phenotype is characterized by the attenuation of the metastatic potential mainly by down regulating MMP-9 and MMP-13 [223]. Similarly, stable overexpression of HO-1 retards hepatocellular carcinoma progression [224]. The antitumorigenic effects of HO-1 have also been demonstrated in human and rat breast cancer, where its overexpression correlates with inhibition of cell proliferation [225] and in pancreatic and prostate cancer, where it is associated with a decrease in cell proliferation and invasiveness by a down regulation of the proangiogenic mediators VEGF and MMP-9 [97,195,226]. In this context, the induction of HO-1 has been proposed to increase conventional cancer therapies, and some "natural" compounds derived from plants have shown interesting properties. In colorectal cancer, it has been demonstrated that treatment with extracts from *Sageretia thea*, a medicinal plant used for treating hepatitis and fevers in Korea and China, decreases cell viability by inducing GSK3β-dependent cyclin D1 degradation and increasing HO-1 expression via activation of Nrf2 [227]. In addition, Ginnalin A, a polyphenolic compound isolated from red maple (*Acer rubrum*), inhibits cell viability and colony formation in colorectal cancer, inducing cell cycle arrest by activating the Nrf2/HO-1 pathway through the up-regulation of p62 and the inhibition of Keap1 [228]. Similarly, treatment with fisetin, a bioactive flavonol molecule abundantly found in strawberries, decreases the level of MMPs and cell migration in metastatic breast cancer with a mechanism depending on Nrf2 nuclear translocation and HO-1 up-regulation [229].

Since ferroptosis may be a way to kill cancer cells, and it can be enhanced by HO-1 overactivation, the pharmacological induction of HO-1 has been proposed. Indeed, HO-1-dependent intracellular Fe^{2+} overload induces lipid peroxidation and triggers a noncanonical ferroptosis [230]. Phytochemicals are often used for this purpose [231]. Neuroblastoma cell treatment with withaferin A, a steroidal lactone derived from *Withania somnifera* (Indian ginseng), directly targets Keap1, leading to Nrf2 release and HO-1 up-regulation and consequently increasing intracellular Fe^{2+} and inducing ferroptosis [232]. Similarly, in human colon cancer cells, a high concentration of extract of *Betula etnensis* extract induces HO-1 leading to ferroptotic cell death through an increase of ROS production and in lipid peroxidation mediated by iron accumulation [233]. Moreover, HO-1 up-regulation has been proved to be the primary factor for curcumin-induced ferroptosis in human breast adenocarcinoma-derived MCF7 cells and in human triple-negative MDA-MB-231 cell line [234]. In addition, β-elemene, a sesquiterpene found in a variety of plants, is able to induce ferroptosis by enhancing HO-1 activity in KRAS mutant colorectal HCT116 cancer cells [235]. In addition, in this work, the presence of possible side effects of β-elemene were tested in the derived orthotopic murine colon cancer model, and no toxicity was found relatively the different organs analyzed (lung, heart, liver, kidney, and spleen) by H&E staining.

Thus, the evaluation of HO-1 expression in cancer samples from patients may help to define a therapeutic strategy where inhibition or induction of HO-1 could improve the efficacy of the standard antineoplastic therapy used.

7. Future Perspectives and Conclusions

The chance to analyze HO-1 expression in cancer patients seems to be a useful tool to improve tumor diagnosis and to better define prognosis and therapy. On the one hand, the analysis of (GT)n length polymorphisms seems a very promising approach to assess the risk of treatment failure as recently proved in ALL patients carrier of short (GT) repeats [60]. On the other hand, the characterization of HO-1 expression in tumors may be a useful tool to improve tumor diagnosis and prognosis because it can correlate with tumor grade/stage, invasiveness, and clinical outcomes. However, contrasting data are reported, and larger analyses need to be performed. Importantly, it has been recently highlighted the role played by the truncated form of HO-1 in favoring cell growth, opening to a new scenario in which HO-1 can be involved in tumor biology [92].

As a future perspective, in order to better assess tumor progression, the correlation between tissue expression of HO-1 and its levels in a blood sample could be taken into consideration, even though no evidence has been reported so far. However, the analysis of HO-1 level may be proposed in other biological fluids such as urine, peritoneal or pleural fluids, if directly related to the tissue bearing neoplastic cells. It is important to remember that in other diseases, HO-1 levels in bio fluids correlate with HO-1 expression levels in tissues [103].

Moreover, a great amount of data support the efficacy of HO-1 modulation in order to improve cancer response to therapies (Figure 2). Different approaches have been proposed, using either pharmacological agents or genetic tools. Unfortunately, concerning HO-1 pharmacological inhibitors, the translational applicability is not completely elucidated, even though both SnPPIX and SnMPIX have been already tested in humans [236] and approved for the treatment of hyperbilirubinemia [237]. Instead, genetic tools have been tested only in experimental animal models. Therefore, HO-1 modulation may represent an important strategy also to prevent cancer immune escape. However, we must consider that, so far, little data in the literature are available on the role played by HO-1 in the function of tumor-related immune cells. This is still an open field of research.

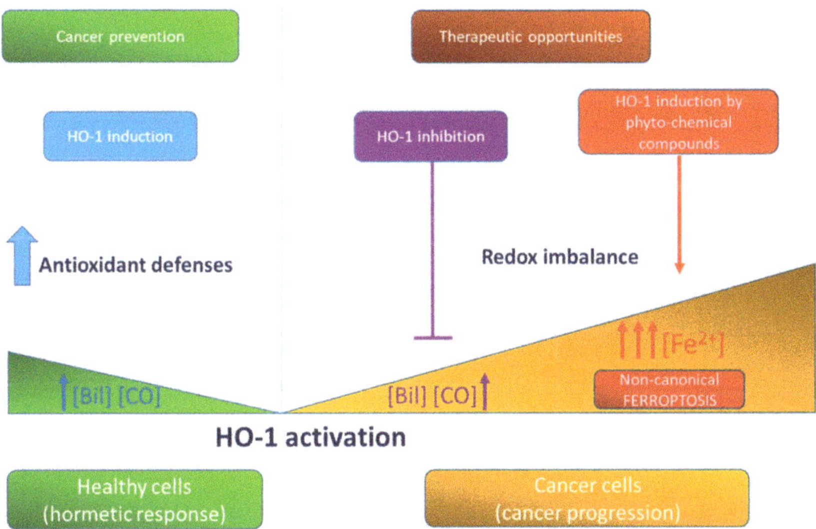

Figure 2. Schematic representation of the effects of HO-1 activation and generation of its metabolic products in healthy and cancer cells. HO-1 activation is involved in antioxidant defenses and in healthy cells promotes the hormetic response and cancer prevention through the generation of bilirubin and CO. In cancer cells, HO-1 favors cancer progression, and its inhibition represents a therapeutic opportunity. However, also HO-1 over-activation can be proposed as a therapeutic option, as it can favor unconventional ferroptosis through the accumulation of pro-oxidant-free iron.

Conversely, molecules able to induce HO-1 may be used in order to favor cancer cell death due to iron imbalance. About this issue, as mentioned before, many natural compounds have been tested and showed their efficacy in this sense, but even in this case, translational applicability in humans seems to be still far away.

In conclusion, a deeper investigation of the specific multifaceted role played by HO-1 in different types of cancers, in the tumor microenvironment and bio fluids is needed in order to customize therapy and improve the outcome of cancer patients. Thus, HO-1 could become in the future an important clinical tool for cancer management.

Author Contributions: M.N. and A.L.F. conceived and wrote the manuscript; M.N., C.I., N.T., and A.L.F. revised the literature; M.N. and N.T. provided financial support. All authors have read and agreed to the published version of the manuscript.

Funding: This research received no external funding.

Acknowledgments: Grants from Genoa University.

Conflicts of Interest: The authors declare no conflict of interest.

Abbreviations

ALL	acute lymphoblastic leukemia
AML	acute myeloid leukemia
BR	bilirubin
BVRA	biliverdin reductase A
ccRCC	clear cell Renal cell carcinoma
CML	chronic myeloid leukemia
DC	dendritic cells
EV	extracellular vescicles
HO-1	heme oxygenase 1
MAPK	mitogen-activated protein kinase pathway
NK	natural killer cells
NMIBC	non-muscle-invasive bladder cancer
NOS	nitric oxide synthase
PMBC	peripheral blood mononuclear cells
ROS	reactive oxygen species
sGC	soluble guanylyl cyclase
SnMPPIX	tin mesoporphyrin IX
SnPPIX	tin protoporphyrin IX
SSP	signal peptide peptidase
VSMC	vascular smooth muscle cells
ZnPPIX	zinc(II) protoporphyrin IX

References

1. Maines, M.D. Heme Oxygenase: Function, Multiplicity, Regulatory Mechanisms, and Clinical Applications. *FASEB J.* **1988**, *2*, 2557–2568. [CrossRef] [PubMed]
2. Waza, A.A.; Hamid, Z.; Ali, S.; Bhat, S.A.; Bhat, M.A. A Review on Heme Oxygenase-1 Induction: Is It a Necessary Evil. *Inflamm. Res.* **2018**, *67*, 579–588. [CrossRef]
3. Kutty, R.K.; Nagineni, C.N.; Kutty, G.; Hooks, J.J.; Chader, G.J.; Wiggert, B. Increased Expression of Heme Oxygenase-1 in Human Retinal Pigment Epithelial Cells by Transforming Growth Factor-Beta. *J. Cell Physiol.* **1994**, *159*, 371–378. [CrossRef]
4. Bian, C.; Zhong, M.; Nisar, M.F.; Wu, Y.; Ouyang, M.; Bartsch, J.W.; Zhong, J.L. A Novel Heme Oxygenase-1 Splice Variant, 14kDa HO-1, Promotes Cell Proliferation and Increases Relative Telomere Length. *Biochem. Biophys Res. Commun.* **2018**, *500*, 429–434. [CrossRef] [PubMed]
5. Furfaro, A.L.; Traverso, N.; Domenicotti, C.; Piras, S.; Moretta, L.; Marinari, U.M.; Pronzato, M.A.; Nitti, M. The Nrf2/HO-1 Axis in Cancer Cell Growth and Chemoresistance. *Oxid. Med. Cell. Longev.* **2016**, *2016*, 1958174. [CrossRef]
6. Keyse, S.M.; Tyrrell, R.M. Heme Oxygenase Is the Major 32-KDa Stress Protein Induced in Human Skin Fibroblasts by UVA Radiation, Hydrogen Peroxide, and Sodium Arsenite. *Proc. Natl. Acad. Sci. USA* **1989**, *86*, 99–103. [CrossRef]
7. Alam, J.; Shibahara, S.; Smith, A. Transcriptional Activation of the Heme Oxygenase Gene by Heme and Cadmium in Mouse Hepatoma Cells. *J. Biol. Chem.* **1989**, *264*, 6371–6375. [CrossRef]

8.	Foresti, R.; Clark, J.E.; Green, C.J.; Motterlini, R. Thiol Compounds Interact with Nitric Oxide in Regulating Heme Oxygenase-1 Induction in Endothelial Cells. Involvement of Superoxide and Peroxynitrite Anions. *J. Biol. Chem.* **1997**, *272*, 18411–18417. [CrossRef]

9.	Ayer, A.; Zarjou, A.; Agarwal, A.; Stocker, R. Heme Oxygenases in Cardiovascular Health and Disease. *Physiol. Rev.* **2016**, *96*, 1449–1508. [CrossRef]

10.	Nitti, M.; Piras, S.; Marinari, U.M.; Moretta, L.; Pronzato, M.A.; Furfaro, A.L. HO-1 Induction in Cancer Progression: A Matter of Cell Adaptation. *Antioxidants* **2017**, *6*, 29. [CrossRef] [PubMed]

11.	Loboda, A.; Jozkowicz, A.; Dulak, J. HO-1/CO System in Tumor Growth, Angiogenesis and Metabolism-Targeting HO-1 as an Anti-Tumor Therapy. *Vasc. Pharmacol.* **2015**, *74*, 11–22. [CrossRef] [PubMed]

12.	Riquelme, S.A.; Carreño, L.J.; Espinoza, J.A.; Mackern-Oberti, J.P.; Alvarez-Lobos, M.M.; Riedel, C.A.; Bueno, S.M.; Kalergis, A.M. Modulation of Antigen Processing by Haem-Oxygenase 1. Implications on Inflammation and Tolerance. *Immunology* **2016**, *149*, 1–12. [CrossRef]

13.	Vijayan, V.; Wagener, F.A.D.T.G.; Immenschuh, S. The Macrophage Heme-Heme Oxygenase-1 System and Its Role in Inflammation. *Biochem. Pharmacol.* **2018**, *153*, 159–167. [CrossRef] [PubMed]

14.	Siow, R.C.; Sato, H.; Mann, G.E. Heme Oxygenase-Carbon Monoxide Signalling Pathway in Atherosclerosis: Anti-Atherogenic Actions of Bilirubin and Carbon Monoxide? *Cardiovasc. Res.* **1999**, *41*, 385–394. [CrossRef]

15.	Otterbein, L.E.; Foresti, R.; Motterlini, R. Heme Oxygenase-1 and Carbon Monoxide in the Heart: The Balancing Act Between Danger Signaling and Pro-Survival. *Circ. Res.* **2016**, *118*, 1940–1959. [CrossRef]

16.	Kishimoto, Y.; Kondo, K.; Momiyama, Y. The Protective Role of Heme Oxygenase-1 in Atherosclerotic Diseases. *Int. J. Mol. Sci.* **2019**, *20*, 3628. [CrossRef]

17.	Stocker, R.; Yamamoto, Y.; McDonagh, A.F.; Glazer, A.N.; Ames, B.N. Bilirubin Is an Antioxidant of Possible Physiological Importance. *Science* **1987**, *235*, 1043–1046. [CrossRef] [PubMed]

18.	Stocker, R. Antioxidant Activities of Bile Pigments. *Antioxid. Redox Signal.* **2004**, *6*, 841–849. [CrossRef]

19.	Tenhunen, R.; Marver, H.S.; Schmid, R. The Enzymatic Conversion of Heme to Bilirubin by Microsomal Heme Oxygenase. *Proc. Natl. Acad. Sci. USA* **1968**, *61*, 748–755. [CrossRef]

20.	Tenhunen, R.; Marver, H.S.; Schmid, R. The Enzymatic Conversion of Hemoglobin to Bilirubin. *Trans. Assoc. Am. Physicians* **1969**, *82*, 363–371.

21.	Neuzil, J.; Stocker, R. Bilirubin Attenuates Radical-Mediated Damage to Serum Albumin. *FEBS Lett.* **1993**, *331*, 281–284. [CrossRef]

22.	Sedlak, T.W.; Saleh, M.; Higginson, D.S.; Paul, B.D.; Juluri, K.R.; Snyder, S.H. Bilirubin and Glutathione Have Complementary Antioxidant and Cytoprotective Roles. *Proc. Natl. Acad. Sci. USA* **2009**, *106*, 5171–5176. [CrossRef]

23.	Wu, T.W.; Fung, K.P.; Yang, C.C. Unconjugated Bilirubin Inhibits the Oxidation of Human Low Density Lipoprotein Better than Trolox. *Life Sci.* **1994**, *54*, P477–P481. [CrossRef]

24.	He, M.; Nitti, M.; Piras, S.; Furfaro, A.L.; Traverso, N.; Pronzato, M.A.; Mann, G.E. Heme Oxygenase-1-Derived Bilirubin Protects Endothelial Cells against High Glucose-Induced Damage. *Free Radic. Biol. Med.* **2015**, *89*, 91–98. [CrossRef]

25.	Wu, B.; Wu, Y.; Tang, W. Heme Catabolic Pathway in Inflammation and Immune Disorders. *Front. Pharmacol.* **2019**, *10*, 825. [CrossRef] [PubMed]

26.	Canesin, G.; Hejazi, S.M.; Swanson, K.D.; Wegiel, B. Heme-Derived Metabolic Signals Dictate Immune Responses. *Front. Immunol.* **2020**, *11*, 66. [CrossRef] [PubMed]

27.	Ryter, S.W. Heme Oxygenase-1/Carbon Monoxide as Modulators of Autophagy and Inflammation. *Arch. Biochem. Biophys.* **2019**, *678*, 108186. [CrossRef]

28.	Keshavan, P.; Deem, T.L.; Schwemberger, S.J.; Babcock, G.F.; Cook-Mills, J.M.; Zucker, S.D. Unconjugated Bilirubin Inhibits VCAM-1-Mediated Transendothelial Leukocyte Migration. *J. Immunol.* **2005**, *174*, 3709–3718. [CrossRef] [PubMed]

29.	Grochot-Przeczek, A.; Dulak, J.; Jozkowicz, A. Haem Oxygenase-1: Non-Canonical Roles in Physiology and Pathology. *Clin. Sci.* **2012**, *122*, 93–103. [CrossRef] [PubMed]

30.	Mazzone, G.L.; Rigato, I.; Ostrow, J.D.; Bossi, F.; Bortoluzzi, A.; Sukowati, C.H.C.; Tedesco, F.; Tiribelli, C. Bilirubin Inhibits the TNFalpha-Related Induction of Three Endothelial Adhesion Molecules. *Biochem. Biophys. Res. Commun.* **2009**, *386*, 338–344. [CrossRef] [PubMed]

31.	Nitti, M.; Furfaro, A.L.; Mann, G.E. Heme Oxygenase Dependent Bilirubin Generation in Vascular Cells: A Role in Preventing Endothelial Dysfunction in Local Tissue Microenvironment? *Front. Physiol.* **2020**, *11*, 23. [CrossRef] [PubMed]

32.	Petrache, I.; Otterbein, L.E.; Alam, J.; Wiegand, G.W.; Choi, A.M. Heme Oxygenase-1 Inhibits TNF-Alpha-Induced Apoptosis in Cultured Fibroblasts. *Am. J. Physiol. Lung Cell Mol. Physiol.* **2000**, *278*, L312–L319. [CrossRef]

33.	Ryter, S.W.; Alam, J.; Choi, A.M. Heme Oxygenase-1/Carbon Monoxide: From Basic Science to Therapeutic Applications. *Physiol. Rev.* **2006**, *86*, 583–650. [CrossRef] [PubMed]

34.	Ryter, S.W.; Ma, K.C.; Choi, A.M.K. Carbon Monoxide in Lung Cell Physiology and Disease. *Am. J. Physiol. Cell Physiol.* **2018**, *314*, C211–C227. [CrossRef]

35.	Motterlini, R.; Foresti, R.; Bassi, R.; Green, C.J. Curcumin, an Antioxidant and Anti-Inflammatory Agent, Induces Heme Oxygenase-1 and Protects Endothelial Cells against Oxidative Stress. *Free Radic. Biol. Med.* **2000**, *28*, 1303–1312. [CrossRef]

36.	Dennery, P.A. Heme Oxygenase in Neonatal Lung Injury and Repair. *Antioxid. Redox Signal.* **2014**, *21*, 1881–1892. [CrossRef] [PubMed]

37. Dulak, J.; Jozkowicz, A.; Foresti, R.; Kasza, A.; Frick, M.; Huk, I.; Green, C.J.; Pachinger, O.; Weidinger, F.; Motterlini, R. Heme Oxygenase Activity Modulates Vascular Endothelial Growth Factor Synthesis in Vascular Smooth Muscle Cells. *Antioxid. Redox Signal.* **2002**, *4*, 229–240. [CrossRef]

38. Jozkowicz, A.; Huk, I.; Nigisch, A.; Weigel, G.; Dietrich, W.; Motterlini, R.; Dulak, J. Heme Oxygenase and Angiogenic Activity of Endothelial Cells: Stimulation by Carbon Monoxide and Inhibition by Tin Protoporphyrin-IX. *Antioxid. Redox Signal.* **2003**, *5*, 155–162. [CrossRef]

39. Loboda, A.; Jazwa, A.; Grochot-Przeczek, A.; Rutkowski, A.J.; Cisowski, J.; Agarwal, A.; Jozkowicz, A.; Dulak, J. Heme Oxygenase-1 and the Vascular Bed: From Molecular Mechanisms to Therapeutic Opportunities. *Antioxid. Redox Signal.* **2008**, *10*, 1767–1812. [CrossRef]

40. Rochette, L.; Cottin, Y.; Zeller, M.; Vergely, C. Carbon Monoxide: Mechanisms of Action and Potential Clinical Implications. *Pharmacol. Ther.* **2013**, *137*, 133–152. [CrossRef]

41. Ryter, S.W.; Choi, A.M.K. Targeting Heme Oxygenase-1 and Carbon Monoxide for Therapeutic Modulation of Inflammation. *Transl. Res.* **2016**, *167*, 7–34. [CrossRef]

42. Pae, H.O.; Choi, B.M.; Oh, G.S.; Lee, M.S.; Ryu, D.G.; Rhew, H.Y.; Kim, Y.M.; Chung, H.T. Roles of Heme Oxygenase-1 in the Antiproliferative and Antiapoptotic Effects of Nitric Oxide on Jurkat T Cells. *Mol. Pharmacol.* **2004**, *66*, 122–128. [CrossRef] [PubMed]

43. Otterbein, L.E.; Bach, F.H.; Alam, J.; Soares, M.; Tao Lu, H.; Wysk, M.; Davis, R.J.; Flavell, R.A.; Choi, A.M. Carbon Monoxide Has Anti-Inflammatory Effects Involving the Mitogen-Activated Protein Kinase Pathway. *Nat. Med.* **2000**, *6*, 422–428. [CrossRef] [PubMed]

44. Balla, G.; Jacob, H.S.; Balla, J.; Rosenberg, M.; Nath, K.; Apple, F.; Eaton, J.W.; Vercellotti, G.M. Ferritin: A Cytoprotective Antioxidant Strategem of Endothelium. *J. Biol. Chem.* **1992**, *267*, 18148–18153. [CrossRef]

45. Baker, H.M.; Anderson, B.F.; Baker, E.N. Dealing with Iron: Common Structural Principles in Proteins That Transport Iron and Heme. *Proc. Natl. Acad. Sci. USA* **2003**, *100*, 3579–3583. [CrossRef]

46. Chiang, S.-K.; Chen, S.-E.; Chang, L.-C. A Dual Role of Heme Oxygenase-1 in Cancer Cells. *Int. J. Mol. Sci.* **2018**, *20*, 39. [CrossRef] [PubMed]

47. Song, X.; Long, D. Nrf2 and Ferroptosis: A New Research Direction for Neurodegenerative Diseases. *Front. Neurosci.* **2020**, *14*, 267. [CrossRef]

48. Di Biase, S.; Longo, V.D. Fasting-Induced Differential Stress Sensitization in Cancer Treatment. *Mol. Cell. Oncol.* **2016**, *3*, e1117701. [CrossRef]

49. Lavrovsky, Y.; Schwartzman, M.L.; Levere, R.D.; Kappas, A.; Abraham, N.G. Identification of Binding Sites for Transcription Factors NF-Kappa B and AP-2 in the Promoter Region of the Human Heme Oxygenase 1 Gene. *Proc. Natl. Acad. Sci. USA* **1994**, *91*, 5987–5991. [CrossRef] [PubMed]

50. Lavrovsky, Y.; Schwartzman, M.L.; Abraham, N.G. Novel Regulatory Sites of the Human Heme Oxygenase-1 Promoter Region. *Biochem. Biophys. Res. Commun.* **1993**, *196*, 336–341. [CrossRef] [PubMed]

51. Medina, M.V.; Sapochnik, D.; Garcia Solá, M.; Coso, O. Regulation of the Expression of Heme Oxygenase-1: Signal Transduction, Gene Promoter Activation, and Beyond. *Antioxid. Redox Signal.* **2020**, *32*, 1033–1044. [CrossRef]

52. Was, H.; Dulak, J.; Jozkowicz, A. Heme Oxygenase-1 in Tumor Biology and Therapy. *Curr. Drug Targets* **2010**, *11*, 1551–1570. [CrossRef]

53. Exner, M.; Minar, E.; Wagner, O.; Schillinger, M. The Role of Heme Oxygenase-1 Promoter Polymorphisms in Human Disease. *Free Radic. Biol. Med.* **2004**, *37*, 1097–1104. [CrossRef] [PubMed]

54. Zhang, M.-M.; Zheng, Y.-Y.; Gao, Y.; Zhang, J.-Z.; Liu, F.; Yang, Y.-N.; Li, X.-M.; Ma, Y.-T.; Xie, X. Heme Oxygenase-1 Gene Promoter Polymorphisms Are Associated with Coronary Heart Disease and Restenosis after Percutaneous Coronary Intervention: A Meta-Analysis. *Oncotarget* **2016**, *7*, 83437–83450. [CrossRef]

55. Daenen, K.E.L.; Martens, P.; Bammens, B. Association of HO-1 (GT)n Promoter Polymorphism and Cardiovascular Disease: A Reanalysis of the Literature. *Can. J. Cardiol.* **2016**, *32*, 160–168. [CrossRef] [PubMed]

56. Zhang, L.; Song, F.-F.; Huang, Y.-B.; Zheng, H.; Song, F.-J.; Chen, K.-X. Association between the (GT)n Polymorphism of the HO-1 Gene Promoter Region and Cancer Risk: A Meta-Analysis. *Asian Pac. J. Cancer Prev.* **2014**, *15*, 4617–4622. [CrossRef] [PubMed]

57. Horio, T.; Morishita, E.; Mizuno, S.; Uchino, K.; Hanamura, I.; Espinoza, J.L.; Morishima, Y.; Kodera, Y.; Onizuka, M.; Kashiwase, K.; et al. Donor Heme Oxygenase-1 Promoter Gene Polymorphism Predicts Survival after Unrelated Bone Marrow Transplantation for High-Risk Patients. *Cancers* **2020**, *12*, 424. [CrossRef] [PubMed]

58. Ono, K.; Goto, Y.; Takagi, S.; Baba, S.; Tago, N.; Nonogi, H.; Iwai, N. A Promoter Variant of the Heme Oxygenase-1 Gene May Reduce the Incidence of Ischemic Heart Disease in Japanese. *Atherosclerosis* **2004**, *173*, 315–319. [CrossRef] [PubMed]

59. Buis, C.I.; van der Steege, G.; Visser, D.S.; Nolte, I.M.; Hepkema, B.G.; Nijsten, M.; Slooff, M.J.H.; Porte, R.J. Heme Oxygenase-1 Genotype of the Donor Is Associated with Graft Survival after Liver Transplantation. *Am. J. Transplant.* **2008**, *8*, 377–385. [CrossRef]

60. Bukowska-Strakova, K.; Włodek, J.; Pitera, E.; Kozakowska, M.; Konturek-Cieśla, A.; Cieśla, M.; Gońka, M.; Nowak, W.; Wieczorek, A.; Pawińska-Wąsikowska, K.; et al. Role of HMOX1 Promoter Genetic Variants in Chemoresistance and Chemotherapy Induced Neutropenia in Children with Acute Lymphoblastic Leukemia. *Int. J. Mol. Sci.* **2021**, *22*, 988. [CrossRef]

61. Paladino, S.; Conte, A.; Caggiano, R.; Pierantoni, G.M.; Faraonio, R. Nrf2 Pathway in Age-Related Neurological Disorders: Insights into MicroRNAs. *Cell Physiol. Biochem.* **2018**, *47*, 1951–1976. [CrossRef]

62. Mitsuishi, Y.; Motohashi, H.; Yamamoto, M. The Keap1-Nrf2 System in Cancers: Stress Response and Anabolic Metabolism. *Front. Oncol.* **2012**, *2*, 200. [CrossRef]

63. Shibata, T.; Kokubu, A.; Gotoh, M.; Ojima, H.; Ohta, T.; Yamamoto, M.; Hirohashi, S. Genetic Alteration of Keap1 Confers Constitutive Nrf2 Activation and Resistance to Chemotherapy in Gallbladder Cancer. *Gastroenterology* **2008**, *135*, 1358–1368.e1-4. [CrossRef]

64. Na, H.K.; Surh, Y.J. Oncogenic Potential of Nrf2 and Its Principal Target Protein Heme Oxygenase-1. *Free Radic. Biol. Med.* **2014**, *67*, 353–365. [CrossRef]

65. Muscarella, L.A.; Parrella, P.; D'Alessandro, V.; la Torre, A.; Barbano, R.; Fontana, A.; Tancredi, A.; Guarnieri, V.; Balsamo, T.; Coco, M.; et al. Frequent Epigenetics Inactivation of KEAP1 Gene in Non-Small Cell Lung Cancer. *Epigenetics* **2011**, *6*, 710–719. [CrossRef] [PubMed]

66. Hanada, N.; Takahata, T.; Zhou, Q.; Ye, X.; Sun, R.; Itoh, J.; Ishiguro, A.; Kijima, H.; Mimura, J.; Itoh, K.; et al. Methylation of the KEAP1 Gene Promoter Region in Human Colorectal Cancer. *BMC Cancer* **2012**, *12*, 66. [CrossRef]

67. Zhao, X.-Q.; Zhang, Y.-F.; Xia, Y.-F.; Zhou, Z.-M.; Cao, Y.-Q. Promoter Demethylation of Nuclear Factor-Erythroid 2-Related Factor 2 Gene in Drug-Resistant Colon Cancer Cells. *Oncol. Lett.* **2015**, *10*, 1287–1292. [CrossRef] [PubMed]

68. van der Wijst, M.G.; Brown, R.; Rots, M.G. Nrf2, the Master Redox Switch: The Achilles' Heel of Ovarian Cancer? *Biochim. Biophys. Acta* **2014**, *1846*, 494–509. [CrossRef]

69. Ogawa, K.; Sun, J.; Taketani, S.; Nakajima, O.; Nishitani, C.; Sassa, S.; Hayashi, N.; Yamamoto, M.; Shibahara, S.; Fujita, H.; et al. Heme Mediates Derepression of Maf Recognition Element through Direct Binding to Transcription Repressor Bach1. *EMBO J.* **2001**, *20*, 2835–2843. [CrossRef] [PubMed]

70. Davudian, S.; Mansoori, B.; Shajari, N.; Mohammadi, A.; Baradaran, B. BACH1, the Master Regulator Gene: A Novel Candidate Target for Cancer Therapy. *Gene* **2016**, *588*, 30–37. [CrossRef] [PubMed]

71. Lignitto, L.; LeBoeuf, S.E.; Homer, H.; Jiang, S.; Askenazi, M.; Karakousi, T.R.; Pass, H.I.; Bhutkar, A.J.; Tsirigos, A.; Ueberheide, B.; et al. Nrf2 Activation Promotes Lung Cancer Metastasis by Inhibiting the Degradation of Bach1. *Cell* **2019**, *178*, 316–329.e18. [CrossRef]

72. Wiel, C.; Le Gal, K.; Ibrahim, M.X.; Jahangir, C.A.; Kashif, M.; Yao, H.; Ziegler, D.V.; Xu, X.; Ghosh, T.; Mondal, T.; et al. BACH1 Stabilization by Antioxidants Stimulates Lung Cancer Metastasis. *Cell* **2019**, *178*, 330–345.e22. [CrossRef]

73. Alam, J.; Wicks, C.; Stewart, D.; Gong, P.; Touchard, C.; Otterbein, S.; Choi, A.M.; Burow, M.E.; Tou, J. Mechanism of Heme Oxygenase-1 Gene Activation by Cadmium in MCF-7 Mammary Epithelial Cells. Role of P38 Kinase and Nrf2 Transcription Factor. *J. Biol. Chem.* **2000**, *275*, 27694–27702. [CrossRef]

74. Liu, Z.-M.; Chen, G.G.; Ng, E.K.W.; Leung, W.-K.; Sung, J.J.Y.; Chung, S.C.S. Upregulation of Heme Oxygenase-1 and P21 Confers Resistance to Apoptosis in Human Gastric Cancer Cells. *Oncogene* **2004**, *23*, 503–513. [CrossRef] [PubMed]

75. Dal-Cim, T.; Molz, S.; Egea, J.; Parada, E.; Romero, A.; Budni, J.; Martín de Saavedra, M.D.; del Barrio, L.; Tasca, C.I.; López, M.G. Guanosine Protects Human Neuroblastoma SH-SY5Y Cells against Mitochondrial Oxidative Stress by Inducing Heme Oxigenase-1 via PI3K/Akt/GSK-3β Pathway. *Neurochem. Int.* **2012**, *61*, 397–404. [CrossRef]

76. Talabnin, C.; Talabnin, K.; Wongkham, S. Enhancement of Piperlongumine Chemosensitivity by Silencing Heme Oxygenase-1 Expression in Cholangiocarcinoma Cell Lines. *Oncol. Lett.* **2020**, *20*, 2483–2492. [CrossRef] [PubMed]

77. Cheng, X.; Ku, C.H.; Siow, R.C. Regulation of the Nrf2 Antioxidant Pathway by MicroRNAs: New Players in Micromanaging Redox Homeostasis. *Free Radic. Biol. Med.* **2013**, *64*, 4–11. [CrossRef]

78. Gu, S.; Lai, Y.; Chen, H.; Liu, Y.; Zhang, Z. MiR-155 Mediates Arsenic Trioxide Resistance by Activating Nrf2 and Suppressing Apoptosis in Lung Cancer Cells. *Sci. Rep.* **2017**, *7*, 12155. [CrossRef] [PubMed]

79. Eades, G.; Yang, M.; Yao, Y.; Zhang, Y.; Zhou, Q. MiR-200a Regulates Nrf2 Activation by Targeting Keap1 MRNA in Breast Cancer Cells. *J. Biol. Chem.* **2011**, *286*, 40725–40733. [CrossRef] [PubMed]

80. Pu, M.; Li, C.; Qi, X.; Chen, J.; Wang, Y.; Gao, L.; Miao, L.; Ren, J. MiR-1254 Suppresses HO-1 Expression through Seed Region-Dependent Silencing and Non-Seed Interaction with TFAP2A Transcript to Attenuate NSCLC Growth. *PLoS Genet.* **2017**, *13*, e1006896. [CrossRef]

81. Yang, Z.; Chen, J.-S.; Wen, J.-K.; Gao, H.-T.; Zheng, B.; Qu, C.-B.; Liu, K.-L.; Zhang, M.-L.; Gu, J.-F.; Li, J.-D.; et al. Silencing of MiR-193a-5p Increases the Chemosensitivity of Prostate Cancer Cells to Docetaxel. *J. Exp. Clin. Cancer Res.* **2017**, *36*, 178. [CrossRef]

82. Piras, S.; Furfaro, A.L.; Caggiano, R.; Brondolo, L.; Garibaldi, S.; Ivaldo, C.; Marinari, U.M.; Pronzato, M.A.; Faraonio, R.; Nitti, M. MicroRNA-494 Favors HO-1 Expression in Neuroblastoma Cells Exposed to Oxidative Stress in a Bach1-Independent Way. *Front. Oncol.* **2018**, *8*, 199. [CrossRef]

83. Durante, W. Targeting Heme Oxygenase-1 in the Arterial Response to Injury and Disease. *Antioxidants* **2020**, *9*, 829. [CrossRef]

84. Huber, W.J.; Backes, W.L. Expression and Characterization of Full-Length Human Heme Oxygenase-1: The Presence of Intact Membrane-Binding Region Leads to Increased Binding Affinity for NADPH Cytochrome P450 Reductase. *Biochemistry* **2007**, *46*, 12212–12219. [CrossRef]

85. Jung, N.-H.; Kim, H.P.; Kim, B.-R.; Cha, S.H.; Kim, G.A.; Ha, H.; Na, Y.E.; Cha, Y.-N. Evidence for Heme Oxygenase-1 Association with Caveolin-1 and -2 in Mouse Mesangial Cells. *IUBMB Life* **2003**, *55*, 525–532. [CrossRef]

86. Slebos, D.J.; Ryter, S.W.; van der Toorn, M.; Liu, F.; Guo, F.; Baty, C.J.; Karlsson, J.M.; Watkins, S.C.; Kim, H.P.; Wang, X.; et al. Mitochondrial Localization and Function of Heme Oxygenase-1 in Cigarette Smoke-Induced Cell Death. *Am. J. Respir. Cell Mol. Biol.* **2007**, *36*, 409–417. [CrossRef] [PubMed]

87. Lin, Q.; Weis, S.; Yang, G.; Weng, Y.H.; Helston, R.; Rish, K.; Smith, A.; Bordner, J.; Polte, T.; Gaunitz, F.; et al. Heme Oxygenase-1 Protein Localizes to the Nucleus and Activates Transcription Factors Important in Oxidative Stress. *J. Biol. Chem.* **2007**, *282*, 20621–20633. [CrossRef]

88. Lin, Q.S.; Weis, S.; Yang, G.; Zhuang, T.; Abate, A.; Dennery, P.A. Catalytic Inactive Heme Oxygenase-1 Protein Regulates Its Own Expression in Oxidative Stress. *Free Radic. Biol. Med.* **2008**, *44*, 847–855. [CrossRef]

89. Hsu, F.F.; Yeh, C.T.; Sun, Y.J.; Chiang, M.T.; Lan, W.M.; Li, F.A.; Lee, W.H.; Chau, L.Y. Signal Peptide Peptidase-Mediated Nuclear Localization of Heme Oxygenase-1 Promotes Cancer Cell Proliferation and Invasion Independent of Its Enzymatic Activity. *Oncogene* **2015**, *34*, 2410–2411. [CrossRef]

90. Biswas, C.; Shah, N.; Muthu, M.; La, P.; Fernando, A.P.; Sengupta, S.; Yang, G.; Dennery, P.A. Nuclear Heme Oxygenase-1 (HO-1) Modulates Subcellular Distribution and Activation of Nrf2, Impacting Metabolic and Anti-Oxidant Defenses. *J. Biol. Chem.* **2014**, *289*, 26882–26894. [CrossRef]

91. Hsu, F.-F.; Chiang, M.-T.; Li, F.-A.; Yeh, C.-T.; Lee, W.-H.; Chau, L.-Y. Acetylation Is Essential for Nuclear Heme Oxygenase-1-Enhanced Tumor Growth and Invasiveness. *Oncogene* **2017**, *36*, 6805–6814. [CrossRef]

92. Mascaró, M.; Alonso, E.N.; Alonso, E.G.; Lacunza, E.; Curino, A.C.; Facchinetti, M.M. Nuclear Localization of Heme Oxygenase-1 in Pathophysiological Conditions: Does It Explain the Dual Role in Cancer? *Antioxidants* **2021**, *10*, 87. [CrossRef]

93. Dennery, P.A. Signaling Function of Heme Oxygenase Proteins. *Antioxid. Redox Signal.* **2014**, *20*, 1743–1753. [CrossRef]

94. Tibullo, D.; Barbagallo, I.; Giallongo, C.; La Cava, P.; Parrinello, N.; Vanella, L.; Stagno, F.; Palumbo, G.A.; Li Volti, G.; Di Raimondo, F. Nuclear Translocation of Heme Oxygenase-1 Confers Resistance to Imatinib in Chronic Myeloid Leukemia Cells. *Curr. Pharm. Des.* **2013**, *19*, 2765–2770. [CrossRef]

95. Elguero, B.; Gueron, G.; Giudice, J.; Toscani, M.A.; De Luca, P.; Zalazar, F.; Coluccio-Leskow, F.; Meiss, R.; Navone, N.; De Siervi, A.; et al. Unveiling the Association of STAT3 and HO-1 in Prostate Cancer: Role beyond Heme Degradation. *Neoplasia* **2012**, *14*, 1043–1056. [CrossRef] [PubMed]

96. Gueron, G.; De Siervi, A.; Ferrando, M.; Salierno, M.; De Luca, P.; Elguero, B.; Meiss, R.; Navone, N.; Vazquez, E.S. Critical Role of Endogenous Heme Oxygenase 1 as a Tuner of the Invasive Potential of Prostate Cancer Cells. *Mol. Cancer Res. MCR* **2009**, *7*, 1745–1755. [CrossRef]

97. Ferrando, M.; Gueron, G.; Elguero, B.; Giudice, J.; Salles, A.; Leskow, F.C.; Jares-Erijman, E.A.; Colombo, L.; Meiss, R.; Navone, N.; et al. Heme Oxygenase 1 (HO-1) Challenges the Angiogenic Switch in Prostate Cancer. *Angiogenesis* **2011**, *14*, 467–479. [CrossRef]

98. Degese, M.S.; Mendizabal, J.E.; Gandini, N.A.; Gutkind, J.S.; Molinolo, A.; Hewitt, S.M.; Curino, A.C.; Coso, O.A.; Facchinetti, M.M. Expression of Heme Oxygenase-1 in Non-Small Cell Lung Cancer (NSCLC) and Its Correlation with Clinical Data. *Lung Cancer* **2012**, *77*, 168–175. [CrossRef]

99. Vanella, L.; Barbagallo, I.; Tibullo, D.; Forte, S.; Zappalà, A.; Li Volti, G. The Non-Canonical Functions of the Heme Oxygenases. *Oncotarget* **2016**, *7*, 69075–69086. [CrossRef]

100. Signorelli, S.S.; Li Volsi, G.; Fiore, V.; Mangiafico, M.; Barbagallo, I.; Parenti, R.; Rizzo, M.; Li Volti, G. Plasma Heme Oxygenase-1 Is Decreased in Peripheral Artery Disease Patients. *Mol. Med. Rep.* **2016**, *14*, 3459–3463. [CrossRef] [PubMed]

101. Serpero, L.D.; Frigiola, A.; Gazzolo, D. Human Milk and Formulae: Neurotrophic and New Biological Factors. *Early Hum. Dev.* **2012**, *88* (Suppl. 1), S9–S12. [CrossRef]

102. Novo, G.; Cappello, F.; Rizzo, M.; Fazio, G.; Zambuto, S.; Tortorici, E.; Marino Gammazza, A.; Gammazza, A.M.; Corrao, S.; Zummo, G.; et al. Hsp60 and Heme Oxygenase-1 (Hsp32) in Acute Myocardial Infarction. *Transl. Res.* **2011**, *157*, 285–292. [CrossRef]

103. Zager, R.A.; Johnson, A.C.M.; Becker, K. Plasma and Urinary Heme Oxygenase-1 in AKI. *J. Am. Soc. Nephrol.* **2012**, *23*, 1048–1057. [CrossRef]

104. van Niel, G.; D'Angelo, G.; Raposo, G. Shedding Light on the Cell Biology of Extracellular Vesicles. *Nat. Rev. Mol. Cell Biol.* **2018**, *19*, 213–228. [CrossRef] [PubMed]

105. Margolis, L.; Sadovsky, Y. The Biology of Extracellular Vesicles: The Known Unknowns. *PLoS Biol.* **2019**, *17*, e3000363. [CrossRef]

106. El-Rifaie, A.-A.A.; Sabry, D.; Doss, R.W.; Kamal, M.A.; Abd El Hassib, D.M. Heme Oxygenase and Iron Status in Exosomes of Psoriasis Patients. *Arch. Dermatol. Res.* **2018**, *310*, 651–656. [CrossRef]

107. Cressatti, M.; Galindez, J.M.; Juwara, L.; Orlovetskie, N.; Velly, A.M.; Eintracht, S.; Liberman, A.; Gornitsky, M.; Schipper, H.M. Characterization and Heme Oxygenase-1 Content of Extracellular Vesicles in Human Biofluids. *J. Neurochem.* **2020**. [CrossRef]

108. Hurwitz, S.N.; Rider, M.A.; Bundy, J.L.; Liu, X.; Singh, R.K.; Meckes, D.G. Proteomic Profiling of NCI-60 Extracellular Vesicles Uncovers Common Protein Cargo and Cancer Type-Specific Biomarkers. *Oncotarget* **2016**, *7*, 86999–87015. [CrossRef]

109. Goodman, A.I.; Choudhury, M.; da Silva, J.L.; Schwartzman, M.L.; Abraham, N.G. Overexpression of the Heme Oxygenase Gene in Renal Cell Carcinoma. *Proc. Soc. Exp. Biol. Med.* **1997**, *214*, 54–61. [CrossRef] [PubMed]

110. Was, H.; Cichon, T.; Smolarczyk, R.; Rudnicka, D.; Stopa, M.; Chevalier, C.; Leger, J.J.; Lackowska, B.; Grochot, A.; Bojkowska, K.; et al. Overexpression of Heme Oxygenase-1 in Murine Melanoma: Increased Proliferation and Viability of Tumor Cells, Decreased Survival of Mice. *Am. J. Pathol.* **2006**, *169*, 2181–2198. [CrossRef]

111. Maines, M.D.; Abrahamsson, P.A. Expression of Heme Oxygenase-1 (HSP32) in Human Prostate: Normal, Hyperplastic, and Tumor Tissue Distribution. *Urology* **1996**, *47*, 727–733. [CrossRef]

112. Schacter, B.A.; Kurz, P. Alterations in Microsomal Drug Metabolism and Heme Oxygenase Activity in Isolated Hepatic Parenchymal and Sinusoidal Cells in Murphy-Sturm Lymphosarcoma-Bearing Rats. *Clin. Investig. Med. Med. Clin. Exp.* **1986**, *9*, 150–155.

113. Sferrazzo, G.; Di Rosa, M.; Barone, E.; Li Volti, G.; Musso, N.; Tibullo, D.; Barbagallo, I. Heme Oxygenase-1 in Central Nervous System Malignancies. *J. Clin. Med.* **2020**, *9*, 1562. [CrossRef]

114. Li Volti, G.; Tibullo, D.; Vanella, L.; Giallongo, C.; Di Raimondo, F.; Forte, S.; Di Rosa, M.; Signorelli, S.S.; Barbagallo, I. The Heme Oxygenase System in Hematological Malignancies. *Antioxid. Redox Signal.* **2017**, *27*, 363–377. [CrossRef] [PubMed]

115. Sunamura, M.; Duda, D.G.; Ghattas, M.H.; Lozonschi, L.; Motoi, F.; Yamauchi, J.; Matsuno, S.; Shibahara, S.; Abraham, N.G. Heme Oxygenase-1 Accelerates Tumor Angiogenesis of Human Pancreatic Cancer. *Angiogenesis* **2003**, *6*, 15–24. [CrossRef]

116. Lee, S.S.; Yang, S.F.; Tsai, C.H.; Chou, M.C.; Chou, M.Y.; Chang, Y.C. Upregulation of Heme Oxygenase-1 Expression in Areca-Quid-Chewing-Associated Oral Squamous Cell Carcinoma. *J. Formos. Med. Assoc.* **2008**, *107*, 355–363. [CrossRef]

117. Halin Bergström, S.; Nilsson, M.; Adamo, H.; Thysell, E.; Jernberg, E.; Stattin, P.; Widmark, A.; Wikström, P.; Bergh, A. Extratumoral Heme Oxygenase-1 (HO-1) Expressing Macrophages Likely Promote Primary and Metastatic Prostate Tumor Growth. *PLoS ONE* **2016**, *11*, e0157280. [CrossRef]

118. Tsai, J.R.; Wang, H.M.; Liu, P.L.; Chen, Y.H.; Yang, M.C.; Chou, S.H.; Cheng, Y.J.; Yin, W.H.; Hwang, J.J.; Chong, I.W. High Expression of Heme Oxygenase-1 Is Associated with Tumor Invasiveness and Poor Clinical Outcome in Non-Small Cell Lung Cancer Patients. *Cell Oncol.* **2012**, *35*, 461–471. [CrossRef]

119. Wang, X.; Ye, T.; Xue, B.; Yang, M.; Li, R.; Xu, X.; Zeng, X.; Tian, N.; Bao, L.; Huang, Y. Mitochondrial GRIM-19 Deficiency Facilitates Gastric Cancer Metastasis through Oncogenic ROS-NRF2-HO-1 Axis via a NRF2-HO-1 Loop. *Gastric. Cancer* **2020**. [CrossRef]

120. Zhao, Z.; Zhao, J.; Xue, J.; Zhao, X.; Liu, P. Autophagy Inhibition Promotes Epithelial-Mesenchymal Transition through ROS/HO-1 Pathway in Ovarian Cancer Cells. *Am. J. Cancer Res.* **2016**, *6*, 2162–2177.

121. Chang, Y.J.; Chen, W.Y.; Huang, C.Y.; Liu, H.H.; Wei, P.L. Glucose-Regulated Protein 78 (GRP78) Regulates Colon Cancer Metastasis through EMT Biomarkers and the NRF-2/HO-1 Pathway. *Tumour Biol.* **2015**, *36*, 1859–1869. [CrossRef]

122. Jozkowicz, A.; Huk, I.; Nigisch, A.; Weigel, G.; Weidinger, F.; Dulak, J. Effect of Prostaglandin-J(2) on VEGF Synthesis Depends on the Induction of Heme Oxygenase-1. *Antioxid. Redox Signal.* **2002**, *4*, 577–585. [CrossRef] [PubMed]

123. Miyake, M.; Fujimoto, K.; Anai, S.; Ohnishi, S.; Kuwada, M.; Nakai, Y.; Inoue, T.; Matsumura, Y.; Tomioka, A.; Ikeda, T.; et al. Heme Oxygenase-1 Promotes Angiogenesis in Urothelial Carcinoma of the Urinary Bladder. *Oncol. Rep.* **2011**, *25*, 653–660. [CrossRef]

124. Kozakowska, M.; Dobrowolska-Glazar, B.; Okoń, K.; Józkowicz, A.; Dobrowolski, Z.; Dulak, J. Preliminary Analysis of the Expression of Selected Proangiogenic and Antioxidant Genes and MicroRNAs in Patients with Non-Muscle-Invasive Bladder Cancer. *J. Clin. Med.* **2016**, *5*, 29. [CrossRef] [PubMed]

125. Shang, F.T.; Hui, L.L.; An, X.S.; Zhang, X.C.; Guo, S.G.; Kui, Z. ZnPPIX Inhibits Peritoneal Metastasis of Gastric Cancer via Its Antiangiogenic Activity. *Biomed. Pharmacother.* **2015**, *71*, 240–246. [CrossRef] [PubMed]

126. Cheng, C.C.; Guan, S.S.; Yang, H.J.; Chang, C.C.; Luo, T.Y.; Chang, J.; Ho, A.S. Blocking Heme Oxygenase-1 by Zinc Protoporphyrin Reduces Tumor Hypoxia-Mediated VEGF Release and Inhibits Tumor Angiogenesis as a Potential Therapeutic Agent against Colorectal Cancer. *J. Biomed. Sci.* **2016**, *23*, 18. [CrossRef] [PubMed]

127. Liu, R.; Peng, J.; Wang, H.; Li, L.; Wen, X.; Tan, Y.; Zhang, L.; Wan, H.; Chen, F.; Nie, X. Oxysophocarpine Retards the Growth and Metastasis of Oral Squamous Cell Carcinoma by Targeting the Nrf2/HO-1 Axis. *Cell Physiol. Biochem.* **2018**, *49*, 1717–1733. [CrossRef]

128. Hao, N.B.; Lu, M.H.; Fan, Y.H.; Cao, Y.L.; Zhang, Z.R.; Yang, S.M. Macrophages in Tumor Microenvironments and the Progression of Tumors. *Clin. Dev. Immunol.* **2012**, *2012*, 948098. [CrossRef]

129. Blancou, P.; Anegon, I. Editorial: Heme Oxygenase-1 and Dendritic Cells: What Else? *J. Leukoc. Biol.* **2010**, *87*, 185–187. [CrossRef]

130. Alaluf, E.; Vokaer, B.; Detavernier, A.; Azouz, A.; Splittgerber, M.; Carrette, A.; Boon, L.; Libert, F.; Soares, M.; Le Moine, A.; et al. Heme Oxygenase-1 Orchestrates the Immunosuppressive Program of Tumor-Associated Macrophages. *JCI Insight* **2020**, *5*. [CrossRef]

131. Nemeth, Z.; Li, M.; Csizmadia, E.; Döme, B.; Johansson, M.; Persson, J.L.; Seth, P.; Otterbein, L.; Wegiel, B. Heme Oxygenase-1 in Macrophages Controls Prostate Cancer Progression. *Oncotarget* **2015**, *6*, 33675–33688. [CrossRef] [PubMed]

132. Deng, R.; Wang, S.M.; Yin, T.; Ye, T.H.; Shen, G.B.; Li, L.; Zhao, J.Y.; Sang, Y.X.; Duan, X.G.; Wei, Y.Q. Inhibition of Tumor Growth and Alteration of Associated Macrophage Cell Type by an HO-1 Inhibitor in Breast Carcinoma-Bearing Mice. *Oncol. Res.* **2013**, *20*, 473–482. [CrossRef]

133. Beyer, M.; Schultze, J.L. Regulatory T Cells in Cancer. *Blood* **2006**, *108*, 804–811. [CrossRef]

134. Chattopadhyay, S.; Chakraborty, N.G.; Mukherji, B. Regulatory T Cells and Tumor Immunity. *Cancer Immunol. Immunother. CII* **2005**, *54*, 1153–1161. [CrossRef] [PubMed]

135. Kohno, T.; Yamada, Y.; Akamatsu, N.; Kamihira, S.; Imaizumi, Y.; Tomonaga, M.; Matsuyama, T. Possible Origin of Adult T-Cell Leukemia/Lymphoma Cells from Human T Lymphotropic Virus Type-1-Infected Regulatory T Cells. *Cancer Sci.* **2005**, *96*, 527–533. [CrossRef]

136. Di Biase, S.; Lee, C.; Brandhorst, S.; Manes, B.; Buono, R.; Cheng, C.W.; Cacciottolo, M.; Martin-Montalvo, A.; de Cabo, R.; Wei, M.; et al. Fasting-Mimicking Diet Reduces HO-1 to Promote T Cell-Mediated Tumor Cytotoxicity. *Cancer Cell* **2016**, *30*, 136–146. [CrossRef]

137. El Andaloussi, A.; Lesniak, M.S. CD4+ CD25+ FoxP3+ T-Cell Infiltration and Heme Oxygenase-1 Expression Correlate with Tumor Grade in Human Gliomas. *J. Neuro-Oncol.* **2007**, *83*, 145–152. [CrossRef]

138. Orange, J.S. Formation and Function of the Lytic NK-Cell Immunological Synapse. *Nat. Rev. Immunol.* **2008**, *8*, 713–725. [CrossRef]

139. Gomez-Lomeli, P.; Bravo-Cuellar, A.; Hernandez-Flores, G.; Jave-Suarez, L.F.; Aguilar-Lemarroy, A.; Lerma-Diaz, J.M.; Dominguez-Rodriguez, J.R.; Sanchez-Reyes, K.; Ortiz-Lazareno, P.C. Increase of IFN-Gamma and TNF-Alpha Production in CD107a + NK-92 Cells Co-Cultured with Cervical Cancer Cell Lines Pre-Treated with the HO-1 Inhibitor. *Cancer Cell Int.* **2014**, *14*, 100. [CrossRef] [PubMed]

140. Furfaro, A.L.; Ottonello, S.; Loi, G.; Cossu, I.; Piras, S.; Spagnolo, F.; Queirolo, P.; Marinari, U.M.; Moretta, L.; Pronzato, M.A.; et al. HO-1 Downregulation Favors BRAFV600 Melanoma Cell Death Induced by Vemurafenib/PLX4032 and Increases NK Recognition. *Int. J. Cancer* **2020**, *146*, 1950–1962. [CrossRef]

141. Wang, Y.; Qi, H.; Liu, Y.; Duan, C.; Liu, X.; Xia, T.; Chen, D.; Piao, H.-L.; Liu, H.-X. The Double-Edged Roles of ROS in Cancer Prevention and Therapy. *Theranostics* **2021**, *11*, 4839–4857. [CrossRef]

142. Aggarwal, V.; Tuli, H.S.; Varol, A.; Thakral, F.; Yerer, M.B.; Sak, K.; Varol, M.; Jain, A.; Khan, M.A.; Sethi, G. Role of Reactive Oxygen Species in Cancer Progression: Molecular Mechanisms and Recent Advancements. *Biomolecules* **2019**, *9*, 735. [CrossRef]

143. Liu, Y.-S.; Li, H.-S.; Qi, D.-F.; Zhang, J.; Jiang, X.-C.; Shi, K.; Zhang, X.-J.; Zhang, X.-H. Zinc Protoporphyrin IX Enhances Chemotherapeutic Response of Hepatoma Cells to Cisplatin. *World J. Gastroenterol.* **2014**, *20*, 8572–8582. [CrossRef]

144. Sun, X.; Wang, S.; Gai, J.; Guan, J.; Li, J.; Li, Y.; Zhao, J.; Zhao, C.; Fu, L.; Li, Q. SIRT5 Promotes Cisplatin Resistance in Ovarian Cancer by Suppressing DNA Damage in a ROS-Dependent Manner via Regulation of the Nrf2/HO-1 Pathway. *Front. Oncol.* **2019**, *9*, 754. [CrossRef] [PubMed]

145. Furfaro, A.L.; Macay, J.R.; Marengo, B.; Nitti, M.; Parodi, A.; Fenoglio, D.; Marinari, U.M.; Pronzato, M.A.; Domenicotti, C.; Traverso, N. Resistance of Neuroblastoma GI-ME-N Cell Line to Glutathione Depletion Involves Nrf2 and Heme Oxygenase-1. *Free Radic. Biol. Med.* **2012**, *52*, 488–496. [CrossRef] [PubMed]

146. Tan, Q.; Wang, H.; Hu, Y.; Hu, M.; Li, X.; Aodengqimuge; Ma, Y.; Wei, C.; Song, L. Src/STAT3-Dependent Heme Oxygenase-1 Induction Mediates Chemoresistance of Breast Cancer Cells to Doxorubicin by Promoting Autophagy. *Cancer Sci.* **2015**, *106*, 1023–1032. [CrossRef] [PubMed]

147. Pei, L.; Kong, Y.; Shao, C.; Yue, X.; Wang, Z.; Zhang, N. Heme Oxygenase-1 Induction Mediates Chemoresistance of Breast Cancer Cells to Pharmorubicin by Promoting Autophagy via PI3K/Akt Pathway. *J. Cell Mol. Med.* **2018**, *22*, 5311–5321. [CrossRef]

148. Furfaro, A.L.; Piras, S.; Passalacqua, M.; Domenicotti, C.; Parodi, A.; Fenoglio, D.; Pronzato, M.A.; Marinari, U.M.; Moretta, L.; Traverso, N.; et al. HO-1 up-Regulation: A Key Point in High-Risk Neuroblastoma Resistance to Bortezomib. *Biochim. Biophys. Acta* **2014**, *1842*, 613–622. [CrossRef] [PubMed]

149. Furfaro, A.L.; Piras, S.; Domenicotti, C.; Fenoglio, D.; De Luigi, A.; Salmona, M.; Moretta, L.; Marinari, U.M.; Pronzato, M.A.; Traverso, N.; et al. Role of Nrf2, HO-1 and GSH in Neuroblastoma Cell Resistance to Bortezomib. *PLoS ONE* **2016**, *11*, e0152465. [CrossRef]

150. Barbagallo, I.; Giallongo, C.; Volti, G.L.; Distefano, A.; Camiolo, G.; Raffaele, M.; Salerno, L.; Pittalà, V.; Sorrenti, V.; Avola, R.; et al. Heme Oxygenase Inhibition Sensitizes Neuroblastoma Cells to Carfilzomib. *Mol. Neurobiol.* **2019**, *56*, 1451–1460. [CrossRef] [PubMed]

151. Abdalla, M.Y.; Ahmad, I.M.; Rachagani, S.; Banerjee, K.; Thompson, C.M.; Maurer, H.C.; Olive, K.P.; Bailey, K.L.; Britigan, B.E.; Kumar, S. Enhancing Responsiveness of Pancreatic Cancer Cells to Gemcitabine Treatment under Hypoxia by Heme Oxygenase-1 Inhibition. *Transl. Res.* **2019**, *207*, 56–69. [CrossRef]

152. Ma, J.; Yu, K.N.; Cheng, C.; Ni, G.; Shen, J.; Han, W. Targeting Nrf2-Mediated Heme Oxygenase-1 Enhances Non-Thermal Plasma-Induced Cell Death in Non-Small-Cell Lung Cancer A549 Cells. *Arch. Biochem. Biophys.* **2018**, *658*, 54–65. [CrossRef]

153. Zhang, W.; Qiao, T.; Zha, L. Inhibition of Heme Oxygenase-1 Enhances the Radiosensitivity in Human Nonsmall Cell Lung Cancer A549 Cells. *Cancer Biother. Radiopharm.* **2011**, *26*, 639–645. [CrossRef] [PubMed]

154. Chen, N.; Wu, L.; Yuan, H.; Wang, J. ROS/Autophagy/Nrf2 Pathway Mediated Low-Dose Radiation Induced Radio-Resistance in Human Lung Adenocarcinoma A549 Cell. *Int. J. Biol. Sci.* **2015**, *11*, 833–844. [CrossRef]

155. Wang, R.; Shen, J.; Yang, R.; Wang, W.-G.; Yuan, Y.; Guo, Z.-H. Association between Heme Oxygenase-1 Gene Promoter Polymorphisms and Cancer Susceptibility: A Meta-Analysis. *Biomed. Rep.* **2018**, *8*, 241–248. [CrossRef] [PubMed]

156. Hu, J.L.; Li, Z.Y.; Liu, W.; Zhang, R.G.; Li, G.L.; Wang, T.; Ren, J.H.; Wu, G. Polymorphism in Heme Oxygenase-1 (HO-1) Promoter and Alcohol Are Related to the Risk of Esophageal Squamous Cell Carcinoma on Chinese Males. *Neoplasma* **2010**, *57*, 86–92. [CrossRef]

157. Kikuchi, A.; Yamaya, M.; Suzuki, S.; Yasuda, H.; Kubo, H.; Nakayama, K.; Handa, M.; Sasaki, T.; Shibahara, S.; Sekizawa, K.; et al. Association of Susceptibility to the Development of Lung Adenocarcinoma with the Heme Oxygenase-1 Gene Promoter Polymorphism. *Hum. Genet.* **2005**, *116*, 354–360. [CrossRef]

158. Murakami, A.; Fujimori, Y.; Yoshikawa, Y.; Yamada, S.; Tamura, K.; Hirayama, N.; Terada, T.; Kuribayashi, K.; Tabata, C.; Fukuoka, K.; et al. Heme Oxygenase-1 Promoter Polymorphism Is Associated with Risk of Malignant Mesothelioma. *Lung* **2012**, *190*, 333–337. [CrossRef]

159. Wu, M.-M.; Lee, C.-H.; Hsu, L.-I.; Cheng, W.-F.; Lee, T.-C.; Wang, Y.-H.; Chiou, H.-Y.; Chen, C.-J. Effect of Heme Oxygenase-1 Gene Promoter Polymorphism on Cancer Risk by Histological Subtype: A Prospective Study in Arseniasis-Endemic Areas in Taiwan. *Int. J. Cancer* **2016**, *138*, 1875–1886. [CrossRef] [PubMed]

160. Zhao, Z.; Xu, Y.; Lu, J.; Xue, J.; Liu, P. High Expression of HO-1 Predicts Poor Prognosis of Ovarian Cancer Patients and Promotes Proliferation and Aggressiveness of Ovarian Cancer Cells. *Clin. Transl. Oncol.* **2018**, *20*, 491–499. [CrossRef]

161. Miyake, M.; Fujimoto, K.; Anai, S.; Ohnishi, S.; Nakai, Y.; Inoue, T.; Matsumura, Y.; Tomioka, A.; Ikeda, T.; Okajima, E.; et al. Inhibition of Heme Oxygenase-1 Enhances the Cytotoxic Effect of Gemcitabine in Urothelial Cancer Cells. *Anticancer Res.* **2010**, *30*, 2145–2152.

162. Kim, J.H.; Park, J. Prognostic Significance of Heme Oxygenase-1, S100 Calcium-Binding Protein A4, and Syndecan-1 Expression in Primary Non-Muscle-Invasive Bladder Cancer. *Hum. Pathol.* **2014**, *45*, 1830–1838. [CrossRef]

163. Gandini, N.A.; Fermento, M.E.; Salomon, D.G.; Obiol, D.J.; Andres, N.C.; Zenklusen, J.C.; Arevalo, J.; Blasco, J.; Lopez Romero, A.; Facchinetti, M.M.; et al. Heme Oxygenase-1 Expression in Human Gliomas and Its Correlation with Poor Prognosis in Patients with Astrocytoma. *Tumour Biol.* **2014**, *35*, 2803–2815. [CrossRef] [PubMed]

164. Wang, J.; Zhang, M.; Zhang, L.; Cai, H.; Zhou, S.; Zhang, J.; Wang, Y. Correlation of Nrf2, HO-1, and MRP3 in Gallbladder Cancer and Their Relationships to Clinicopathologic Features and Survival. *J. Surg. Res.* **2010**, *164*, e99–e105. [CrossRef]

165. Deng, Y.; Wu, Y.; Zhao, P.; Weng, W.; Ye, M.; Sun, H.; Xu, M.; Wang, C. The Nrf2/HO-1 Axis Can Be a Prognostic Factor in Clear Cell Renal Cell Carcinoma. *Cancer Manag. Res.* **2019**, *11*, 1221–1230. [CrossRef]

166. Kongpetch, S.; Puapairoj, A.; Ong, C.K.; Senggunprai, L.; Prawan, A.; Kukongviriyapan, U.; Chan-On, W.; Siew, E.Y.; Khuntikeo, N.; Teh, B.T.; et al. Haem Oxygenase 1 Expression Is Associated with Prognosis in Cholangiocarcinoma Patients and with Drug Sensitivity in Xenografted Mice. *Cell Prolif.* **2016**, *49*, 90–101. [CrossRef] [PubMed]

167. Yu, M.; Wang, J.; Ma, D.; Chen, S.; Lin, X.; Fang, Q.; Zhe, N. HO-1, RET and PML as Possible Markers for Risk Stratification of Acute Myelocytic Leukemia and Prognostic Evaluation. *Oncol. Lett.* **2015**, *10*, 3137–3144. [CrossRef] [PubMed]

168. Fest, S.; Soldati, R.; Christiansen, N.M.; Zenclussen, M.L.; Kilz, J.; Berger, E.; Starke, S.; Lode, H.N.; Engel, C.; Zenclussen, A.C.; et al. Targeting of Heme Oxygenase-1 as a Novel Immune Regulator of Neuroblastoma. *Int. J. Cancer* **2016**, *138*, 2030–2042. [CrossRef]

169. Wei, S.; Wang, Y.; Chai, Q.; Fang, Q.; Zhang, Y.; Lu, Y.; Wang, J. Over-Expression of Heme Oxygenase-1 in Peripheral Blood Predicts the Progression and Relapse Risk of Chronic Myeloid Leukemia. *Chin. Med. J.* **2014**, *127*, 2795–2801. [PubMed]

170. Zhe, N.; Wang, J.; Chen, S.; Lin, X.; Chai, Q.; Zhang, Y.; Zhao, J.; Fang, Q. Heme Oxygenase-1 Plays a Crucial Role in Chemoresistance in Acute Myeloid Leukemia. *Hematology* **2015**, *20*, 384–391. [CrossRef]

171. He, Z.; Zhang, S.; Ma, D.; Fang, Q.; Yang, L.; Shen, S.; Chen, Y.; Ren, L.; Wang, J. HO-1 Promotes Resistance to an EZH2 Inhibitor through the PRB-E2F Pathway: Correlation with the Progression of Myelodysplastic Syndrome into Acute Myeloid Leukemia. *J. Transl. Med.* **2019**, *17*, 366. [CrossRef] [PubMed]

172. Wang, T.-Y.; Liu, C.-L.; Chen, M.-J.; Lee, J.-J.; Pun, P.C.; Cheng, S.-P. Expression of Haem Oxygenase-1 Correlates with Tumour Aggressiveness and BRAF V600E Expression in Thyroid Cancer. *Histopathology* **2015**, *66*, 447–456. [CrossRef] [PubMed]

173. Li, Y.; Su, J.; DingZhang, X.; Zhang, J.; Yoshimoto, M.; Liu, S.; Bijian, K.; Gupta, A.; Squire, J.A.; Alaoui Jamali, M.A.; et al. PTEN Deletion and Heme Oxygenase-1 Overexpression Cooperate in Prostate Cancer Progression and Are Associated with Adverse Clinical Outcome. *J. Pathol.* **2011**, *224*, 90–100. [CrossRef]

174. Xu, Y.; Yang, Y.; Huang, Y.; Ma, Q.; Shang, J.; Guo, J.; Cao, X.; Wang, X.; Li, M. Inhibition of Nrf2/HO-1 Signaling Pathway by Dextran Sulfate Suppresses Angiogenesis of Gastric Cancer. *J. Cancer* **2021**, *12*, 1042–1060. [CrossRef] [PubMed]

175. Park, C.-S.; Eom, D.-W.; Ahn, Y.; Jang, H.J.; Hwang, S.; Lee, S.-G. Can Heme Oxygenase-1 Be a Prognostic Factor in Patients with Hepatocellular Carcinoma? *Medicine* **2019**, *98*, e16084. [CrossRef] [PubMed]

176. Liu, L.; Wu, Y.; Bian, C.; Nisar, M.F.; Wang, M.; Hu, X.; Diao, Q.; Nian, W.; Wang, E.; Xu, W.; et al. Heme Oxygenase 1 Facilitates Cell Proliferation via the B-Raf-ERK Signaling Pathway in Melanoma. *Cell Commun. Signal.* **2019**, *17*, 3. [CrossRef] [PubMed]

177. Matsuo, T.; Miyata, Y.; Mitsunari, K.; Yasuda, T.; Ohba, K.; Sakai, H. Pathological Significance and Prognostic Implications of Heme Oxygenase 1 Expression in Non-Muscle-Invasive Bladder Cancer: Correlation with Cell Proliferation, Angiogenesis, Lymphangiogenesis and Expression of VEGFs and COX-2. *Oncol. Lett.* **2017**, *13*, 275–280. [CrossRef]

178. Nishie, A.; Ono, M.; Shono, T.; Fukushi, J.; Otsubo, M.; Onoue, H.; Ito, Y.; Inamura, T.; Ikezaki, K.; Fukui, M.; et al. Macrophage Infiltration and Heme Oxygenase-1 Expression Correlate with Angiogenesis in Human Gliomas. *Clin. Cancer Res.* **1999**, *5*, 1107–1113.

179. Kimura, S.; Aung, N.Y.; Ohe, R.; Yano, M.; Hashimoto, T.; Fujishima, T.; Kimura, W.; Yamakawa, M. Increasing Heme Oxygenase-1-Expressing Macrophages Indicates a Tendency of Poor Prognosis in Advanced Colorectal Cancer. *Digestion* **2020**, *101*, 401–410. [CrossRef]

180. Becker, J.C.; Fukui, H.; Imai, Y.; Sekikawa, A.; Kimura, T.; Yamagishi, H.; Yoshitake, N.; Pohle, T.; Domschke, W.; Fujimori, T. Colonic Expression of Heme Oxygenase-1 Is Associated with a Better Long-Term Survival in Patients with Colorectal Cancer. *Scand. J. Gastroenterol.* **2007**, *42*, 852–858. [CrossRef]

181. Andrés, N.C.; Fermento, M.E.; Gandini, N.A.; Romero, A.L.; Ferro, A.; Donna, L.G.; Curino, A.C.; Facchinetti, M.M. Heme Oxygenase-1 Has Antitumoral Effects in Colorectal Cancer: Involvement of P53. *Exp. Mol. Pathol.* **2014**, *97*, 321–331. [CrossRef]

182. Yin, Y.; Liu, Q.; Wang, B.; Chen, G.; Xu, L.; Zhou, H. Expression and Function of Heme Oxygenase-1 in Human Gastric Cancer. *Exp Biol. Med.* **2012**, *237*, 362–371. [CrossRef] [PubMed]

183. Jun, S.-Y.; Hong, S.-M.; Bae, Y.K.; Kim, H.K.; Jang, K.Y.; Eom, D.W. Clinicopathological and Prognostic Significance of Heme Oxygenase-1 Expression in Small Intestinal Adenocarcinomas. *Pathol. Int.* **2018**, *68*, 294–300. [CrossRef]

184. Tsuji, M.H.; Yanagawa, T.; Iwasa, S.; Tabuchi, K.; Onizawa, K.; Bannai, S.; Toyooka, H.; Yoshida, H. Heme Oxygenase-1 Expression in Oral Squamous Cell Carcinoma as Involved in Lymph Node Metastasis. *Cancer Lett.* **1999**, *138*, 53–59. [CrossRef]

185. Gandini, N.A.; Fermento, M.E.; Salomon, D.G.; Blasco, J.; Patel, V.; Gutkind, J.S.; Molinolo, A.A.; Facchinetti, M.M.; Curino, A.C. Nuclear Localization of Heme Oxygenase-1 Is Associated with Tumor Progression of Head and Neck Squamous Cell Carcinomas. *Exp. Mol. Pathol.* **2012**, *93*, 237–245. [CrossRef]

186. Gandini, N.A.; Alonso, E.N.; Fermento, M.E.; Mascaró, M.; Abba, M.C.; Coló, G.P.; Arévalo, J.; Ferronato, M.J.; Guevara, J.A.; Núñez, M.; et al. Heme Oxygenase-1 Has an Antitumor Role in Breast Cancer. *Antioxid. Redox Signal.* **2019**, *30*, 2030–2049. [CrossRef] [PubMed]

187. Yin, H.; Fang, J.; Liao, L.; Maeda, H.; Su, Q. Upregulation of Heme Oxygenase-1 in Colorectal Cancer Patients with Increased Circulation Carbon Monoxide Levels, Potentially Affects Chemotherapeutic Sensitivity. *BMC Cancer* **2014**, *14*, 436. [CrossRef]

188. Alaoui-Jamali, M.A.; Bismar, T.A.; Gupta, A.; Szarek, W.A.; Su, J.; Song, W.; Xu, Y.; Xu, B.; Liu, G.; Vlahakis, J.Z.; et al. A Novel Experimental Heme Oxygenase-1-Targeted Therapy for Hormone-Refractory Prostate Cancer. *Cancer Res.* **2009**, *69*, 8017–8024. [CrossRef] [PubMed]

189. Li, C.; Wu, H.; Wang, S.; Zhu, J. Expression and Correlation of NRF2, KEAP1, NQO-1 and HO-1 in Advanced Squamous Cell Carcinoma of the Larynx and Their Association with Clinicopathologic Features. *Mol. Med. Rep.* **2016**, *14*, 5171–5179. [CrossRef] [PubMed]

190. Wei, D.; Lu, T.; Ma, D.; Yu, K.; Li, X.; Chen, B.; Xiong, J.; Zhang, T.; Wang, J. Heme Oxygenase-1 Reduces the Sensitivity to Imatinib through Nonselective Activation of Histone Deacetylases in Chronic Myeloid Leukemia. *J. Cell Physiol.* **2019**, *234*, 5252–5263. [CrossRef]

191. Cheng, B.; Tang, S.; Zhe, N.; Ma, D.; Yu, K.; Wei, D.; Zhou, Z.; Lu, T.; Wang, J.; Fang, Q. Low Expression of GFI-1 Gene Is Associated with Panobinostat-Resistance in Acute Myeloid Leukemia through Influencing the Level of HO-1. *Biomed. Pharmacother.* **2018**, *100*, 509–520. [CrossRef] [PubMed]

192. Lin, X.; Fang, Q.; Chen, S.; Zhe, N.; Chai, Q.; Yu, M.; Zhang, Y.; Wang, Z.; Wang, J. Heme Oxygenase-1 Suppresses the Apoptosis of Acute Myeloid Leukemia Cells via the JNK/c-JUN Signaling Pathway. *Leuk. Res.* **2015**, *39*, 544–552. [CrossRef] [PubMed]

193. Ma, D.; Fang, Q.; Wang, P.; Gao, R.; Sun, J.; Li, Y.; Hu, X.Y.; Wang, J.S. Downregulation of HO-1 Promoted Apoptosis Induced by Decitabine via Increasing P15INK4B Promoter Demethylation in Myelodysplastic Syndrome. *Gene Ther.* **2015**, *22*, 287–296. [CrossRef]

194. Barrera, L.N.; Rushworth, S.A.; Bowles, K.M.; MacEwan, D.J. Bortezomib Induces Heme Oxygenase-1 Expression in Multiple Myeloma. *Cell Cycle* **2012**, *11*, 2248–2252. [CrossRef]

195. Podkalicka, P.; Mucha, O.; Józkowicz, A.; Dulak, J.; Łoboda, A. Heme Oxygenase Inhibition in Cancers: Possible Tools and Targets. *Contemp. Oncol.* **2018**, *22*, 23–32. [CrossRef] [PubMed]

196. Vreman, H.J.; Ekstrand, B.C.; Stevenson, D.K. Selection of Metalloporphyrin Heme Oxygenase Inhibitors Based on Potency and Photoreactivity. *Pediatr. Res.* **1993**, *33*, 195–200. [CrossRef]

197. Schulz, S.; Wong, R.J.; Vreman, H.J.; Stevenson, D.K. Metalloporphyrins—An Update. *Front. Pharmacol.* **2012**, *3*, 68. [CrossRef]

198. Shi, L.; Fang, J. Implication of Heme Oxygenase-1 in the Sensitivity of Nasopharyngeal Carcinomas to Radiotherapy. *J. Exp. Clin. Cancer Res.* **2008**, *27*, 13. [CrossRef] [PubMed]

199. Nowis, D.; Legat, M.; Grzela, T.; Niderla, J.; Wilczek, E.; Wilczynski, G.M.; Glodkowska, E.; Mrowka, P.; Issat, T.; Dulak, J.; et al. Heme Oxygenase-1 Protects Tumor Cells against Photodynamic Therapy-Mediated Cytotoxicity. *Oncogene* **2006**, *25*, 3365–3374. [CrossRef] [PubMed]

200. Frank, J.; Lornejad-Schäfer, M.R.; Schöffl, H.; Flaccus, A.; Lambert, C.; Biesalski, H.K. Inhibition of Heme Oxygenase-1 Increases Responsiveness of Melanoma Cells to ALA-Based Photodynamic Therapy. *Int. J. Oncol.* **2007**, *31*, 1539–1545. [CrossRef]

201. Appleton, S.D.; Chretien, M.L.; McLaughlin, B.E.; Vreman, H.J.; Stevenson, D.K.; Brien, J.F.; Nakatsu, K.; Maurice, D.H.; Marks, G.S. Selective Inhibition of Heme Oxygenase, without Inhibition of Nitric Oxide Synthase or Soluble Guanylyl Cyclase, by Metalloporphyrins at Low Concentrations. *Drug Metab. Dispos.* **1999**, *27*, 1214–1219.

202. Kinobe, R.T.; Dercho, R.A.; Nakatsu, K. Inhibitors of the Heme Oxygenase-Carbon Monoxide System: On the Doorstep of the Clinic? *Can. J. Physiol. Pharmacol.* **2008**, *86*, 577–599. [CrossRef] [PubMed]

203. Sardana, M.K.; Kappas, A. Dual Control Mechanism for Heme Oxygenase: Tin(IV)-Protoporphyrin Potently Inhibits Enzyme Activity While Markedly Increasing Content of Enzyme Protein in Liver. *Proc. Natl. Acad. Sci. USA* **1987**, *84*, 2464–2468. [CrossRef] [PubMed]

204. Yang, G.; Nguyen, X.; Ou, J.; Rekulapelli, P.; Stevenson, D.K.; Dennery, P.A. Unique Effects of Zinc Protoporphyrin on HO-1 Induction and Apoptosis. *Blood* **2001**, *97*, 1306–1313. [CrossRef]

205. Kwok, S.C.M. Zinc Protoporphyrin Upregulates Heme Oxygenase-1 in PC-3 Cells via the Stress Response Pathway. *Int. J. Cell Biol.* **2013**, *2013*, 162094. [CrossRef]

206. Sahoo, S.K.; Sawa, T.; Fang, J.; Tanaka, S.; Miyamoto, Y.; Akaike, T.; Maeda, H. Pegylated Zinc Protoporphyrin: A Water-Soluble Heme Oxygenase Inhibitor with Tumor-Targeting Capacity. *Bioconjugate Chem.* **2002**, *13*, 1031–1038. [CrossRef]

207. Iyer, A.K.; Greish, K.; Seki, T.; Okazaki, S.; Fang, J.; Takeshita, K.; Maeda, H. Polymeric Micelles of Zinc Protoporphyrin for Tumor Targeted Delivery Based on EPR Effect and Singlet Oxygen Generation. *J. Drug Target.* **2007**, *15*, 496–506. [CrossRef] [PubMed]

208. Iyer, A.K.; Greish, K.; Fang, J.; Murakami, R.; Maeda, H. High-Loading Nanosized Micelles of Copoly(Styrene-Maleic Acid)-Zinc Protoporphyrin for Targeted Delivery of a Potent Heme Oxygenase Inhibitor. *Biomaterials* **2007**, *28*, 1871–1881. [CrossRef]
209. Fang, J.; Sawa, T.; Akaike, T.; Akuta, T.; Sahoo, S.K.; Khaled, G.; Hamada, A.; Maeda, H. In Vivo Antitumor Activity of Pegylated Zinc Protoporphyrin: Targeted Inhibition of Heme Oxygenase in Solid Tumor. *Cancer Res.* **2003**, *63*, 3567–3574. [PubMed]
210. Herrmann, H.; Kneidinger, M.; Cerny-Reiterer, S.; Rülicke, T.; Willmann, M.; Gleixner, K.V.; Blatt, K.; Hörmann, G.; Peter, B.; Samorapoompichit, P.; et al. The Hsp32 Inhibitors SMA-ZnPP and PEG-ZnPP Exert Major Growth-Inhibitory Effects on D34+/CD38+ and CD34+/CD38- AML Progenitor Cells. *Curr. Cancer Drug Targets* **2012**, *12*, 51–63. [CrossRef]
211. Kinobe, R.T.; Vlahakis, J.Z.; Vreman, H.J.; Stevenson, D.K.; Brien, J.F.; Szarek, W.A.; Nakatsu, K. Selectivity of Imidazole-Dioxolane Compounds for in Vitro Inhibition of Microsomal Haem Oxygenase Isoforms. *Br. J. Pharmacol.* **2006**, *147*, 307–315. [CrossRef]
212. Pittala, V.; Salerno, L.; Romeo, G.; Modica, M.N.; Siracusa, M.A. A Focus on Heme Oxygenase-1 (HO-1) Inhibitors. *Curr. Med. Chem.* **2013**, *20*, 3711–3732. [CrossRef] [PubMed]
213. Vlahakis, J.Z.; Kinobe, R.T.; Bowers, R.J.; Brien, J.F.; Nakatsu, K.; Szarek, W.A. Synthesis and Evaluation of Azalanstat Analogues as Heme Oxygenase Inhibitors. *Bioorg. Med. Chem. Lett.* **2005**, *15*, 1457–1461. [CrossRef]
214. Salerno, L.; Floresta, G.; Ciaffaglione, V.; Gentile, D.; Margani, F.; Turnaturi, R.; Rescifina, A.; Pittalà, V. Progress in the Development of Selective Heme Oxygenase-1 Inhibitors and Their Potential Therapeutic Application. *Eur. J. Med. Chem.* **2019**, *167*, 439–453. [CrossRef] [PubMed]
215. Ciaffaglione, V.; Intagliata, S.; Pittalà, V.; Marrazzo, A.; Sorrenti, V.; Vanella, L.; Rescifina, A.; Floresta, G.; Sultan, A.; Greish, K.; et al. New Arylethanolimidazole Derivatives as HO-1 Inhibitors with Cytotoxicity against MCF-7 Breast Cancer Cells. *Int. J. Mol. Sci.* **2020**, *21*, 1923. [CrossRef]
216. Salerno, L.; Pittalà, V.; Romeo, G.; Modica, M.N.; Marrazzo, A.; Siracusa, M.A.; Sorrenti, V.; Di Giacomo, C.; Vanella, L.; Parayath, N.N.; et al. Novel Imidazole Derivatives as Heme Oxygenase-1 (HO-1) and Heme Oxygenase-2 (HO-2) Inhibitors and Their Cytotoxic Activity in Human-Derived Cancer Cell Lines. *Eur. J. Med. Chem.* **2015**, *96*, 162–172. [CrossRef] [PubMed]
217. Berberat, P.O.; Dambrauskas, Z.; Gulbinas, A.; Giese, T.; Giese, N.; Kunzli, B.; Autschbach, F.; Meuer, S.; Buchler, M.W.; Friess, H. Inhibition of Heme Oxygenase-1 Increases Responsiveness of Pancreatic Cancer Cells to Anticancer Treatment. *Clin. Cancer Res.* **2005**, *11*, 3790–3798. [CrossRef]
218. Sass, G.; Leukel, P.; Schmitz, V.; Raskopf, E.; Ocker, M.; Neureiter, D.; Meissnitzer, M.; Tasika, E.; Tannapfel, A.; Tiegs, G. Inhibition of Heme Oxygenase 1 Expression by Small Interfering RNA Decreases Orthotopic Tumor Growth in Livers of Mice. *Int. J. Cancer* **2008**, *123*, 1269–1277. [CrossRef] [PubMed]
219. Cerny-Reiterer, S.; Meyer, R.A.; Herrmann, H.; Peter, B.; Gleixner, K.V.; Stefanzl, G.; Hadzijusufovic, E.; Pickl, W.F.; Sperr, W.R.; Melo, J.V.; et al. Identification of Heat Shock Protein 32 (Hsp32) as a Novel Target in Acute Lymphoblastic Leukemia. *Oncotarget* **2014**, *5*, 1198–1211. [CrossRef]
220. Zhou, Z.; Fang, Q.; Li, P.; Ma, D.; Zhe, N.; Ren, M.; Chen, B.; He, Z.; Wang, J.; Zhong, Q.; et al. Entinostat Combined with Fludarabine Synergistically Enhances the Induction of Apoptosis in TP53 Mutated CLL Cells via the HDAC1/HO-1 Pathway. *Life Sci.* **2019**, *232*, 116583. [CrossRef]
221. Zhou, Z.; Ma, D.; Liu, P.; Wang, P.; Wei, D.; Yu, K.; Li, P.; Fang, Q.; Wang, J. Deletion of HO-1 Blocks Development of B Lymphocytes in Mice. *Cell Signal.* **2019**, *63*, 109378. [CrossRef] [PubMed]
222. Skrzypek, K.; Tertil, M.; Golda, S.; Ciesla, M.; Weglarczyk, K.; Collet, G.; Guichard, A.; Kozakowska, M.; Boczkowski, J.; Was, H.; et al. Interplay between Heme Oxygenase-1 and MiR-378 Affects Non-Small Cell Lung Carcinoma Growth, Vascularization, and Metastasis. *Antioxid. Redox Signal.* **2013**, *19*, 644–660. [CrossRef] [PubMed]
223. Tertil, M.; Golda, S.; Skrzypek, K.; Florczyk, U.; Weglarczyk, K.; Kotlinowski, J.; Maleszewska, M.; Czauderna, S.; Pichon, C.; Kieda, C.; et al. Nrf2-Heme Oxygenase-1 Axis in Mucoepidermoid Carcinoma of the Lung: Antitumoral Effects Associated with down-Regulation of Matrix Metalloproteinases. *Free Radic. Biol. Med.* **2015**, *89*, 147–157. [CrossRef] [PubMed]
224. Zou, C.; Cheng, W.; Li, Q.; Han, Z.; Wang, X.; Jin, J.; Zou, J.; Liu, Z.; Zhou, Z.; Zhao, W.; et al. Heme Oxygenase-1 Retards Hepatocellular Carcinoma Progression through the MicroRNA Pathway. *Oncol. Rep.* **2016**, *36*, 2715–2722. [CrossRef] [PubMed]
225. Hill, M.; Pereira, V.; Chauveau, C.; Zagani, R.; Remy, S.; Tesson, L.; Mazal, D.; Ubillos, L.; Brion, R.; Asghar, K.; et al. Heme Oxygenase-1 Inhibits Rat and Human Breast Cancer Cell Proliferation: Mutual Cross Inhibition with Indoleamine 2,3-Dioxygenase. *FASEB J.* **2005**, *19*, 1957–1968. [CrossRef]
226. Gueron, G.; Giudice, J.; Valacco, P.; Paez, A.; Elguero, B.; Toscani, M.; Jaworski, F.; Leskow, F.C.; Cotignola, J.; Marti, M.; et al. Heme-Oxygenase-1 Implications in Cell Morphology and the Adhesive Behavior of Prostate Cancer Cells. *Oncotarget* **2014**, *5*, 4087–4102. [CrossRef] [PubMed]
227. Kim, H.N.; Park, G.H.; Park, S.B.; Kim, J.D.; Eo, H.J.; Son, H.-J.; Song, J.H.; Jeong, J.B. Extracts from Sageretia Thea Reduce Cell Viability through Inducing Cyclin D1 Proteasomal Degradation and HO-1 Expression in Human Colorectal Cancer Cells. *BMC Complement. Altern. Med.* **2019**, *19*, 43. [CrossRef]
228. Bi, W.; He, C.-N.; Li, X.-X.; Zhou, L.-Y.; Liu, R.-J.; Zhang, S.; Li, G.-Q.; Chen, Z.-C.; Zhang, P.-F. Ginnalin A from Kujin Tea (*Acer tataricum* subsp. *ginnala*) Exhibits a Colorectal Cancer Chemoprevention Effect via Activation of the Nrf2/HO-1 Signaling Pathway. *Food Funct.* **2018**, *9*, 2809–2819. [CrossRef]
229. Tsai, C.-F.; Chen, J.-H.; Chang, C.-N.; Lu, D.-Y.; Chang, P.-C.; Wang, S.-L.; Yeh, W.-L. Fisetin Inhibits Cell Migration via Inducing HO-1 and Reducing MMPs Expression in Breast Cancer Cell Lines. *Food Chem. Toxicol.* **2018**, *120*, 528–535. [CrossRef]
230. Kim, M.J.; Yun, G.J.; Kim, S.E. Metabolic Regulation of Ferroptosis in Cancer. *Biology* **2021**, *10*, 83. [CrossRef]

231. Greco, G.; Catanzaro, E.; Fimognari, C. Natural Products as Inducers of Non-Canonical Cell Death: A Weapon against Cancer. *Cancers* **2021**, *13*, 304. [CrossRef] [PubMed]

232. Hassannia, B.; Wiernicki, B.; Ingold, I.; Qu, F.; Van Herck, S.; Tyurina, Y.Y.; Bayır, H.; Abhari, B.A.; Angeli, J.P.F.; Choi, S.M.; et al. Nano-Targeted Induction of Dual Ferroptotic Mechanisms Eradicates High-Risk Neuroblastoma. *J. Clin. Investig.* **2018**, *128*, 3341–3355. [CrossRef]

233. Malfa, G.A.; Tomasello, B.; Acquaviva, R.; Genovese, C.; La Mantia, A.; Cammarata, F.P.; Ragusa, M.; Renis, M.; Di Giacomo, C. Betula Etnensis Raf. (Betulaceae) Extract Induced HO-1 Expression and Ferroptosis Cell Death in Human Colon Cancer Cells. *Int. J. Mol. Sci.* **2019**, *20*, 2723. [CrossRef] [PubMed]

234. Li, R.; Zhang, J.; Zhou, Y.; Gao, Q.; Wang, R.; Fu, Y.; Zheng, L.; Yu, H. Transcriptome Investigation and In Vitro Verification of Curcumin-Induced HO-1 as a Feature of Ferroptosis in Breast Cancer Cells. *Oxid. Med. Cell. Longev.* **2020**, *2020*, 3469840. [CrossRef]

235. Chen, P.; Li, X.; Zhang, R.; Liu, S.; Xiang, Y.; Zhang, M.; Chen, X.; Pan, T.; Yan, L.; Feng, J.; et al. Combinative Treatment of β-Elemene and Cetuximab Is Sensitive to KRAS Mutant Colorectal Cancer Cells by Inducing Ferroptosis and Inhibiting Epithelial-Mesenchymal Transformation. *Theranostics* **2020**, *10*, 5107–5119. [CrossRef] [PubMed]

236. Zager, R.A.; Johnson, A.C.M.; Guillem, A.; Keyser, J.; Singh, B. A Pharmacologic "Stress Test" for Assessing Select Antioxidant Defenses in Patients with CKD. *Clin. J. Am. Soc. Nephrol.* **2020**, *15*, 633–642. [CrossRef]

237. Bhutani, V.K.; Poland, R.; Meloy, L.D.; Hegyi, T.; Fanaroff, A.A.; Maisels, M.J. Clinical Trial of Tin Mesoporphyrin to Prevent Neonatal Hyperbilirubinemia. *J. Perinatol.* **2016**, *36*, 533–539. [CrossRef]

Perspective

Opportunities for Ferroptosis in Cancer Therapy

Kenji M. Fujihara [1,2,*], Bonnie Z. Zhang [1,2] and Nicholas J. Clemons [1,2,*]

[1] Gastrointestinal Cancer Program, Cancer Research Division, Peter MacCallum Cancer Centre, Melbourne, VIC 3000, Australia; bonnie.zhang@unimelb.edu.au
[2] Sir Peter MacCallum Department of Oncology, The University of Melbourne, Parkville, VIC 3010, Australia
* Correspondence: authors: kenji.fujihara@petermac.org (K.M.F.); nicholas.clemons@petermac.org (N.J.C.); Tel.: +61-3-8559-8294 (K.M.F.); +61-3-8559-5273 (N.J.C.)

Abstract: A critical hallmark of cancer cells is their ability to evade programmed apoptotic cell death. Consequently, resistance to anti-cancer therapeutics is a hurdle often observed in the clinic. Ferroptosis, a non-apoptotic form of cell death distinguished by toxic lipid peroxidation and iron accumulation, has garnered substantial attention as an alternative therapeutic strategy to selectively destroy tumours. Although there is a plethora of research outlining the molecular mechanisms of ferroptosis, these findings are yet to be translated into clinical compounds inducing ferroptosis. In this perspective, we elaborate on how ferroptosis can be leveraged in the clinic. We discuss a therapeutic window for compounds inducing ferroptosis, the subset of tumour types that are most sensitive to ferroptosis, conventional therapeutics that induce ferroptosis, and potential strategies for lowering the threshold for ferroptosis.

Keywords: ferroptosis; Eprenetapopt; Erastin; GPX4; glutathione (GSH); SLC7A11; iron; oxidative stress; NRF2

Citation: Fujihara, K.M.; Zhang, B.Z.; Clemons, N.J. Opportunities for Ferroptosis in Cancer Therapy. *Antioxidants* **2021**, *10*, 986. https://doi.org/10.3390/antiox10060986

Academic Editors: Cinzia Domenicotti and Barbara Marengo

Received: 2 June 2021
Accepted: 16 June 2021
Published: 21 June 2021

Publisher's Note: MDPI stays neutral with regard to jurisdictional claims in published maps and institutional affiliations.

1. Introduction

A hallmark of cancer is the development of resistance to apoptosis, often through genetic loss of the molecular machinery involved in programmed cell death [1]. Furthermore, resistance to chemotherapeutics and molecular targeted therapies are major challenges in oncology [2]. As a result, harnessing our understanding of non-apoptotic cell death pathways, such as ferroptosis, has substantial therapeutic potential for patients, especially in the metastatic setting where effective therapeutic strategies remain limited [3]. Ferroptosis is an iron-dependent, non-apoptotic form of regulated cell death characterised by aberrant lipid membrane peroxidation [4]. As such, the induction of ferroptosis is experimentally verified by the restoration of cell viability by iron chelators and lipophilic antioxidants, and by lack of cell death rescue by pan-caspase inhibitors (Figure 1). Given that dysregulated iron metabolism and iron accumulation have been frequently observed across both solid tumours and haematological malignancies [5], selectively inducing ferroptosis is an attractive potential anti-cancer strategy with broad clinical implications. In this perspective, we discuss the potential for weaponising ferroptosis in the clinic through two therapeutic avenues: (1) triggering ferroptosis in cancer cells directly with targeted agents and (2) lowering the threshold at which cancer cells undergo ferroptosis to enhance the efficacy of conventional therapies, including chemotherapy, radiotherapy, and immunotherapy.

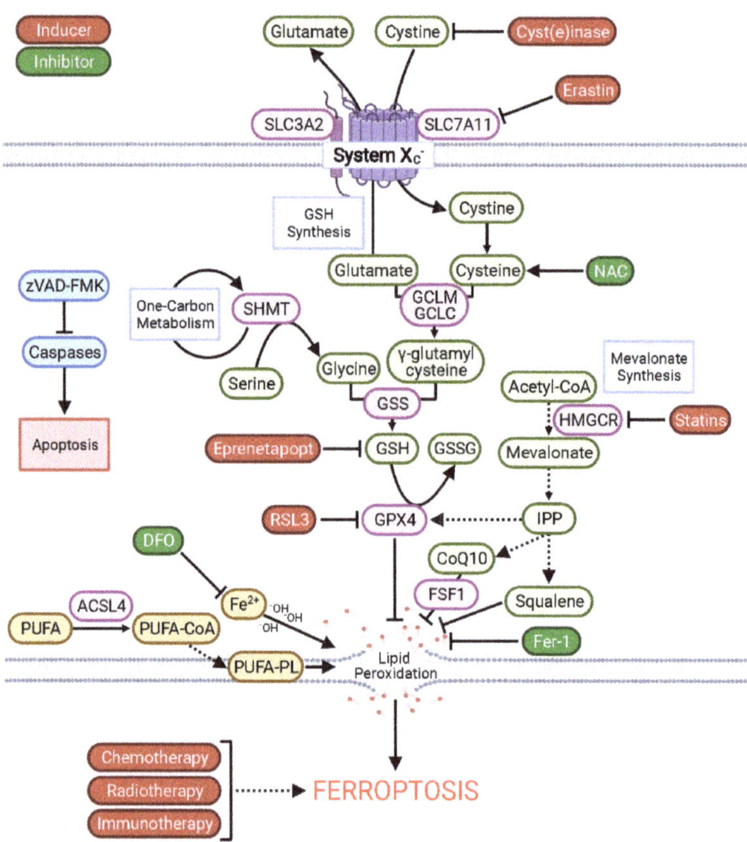

Figure 1. Mechanisms of Ferroptosis Inducers and Inhibitors. Ferroptosis is triggered following the accumulation of iron-catalysed damage to phospholipid-bound polyunsaturated fatty acids (PUFA-PLs). Glutathione peroxidase 4 (GPX4) detoxifies lipid peroxides at the expense of glutathione oxidation (GSH to GSSG). GSH is a tripeptide containing cysteine, glutamate, and glycine and is synthesised through a stepwise pathway catalysed by a glutamate–cysteine ligase catalytic subunit (GCLC), glutamate–cysteine ligase modifier subunit (GCLM), and glutathione synthetase (GSS). Cystine imported in exchange for glutamate by system x$_c^-$ (encoded by *SLC7A11*) provides the main source of cysteine for GSH synthesis. Glycine for GSH synthesis can be sourced from serine catabolism by serine hydroxymethyltransferase (SHMT). Acyl-CoA synthetase long-chain family member 4 (ACSL4) acylates PUFAs, which are incorporated into plasma membranes and are vulnerable to peroxidation. Ferroptosis can be triggered by the GSH depletion (e.g., Cyst(e)inase, Erastin, Eprenetapopt) or direct inhibition of GPX4 (e.g., RSL3). Downstream products of the mevalonate pathway suppress ferroptosis, including isopentenyl pyrophosphate (IPP), which is utilised for selenoprotein synthesis (e.g., GPX4), coenzyme-Q10 (CoQ10) synthesis, which is a co-factor of ferroptosis suppressor protein 1 (FSP1, encoded by *AIFM2*), and squalene synthesis, which is a lipophilic antioxidant. Inhibiting hydroxymethylglutaryl-coenzyme A reductase (HMGCR) with statins amplifies the activity of ferroptosis inducers. Supplementation of exogenous antioxidants (e.g., NAC, N-acetyl-cysteine) to simulate GSH synthesis, lipophilic antioxidants (e.g., Ferrostatin-1, Fer-1) to detoxify lipid peroxides, and iron chelation (e.g., DFO, deferoxamine) blocks the induction of ferroptosis. Pan-caspase inhibitors fail to rescue the cell death (e.g., zVAD-FMK) induced by ferroptosis inducers. Ferroptosis induction by traditional therapies (chemotherapy, radiotherapy, and immunotherapy) contributes to their anti-cancer activity. Figure generated using BioRender.com (21 June 2021).

2. The Development of Ferroptosis Inducers

Ferroptosis is triggered through two mechanisms, either through the depletion of the cellular antioxidant glutathione (GSH), or through direct inhibition of the enzyme responsible for reversing lipid oxidation, glutathione peroxidase 4 (GPX4). While more detailed reviews of ferroptosis can be found elsewhere [3,6,7], here, we highlight the key compounds used in the elucidation of the mechanisms of ferroptosis. The first chemical agent found to trigger ferroptosis, Erastin, was originally identified in a high-throughput chemical library screen to identify compounds that were selectively lethal in oncogenic mutant $HRAS^{V12}$ cells [8]. Later, the protein target of Erastin was elucidated as system x_c^- (encoded by *SLC7A11* and *SLC3A2*), a cell surface cystine–glutamate antiporter [4,9]. Erastin was found to inhibit the activity of system x_c^-, limiting the cellular supply of cystine, which critically leads to the depletion of intracellular GSH. Likewise, cystine deprivation in vitro also induces ferroptosis and phenocopies many of the cell death features induced by Erastin [10]. Furthermore, restricting cystine/cysteine availability to cancer cells through enzymatic degradation with cyst(e)inase triggers ferroptosis and inhibits tumour growth in vivo [11,12]. Moreover, recent work by us and others showed that Eprenetapopt (APR-246, PRIMA-1^{MET}), previously identified as a mutant-p53 reactivator, can also induce ferroptosis and has demonstrated capacity to conjugate to free cysteine and deplete GSH [13–15].

Following on from the discovery of Erastin, 1*S*,3*R*-RSL3 (Ras synthetic lethal-3, RSL3) was identified in an analogous fashion [16]. Here, RSL3, as well as Erastin, were shown to induce cell death through a non-apoptotic, iron-dependent mechanism, and cells transformed with oncogenic RAS had increased levels of iron accumulation due to upregulation of transferrin receptor 1 [16]. Unlike Erastin, however, RSL3 was found to act independently of system x_c^- inhibition [4], and to instead covalently inhibit GPX4 [10], a unique cellular selenoenzyme that reduces phospholipid hydroperoxides to lipid alcohols using GSH as a cofactor [17,18]. As a result, inhibiting GPX4 activity, either directly or indirectly through GSH depletion, triggers unrestricted lipid peroxide accumulation in the presence of iron and subsequently results in the rupture of the plasma membrane [19]. These observations are consistent with our analyses of the Cancer Dependency Map (DepMap) and Cancer Therapeutics Response Portal v2 (CTRPv2) datasets [20–22], which highlight that *GPX4* gene dependency correlates with cancer cell line sensitivity to GPX4 inhibitors (including RSL3), Erastin, and APR-017 (analogue of Eprenetapopt) (Figure 2A,B). The DepMap dataset contains gene dependency data generated from pooled genome-wide CRISPR knockout screening of over 1000 cancer cell lines, whilst the CTRPv2 dataset contains compound activity data from 481 compounds across ~700 cancer cell lines. Correlating these datasets can reveal insights into compound mechanisms of action, as demonstrated.

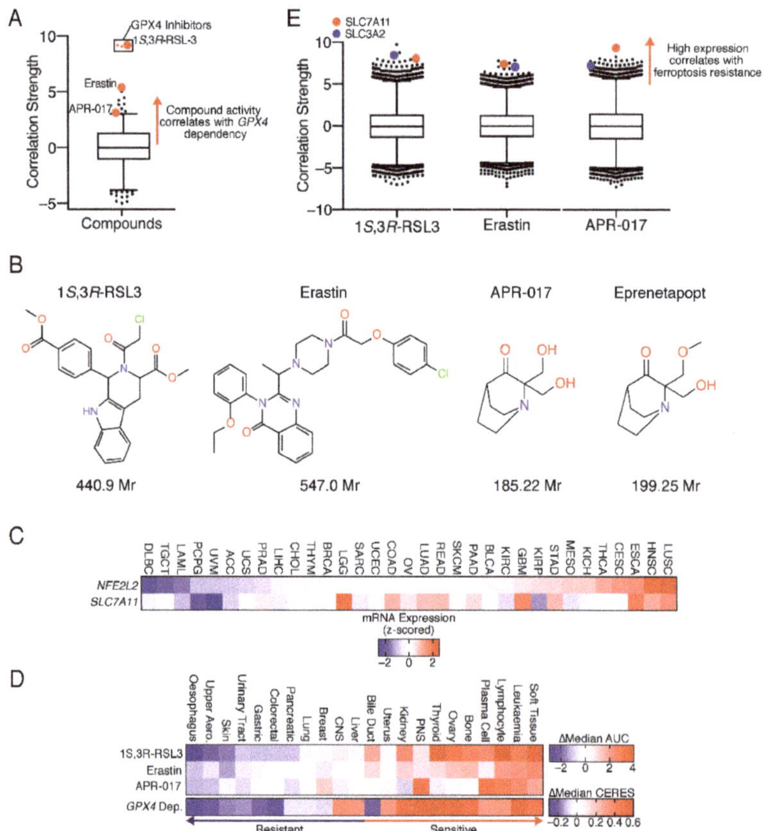

Figure 2. DepMap and TCGA data (**A**) Box-and-whisker plot (1st–99th percentile) of Fischer's transformed z-scored Pearson correlation strength of *GPX4* dependency and the 481 Cancer Therapeutic Response Portal v2 (CTRPv2) compound activity data across ~700 cancer cell lines. Red dots indicate examples of ferroptosis inducers. (**B**) Chemical structures and molecular weights (Mr) of ferroptosis inducers. (**C**) Heatmap of gene expression of *NFE2L2* and *SLC7A11* in patients with cancer were analysed from the TCGA, cancer type ordered by *NFE2L2* expression. ACC, adrenocortical carcinoma; BLCA, bladder urothelial carcinoma; BRCA, breast invasive carcinoma; CESC, cervical squamous cell carcinoma and endocervical adenocarcinoma; CHOL, cholangiocarcinoma; COAD, colon adenocarcinoma; DLBC, lymphoid neoplasm diffuse large B- cell lymphoma; ESCA, oesophageal carcinoma; GBM, glioblastoma multiforme; HNSC, head and neck squamous cell carcinoma; KICH, kidney chromophobe; KIRC, kidney renal clear cell carcinoma; KIRP, kidney renal papillary cell carcinoma; LAML, acute myeloid leukaemia; LGG, brain lower grade glioma; LIHC, liver hepatocellular carcinoma; LUAD, lung adenocarcinoma; LUSC, lung squamous cell carcinoma; MESO, mesothelioma; OV, ovarian serous cystadenocarcinoma; PAAD, pancreatic adenocarcinoma; PCPG, pheochromocytoma and paraganglioma; PRAD, prostate adenocarcinoma; READ, rectum adenocarcinoma; SARC, sarcoma; SKCM, skin cutaneous melanoma; STAD, stomach adenocarcinoma; TGCT, testicular germ cell tumours; THCA, thyroid carcinoma; THYM, thymoma; TPM, transcripts per million; UCEC, uterine corpus endometrial carcinoma; UCS, uterine carcinosarcoma; UVM, uveal melanoma. (**D**) Heatmap of sensitivity to ferroptosis inducers, RSL3, Erastin, Eprenetapopt analogue (APR-017), and *GPX4* dependency across ~700 cancer cell lines. Cancer lineages ordered by sensitivity. AUC, area under the curve (compound activity); CERES, copy-number adjusted gene dependency score. (**E**) Box-and-whisker plot (1st-99th percentile) of Fischer's transformed z-scored Pearson correlation strength of RSL3, Erastin and Eprenetapopt analogue (APR-017) activity and genome-wide expression data across ~700 cancer cell lines. Red dot indicates SLC7A11 and blue dot indicates SLC3A2, which encode system x_c^-. Data accessed from www.depmap.org (accessed on 30 March 2021) and www.cbioportal.org (accessed on 30 March 2021).

3. Therapeutic Index for Ferroptosis

All anti-cancer therapeutics principally rely on the selective targeting and destruction of tumour cells over normal cells, known as the therapeutic index. Understanding the differences in the threshold at which cancer cells undergo ferroptosis compared to normal cells is vital for the clinical deployment of ferroptosis inducers, to both mitigate unwanted toxicities and maximise therapeutic benefit [23]. Ferroptosis inducers could be used to leverage the increased levels of oxidative stress and iron in cancer cells to drive their therapeutic index (Figure 3A). Ultimately, the efficacy and safety profiles of ferroptosis inducers can only be established through clinical trials of compounds that have appropriate pharmacodynamics and pharmacokinetics. Whilst Erastin and RSL3 are not readily bioavailable, Erastin analogues (e.g., PRLX 93936), other system x_c^- inhibitors (e.g., Sorafenib, Sulfasalazine), and GPX4 inhibitors (e.g., Altretamine, Withaferin A) are under clinical investigation across various tumour streams [3]. Eprenetapopt is also being tested in a phase III clinical trial in *TP53*-mutated myeloid dysplastic syndromes (NCT03745716). Nevertheless, lessons can be gleaned from the development and usage of cytotoxic chemotherapeutics and targeted therapies in order to achieve the greatest clinical benefit for patients. These include considerations of cancer type and setting, predictive biomarkers, and the use of rescue compounds to mitigate on-target toxicities.

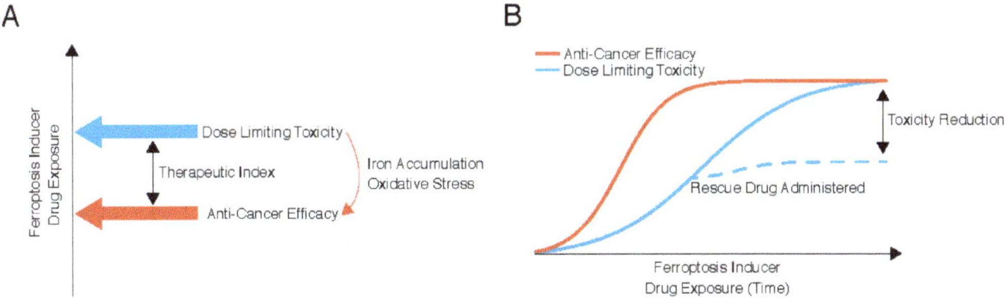

Figure 3. Therapeutic Index for Ferroptosis Inducers. (**A**) Schematic depiction of the proposed therapeutic index for ferroptosis inducers as anti-cancer agents. (**B**) The potential use of rescue drugs following ferroptosis inducer treatment. Rescue drugs could be administered once the maximal efficacy of ferroptosis inducers has been reached to reduce dose limiting toxicities without diminishing the effects on tumours.

4. Oxidative Stress and Iron

Cancer cells experience elevated levels of oxidative stress due to increased reactive oxygen species (ROS) production arising from the augmented metabolic demands to support biomass accumulation and proliferation compared to non-transformed cells [24]. In response, some cancer cells restrict ROS by elevating antioxidant pathways in order to avoid the deleterious effects of oxidative stress [25]. For example, lung cancers frequently harbour mutations in *NFE2L2* (nuclear factor, erythroid 2-like 2, encodes for NRF2) or *KEAP1* (encodes for KEAP1, a negative regulator of NRF2), which results in the activation of antioxidant pathways, including *SLC7A11* and *NQO1* upregulation [26,27]. Furthermore, there is direct evidence supporting the role of NRF2 as a negative regulator of ferroptosis by promoting antioxidant pathways [28,29]. Conversely, NRF2 acts as a guardrail to unchecked cell cycle and proliferation in haematopoietic stem cells [30], and as a result NRF2 and its target genes are found at low levels in haematological malignancies compared to solid tumours (Figure 2C). Iron accumulation is found frequently in several cancer types, especially haematological malignancies [5]. Iron is an important heavy metal required for a multitude of biological processes, including iron–sulphur cluster biogenesis to support mitochondrial metabolism and DNA synthesis, and heme synthesis to support cellular oxygen trafficking. Furthermore, iron participates in ROS generating reactions and lipid peroxidation formation via Fenton chemistry [31]. As a result, the levels of iron accumulation in cancer cells compared to healthy tissues provides a therapeutic index for ferroptosis inducers.

5. Cancer Type

As a result of these factors, it is clear that cancer-type specific factors play a role in the sensitivity of tumour cells to ferroptosis, and identifying which tumour types are most likely to benefit will be a key factor in the successful development of ferroptosis inducers as therapeutics. As such, cancer cells of mesenchymal origin were found to be selectively sensitive to ferroptosis inducers compared to epithelial-derived cancer cells [32]. For example, cancers that arise from soft tissue, bone, haematological, and lymphoid tissues (i.e., mesenchymal origin) display high dependency on GPX4 for survival and high sensitivity to ferroptosis activators compared to epithelial-derived cancer cell lines (e.g., oesophageal, upper aerodigestive, and skin; Figure 2D). Further, cancer cells of epithelial origin that have undergone epithelial-to-mesenchymal transition (EMT) are more susceptible to ferroptosis [32]. Increased polyunsaturated fatty acid (PUFA) synthesis in mesenchymal-like state cells likely drives the increased dependency on GPX4 to dissipate reactive lipid peroxides [32,33]. Moreover, breast cancer cells that enter a mesenchymal-like state following Lapatinib treatment become highly sensitive to GPX4 inhibition [29]. Importantly, EMT drives the metastatic potential of cancer cells [34], which suggests that metastatic cells may be more vulnerable to ferroptosis. Cell-to-cell interactions also play a major role in ferroptosis sensitivity; cells plated at a lower density display increased sensitivity to ferroptosis compared to identical but confluent cells in culture [35]. Mechanistically, cell-to-cell contacts rely on E-cadherin, which suppresses ferroptosis through activation of the NF2 and Hippo signalling pathways [35]. This matches the findings relating to EMT and ferroptosis as mesenchymal-like cells lose E-cadherin expression in order to diminish cell-to-cell interactions. Collectively, these findings suggest that cancers of mesenchymal origin, especially haematological cancers, and those prone to EMT and metastasis are likely to be strong candidates for therapeutically leveraging ferroptosis inducers.

6. Therapeutic Biomarker

Predictive therapeutic biomarkers could also be utilised to screen and select for patients with a higher probability of response to ferroptosis inducers. A likely beneficial approach would be to screen patient tumours for low SLC7A11 expression, as high expression of SLC7A11 correlates with resistance to ferroptosis inducers (Figure 2E). This further highlights the strong likelihood that patients with haematological malignancies would benefit from treatment with ferroptosis inducers as SLC7A11 expression in haematological cancers is low, correlating with their low NRF2 expression (Figure 2C).

7. Use of Rescue Agents to Mitigate Toxicity

Rescue interventions could be used to mitigate the on-target side effects of ferroptosis. For example, chemotherapeutic dosing with methotrexate is often followed by folinic acid supplementation in order to limit haematological and hepatic toxicities [36]. To date, while it is possible to rescue the cell death induced by ferroptosis inducers, no attempts have been made to establish whether the selectivity of ferroptosis inducers for tumour cells could be improved by selectively blocking ferroptosis in normal cells. One could hypothesise that high-dose N-acetyl-cysteine, which is routinely used to treat paracetamol poisoning [37], could be used to rescue normal cells and limit side-effects following treatments with ferroptosis inducers (Figure 3B). Furthermore, iron chelators (e.g., deferoxamine) are commonly used to treat patients with iron overload and could also be utilised to limit deleterious side-effects of ferroptosis inducers.

8. Conventional Therapeutics That Induce Ferroptosis

Several recent studies have identified that cancer cells undergo ferroptosis in response to conventional therapeutics, including chemotherapy, radiotherapy, and immunotherapy. Cytotoxic chemotherapies typically target rapidly proliferating cells by interfering with cellular processes involved in cell division and DNA replication. Whilst the evidence indicating that ferroptosis induction directly by conventional chemotherapeutics, such as cisplatin and gemcitabine, is limited [38,39], there is pre-clinical evidence that ferroptosis inducers can synergise with traditional chemotherapeutics [40]. In the case of Eprenetapopt, significant pre-clinical evidence demonstrates the chemosensitisation capacity of ferroptosis activation, including in oesophageal, ovarian, and haematological malignancies [41–44]. Meanwhile, sorafenib, a multi-tyrosine kinase inhibitor used in the treatment of advanced liver cancer, and sulfasalazine, an anti-inflammatory drug often used to treat rheumatoid arthritis, have both been shown to induce ferroptosis through inhibition of system x_c^- [9,45,46].

Radiotherapy uses high-dose ionising radiation delivered locally to tumour-affected tissues to kill cancer cells, predominantly by causing DNA damage. Radiotherapy is also known to induce oxidative stress in cancer cells by generating reactive oxygen species [47]. Radiotherapy was recently shown to trigger ferroptosis through ATM-mediated suppression of SLC7A11 [48]. Furthermore, upregulation of acyl-CoA synthetase long-chain family member 4 (ACSL4) also contributes to the promotion of ferroptosis by radiotherapy [49]. ACLS4 preferentially acylates long-chain PUFAs [50], which are incorporated into plasma membranes, increasing the membrane-resident pool of oxidation-sensitive lipids [51]. Interestingly, whilst ferroptosis inhibitors partially block the cell death induced by radiotherapy, they do not block the DNA damage triggered by ionising radiation [49]. Importantly, strong synergy was reported between radiotherapy and ferroptosis activators, including cyst(e)inase and sulfasalazine [48,49].

Immunotherapy with immune checkpoint inhibitors (ICI) has revolutionised clinical care of cancer patients, providing an additional pillar to the suite of cancer treatment modalities. ICI predominantly elicit their anti-cancer effects by inhibiting tumour cell capacity to dampen cytotoxic T cell-mediated killing. ICI have also been shown to induce CD8+ T cell-mediated ferroptosis through suppression of SLC7A11 and SLC3A2 by interferon gamma (IFNγ) released by the T-cells [52]. Further, PD-L1 blockage therapy synergised with ferroptosis inducers, Erastin, RSL-3, and cyst(e)ine, both in vitro and in vivo [52], as well as with radiotherapy [48]. More recently, high expression of the receptor tyrosine kinase, TYRO3, was shown to correlate with resistance to anti-PD-1 therapy and suppress the induction of ferroptosis in tumour cells by activating NRF2 [53]. Conversely, ferroptosis inducers have also been shown to induce ferroptosis in CD8+ T cells, limiting the anti-cancer efficacy of ICI [54]. Preliminary results also reported that Eprenetapopt and anti-PD-1 therapy synergise in murine solid tumour models [55]. This has prompted the initiation of a phase I clinical trial to test the safety of Eprenetapopt and anti-PD-1 therapy, Pembrolizumab, in solid tumour malignancies (NCT04383938). Given the tension between how ferroptosis inducers affect ICI and T cell killing, trials like this will provide pivotal insights into the role of ferroptosis inducers as adjuvants to ICI.

It is likely that a portion of the cell death induced by most conventional therapeutic regimens is ferroptotic. However, to date, there has been no systematic attempt to quantify the contribution of cell death driven by ferroptosis in an in vivo or clinical setting with any therapeutic regimen. Such a study could potentially provide rationale for when and how ferroptosis inhibitors or sensitisers should be applied in combination with conventional chemotherapies to maximise therapeutic gain, particularly in the setting of tumour resistance to apoptosis.

9. Lowering the Threshold for Ferroptosis

Given that ferroptotic cell death is in part responsible for the tumour killing achieved by chemotherapy, radiotherapy, and immunotherapy, reducing cancer cell capacity to evade ferroptotic cell death would be a powerful therapeutic strategy. Following on from the study investigating ferroptosis sensitivity and mesenchymal-like cell state, statins, common anti-cholesterol drugs, were also identified as modulators of ferroptosis sensitivity in mesenchymal cells [32]. Statins inhibit the rate-limiting enzyme in the mevalonate synthesis pathway, HMG-CoA reductase (HMGCR), which decreases cholesterol abundance (Figure 1). Previous reports detailed that statins inhibit isopentenylation of the selenocysteine-charged transfer RNA, which is required for synthesis of selenoproteins like GPX4 [56]. In keeping with this, statins synergise with GPX4 inhibitors through decreasing the abundance of the GPX4 protein and inducing lipid peroxidation [32]. The mevalonate synthesis pathway also directs several other downstream pathways involved in ferroptosis, including CoQ10 and squalene synthesis—GSH-independent mechanisms of protection against ferroptosis (Figure 1) [57–59]. As a result, targeting the mevalonate synthesis pathway with statins could be utilised to lower the threshold at which cancer cells undergo ferroptosis.

Limiting endogenous supply of other key nutrients may provide an alternative strategy to sensitise tumours to undergoing ferroptosis. Chronic activation of the antioxidant response induced by NRF2 activation increases the demand for the supply of glutamine and other non-essential amino acids, due to the increased efflux of glutamate from SLC7A11 to supply cyst(e)ine for GSH synthesis [60,61]. In addition, we recently demonstrated that limiting the availability of serine and glycine (SG) through dietary restriction significantly enhanced the efficacy of Eprenetapopt in vivo by limiting the availability of glycine required for de novo GSH synthesis [13]. However, this differs from the effects of ferroptosis induction by Erastin under SG restricted conditions seen in other studies, where Erastin treatment was found to reverse the sensitivity of *KEAP1* mutant tumour limits to SG restriction [60]. This is likely explained by how Eprenetapopt and Erastin differ in their mechanisms of GSH depletion and effects on cyst(e)ine and glutamate availability (Table 1). Recently, arginine deprivation also demonstrated protection against Erastin and cystine depletion but not against GPX4 inhibition with RSL3 [62]. Given that cyst(e)ine are considered non-essential amino acids, dietary restriction of cystine could provide improved therapeutic benefit for Erastin or other SLC7A11 inhibitors in vivo.

Table 1. Effect of ferroptosis inducers of GSH, cyst(e)ine uptake, glutamate release, and serine/glycine (SG) restriction.

Inhibitor	GSH Depletion	Cyst(e)ine Uptake	Glutamate Release	SG Restriction
Eprenetapopt	Yes	Increases	Increases	Increases activity
Erastin	Yes	Decreases	Decreases	Decreases activity
1S,3R-RSL3	No	Unknown	Unknown	Unknown

10. Conclusions

The identification of ferroptosis as a non-redundant, regulated cell death pathway opens up opportunities for circumventing tumour cell resistance to other forms of regulated cell death, such as apoptosis. Exploiting ferroptosis in cancer therapy requires continued building of our understanding of the mechanisms underlying this cell death pathway; in particular, identification of molecular regulators of sensitivity and resistance to ferroptosis will be crucial for future clinical application. Complementary to this will be the identification of certain tumour types or cell states (such as a mesenchymal phenotype) that are particularly amenable to induction of ferroptosis. Whilst there is active interest in the development of specific ferroptosis inducers as novel therapeutics, the recognition that ferroptosis can be engaged by many current treatment modalities opens up the potential for strategies to leverage this activity indirectly by lowering the threshold for activation of ferroptotic cell death. Given the dependence on specific metabolic pathways to protect tumour cells from ferroptosis, the application of specific diets concurrent with anti-cancer treatments holds much interest and future potential.

Author Contributions: Conceptualization, K.M.F. and N.J.C.; data curation and formal analysis, K.M.F.; writing—original draft preparation, K.M.F., B.Z.Z. and N.J.C.; writing—review and editing, K.M.F., B.Z.Z. and N.J.C.; visualization, K.M.F., B.Z.Z., N.J.C.; supervision, N.J.C.; project administration, N.J.C.; funding acquisition, N.J.C. All authors have read and agreed to the published version of the manuscript.

Funding: This work was supported by a National Health and Medical Research Council (NMHRC) Project Grant (APP1120293) to N.J.C. N.J.C. is supported by a Fellowship (MCRF16002) from the Department of Health and Human Services acting through the Victorian Cancer Agency Fellowship. K.M.F. is supported by an Australian Research Training Program (RTP) Scholarship.

Institutional Review Board Statement: Not applicable.

Informed Consent Statement: Not applicable.

Data Availability Statement: Data used in Figure 2 is publically accessible at www.depmap.org and www.cbioportal.org.

Conflicts of Interest: The authors disclose that there are no conflict of interest in the work that contributed towards this manuscript.

References

1. Hanahan, D.; Weinberg, R.A. Hallmarks of cancer: The next generation. *Cell* **2011**, *144*, 646–674. [CrossRef]
2. Ward, R.A.; Fawell, S.; Floc'h, N.; Flemington, V.; McKerrecher, D.; Smith, P.D. Challenges and Opportunities in Cancer Drug Resistance. *Chem. Rev.* **2021**, *121*, 3297–3351. [CrossRef]
3. Chen, X.; Kang, R.; Kroemer, G.; Tang, D. Broadening horizons: The role of ferroptosis in cancer. *Nat. Rev. Clin. Oncol.* **2021**, *18*, 280–296. [CrossRef]
4. Dixon, S.J.; Lemberg, K.M.; Lamprecht, M.R.; Skouta, R.; Zaitsev, E.M.; Gleason, C.E.; Patel, D.N.; Bauer, A.J.; Cantley, A.M.; Yang, W.S.; et al. Ferroptosis: An iron-dependent form of nonapoptotic cell death. *Cell* **2012**, *149*, 1060–1072. [CrossRef]
5. Torti, S.V.; Manz, D.H.; Paul, B.T.; Blanchette-Farra, N.; Torti, F.M. Iron and Cancer. *Annu. Rev. Nutr.* **2018**, *38*, 97–125. [CrossRef] [PubMed]
6. Zou, Y.; Schreiber, S.L. Progress in Understanding Ferroptosis and Challenges in Its Targeting for Therapeutic Benefit. *Cell Chem. Biol.* **2020**, *27*, 463–471. [CrossRef] [PubMed]
7. Armenta, D.A.; Dixon, S.J. Investigating Nonapoptotic Cell Death Using Chemical Biology Approaches. *Cell Chem. Biol.* **2020**, *27*, 376–386. [CrossRef] [PubMed]
8. Dolma, S.; Lessnick, S.L.; Hahn, W.C.; Stockwell, B.R. Identification of genotype-selective antitumor agents using synthetic lethal chemical screening in engineered human tumor cells. *Cancer Cell* **2003**, *3*, 285–296. [CrossRef]
9. Dixon, S.J.; Patel, D.N.; Welsch, M.; Skouta, R.; Lee, E.D.; Hayano, M.; Thomas, A.G.; Gleason, C.E.; Tatonetti, N.P.; Slusher, B.S.; et al. Pharmacological inhibition of cystine-glutamate exchange induces endoplasmic reticulum stress and ferroptosis. *eLife* **2014**, *3*, e02523. [CrossRef]
10. Yang, W.S.; SriRamaratnam, R.; Welsch, M.E.; Shimada, K.; Skouta, R.; Viswanathan, V.S.; Cheah, J.H.; Clemons, P.A.; Shamji, A.F.; Clish, C.B.; et al. Regulation of ferroptotic cancer cell death by GPX4. *Cell* **2014**, *156*, 317–331. [CrossRef]
11. Badgley, M.A.; Kremer, D.M.; Maurer, H.C.; DelGiorno, K.E.; Lee, H.J.; Purohit, V.; Sagalovskiy, I.R.; Ma, A.; Kapilian, J.; Firl, C.E.M.; et al. Cysteine depletion induces pancreatic tumor ferroptosis in mice. *Science* **2020**, *368*, 85–89. [CrossRef] [PubMed]
12. Cramer, S.L.; Saha, A.; Liu, J.; Tadi, S.; Tiziani, S.; Yan, W.; Triplett, K.; Lamb, C.; Alters, S.E.; Rowlinson, S.; et al. Systemic depletion of L-cyst(e)ine with cyst(e)inase increases reactive oxygen species and suppresses tumor growth. *Nat. Med.* **2017**, *23*, 120–127. [CrossRef]
13. Fujihara, K.M.; Zhang, B.; Jackson, T.D.; Nijiagel, B.; Ang, C.-S.; Nikolic, I.; Sutton, V.; Trapani, J.; Simpson, K.J.; Stojanovski, D.; et al. Genome-wide CRISPR screens reveal APR-246 (Eprenetapopt) triggers ferroptosis and inhibits iron-sulfur cluster biogenesis. *bioRxiv* **2020**. [CrossRef]
14. Birsen, R.; Larrue, C.; Decroocq, J.; Johnson, N.; Guiraud, N.; Gotanegre, M.; Cantero-Aguilar, L.; Grignano, E.; Huynh, T.; Fontenay, M.; et al. APR-246 induces early cell death by ferroptosis in acute myeloid leukemia. *Haematologica* **2021**. online ahead of print Jan 7. [CrossRef] [PubMed]
15. Ceder, S.; Eriksson, S.E.; Cheteh, E.H.; Dawar, S.; Corrales Benitez, M.; Bykov, V.J.N.; Fujihara, K.M.; Grandin, M.; Li, X.; Ramm, S.; et al. A thiol-bound drug reservoir enhances APR-246-induced mutant p53 tumor cell death. *EMBO Mol. Med.* **2021**, *13*, e10852. [CrossRef]
16. Yang, W.S.; Stockwell, B.R. Synthetic lethal screening identifies compounds activating iron-dependent, nonapoptotic cell death in oncogenic-RAS-harboring cancer cells. *Chem. Biol.* **2008**, *15*, 234–245. [CrossRef]
17. Ursini, F.; Maiorino, M.; Gregolin, C. The selenoenzyme phospholipid hydroperoxide glutathione peroxidase. *Biochim. Biophys. Acta* **1985**, *839*, 62–70. [CrossRef]

18. Schuckelt, R.; Brigelius-Flohe, R.; Maiorino, M.; Roveri, A.; Reumkens, J.; Strassburger, W.; Ursini, F.; Wolf, B.; Flohe, L. Phospholipid hydroperoxide glutathione peroxidase is a selenoenzyme distinct from the classical glutathione peroxidase as evident from cDNA and amino acid sequencing. *Free Radic Res. Commun.* **1991**, *14*, 343–361. [CrossRef]
19. Jiang, X.; Stockwell, B.R.; Conrad, M. Ferroptosis: Mechanisms, biology and role in disease. *Nat. Rev. Mol. Cell Biol.* **2021**, *22*, 266–282. [CrossRef]
20. Meyers, R.M.; Bryan, J.G.; McFarland, J.M.; Weir, B.A.; Sizemore, A.E.; Xu, H.; Dharia, N.V.; Montgomery, P.G.; Cowley, G.S.; Pantel, S.; et al. Computational correction of copy number effect improves specificity of CRISPR-Cas9 essentiality screens in cancer cells. *Nat. Genet.* **2017**, *49*, 1779–1784. [CrossRef]
21. Seashore-Ludlow, B.; Rees, M.G.; Cheah, J.H.; Cokol, M.; Price, E.V.; Coletti, M.E.; Jones, V.; Bodycombe, N.E.; Soule, C.K.; Gould, J.; et al. Harnessing Connectivity in a Large-Scale Small-Molecule Sensitivity Dataset. *Cancer Discov.* **2015**, *5*, 1210–1223. [CrossRef] [PubMed]
22. Rees, M.G.; Seashore-Ludlow, B.; Cheah, J.H.; Adams, D.J.; Price, E.V.; Gill, S.; Javaid, S.; Coletti, M.E.; Jones, V.L.; Bodycombe, N.E.; et al. Correlating chemical sensitivity and basal gene expression reveals mechanism of action. *Nat. Chem. Biol.* **2016**, *12*, 109–116. [CrossRef]
23. Chang, L.; Ruiz, P.; Ito, T.; Sellers, W.R. Targeting pan-essential genes in cancer: Challenges and opportunities. *Cancer Cell* **2021**, *39*, 466–479. [CrossRef]
24. Trachootham, D.; Zhou, Y.; Zhang, H.; Demizu, Y.; Chen, Z.; Pelicano, H.; Chiao, P.J.; Achanta, G.; Arlinghaus, R.B.; Liu, J.; et al. Selective killing of oncogenically transformed cells through a ROS-mediated mechanism by beta-phenylethyl isothiocyanate. *Cancer Cell* **2006**, *10*, 241–252. [CrossRef]
25. Harris, I.S.; DeNicola, G.M. The Complex Interplay between Antioxidants and ROS in Cancer. *Trends Cell Biol.* **2020**, *30*, 440–451. [CrossRef] [PubMed]
26. DeNicola, G.M.; Karreth, F.A.; Humpton, T.J.; Gopinathan, A.; Wei, C.; Frese, K.; Mangal, D.; Yu, K.H.; Yeo, C.J.; Calhoun, E.S.; et al. Oncogene-induced Nrf2 transcription promotes ROS detoxification and tumorigenesis. *Nature* **2011**, *475*, 106–109. [CrossRef] [PubMed]
27. de la Vega, M.R.; Chapman, E.; Zhang, D.D. NRF2 and the Hallmarks of Cancer. *Cancer Cell* **2018**, *34*, 21–43. [CrossRef] [PubMed]
28. Sun, X.; Ou, Z.; Chen, R.; Niu, X.; Chen, D.; Kang, R.; Tang, D. Activation of the p62-Keap1-NRF2 pathway protects against ferroptosis in hepatocellular carcinoma cells. *Hepatology* **2016**, *63*, 173–184. [CrossRef]
29. Hangauer, M.J.; Viswanathan, V.S.; Ryan, M.J.; Bole, D.; Eaton, J.K.; Matov, A.; Galeas, J.; Dhruv, H.D.; Berens, M.E.; Schreiber, S.L.; et al. Drug-tolerant persister cancer cells are vulnerable to GPX4 inhibition. *Nature* **2017**, *551*, 247–250. [CrossRef]
30. Tsai, J.J.; Dudakov, J.A.; Takahashi, K.; Shieh, J.H.; Velardi, E.; Holland, A.M.; Singer, N.V.; West, M.L.; Smith, O.M.; Young, L.F.; et al. Nrf2 regulates haematopoietic stem cell function. *Nat. Cell Biol.* **2013**, *15*, 309–316. [CrossRef]
31. Dixon, S.J.; Stockwell, B.R. The role of iron and reactive oxygen species in cell death. *Nat. Chem. Biol.* **2014**, *10*, 9–17. [CrossRef]
32. Viswanathan, V.S.; Ryan, M.J.; Dhruv, H.D.; Gill, S.; Eichhoff, O.M.; Seashore-Ludlow, B.; Kaffenberger, S.D.; Eaton, J.K.; Shimada, K.; Aguirre, A.J.; et al. Dependency of a therapy-resistant state of cancer cells on a lipid peroxidase pathway. *Nature* **2017**, *547*, 453–457. [CrossRef] [PubMed]
33. Lee, J.Y.; Nam, M.; Son, H.Y.; Hyun, K.; Jang, S.Y.; Kim, J.W.; Kim, M.W.; Jung, Y.; Jang, E.; Yoon, S.J.; et al. Polyunsaturated fatty acid biosynthesis pathway determines ferroptosis sensitivity in gastric cancer. *Proc. Natl. Acad. Sci. USA* **2020**, *117*, 32433–32442. [CrossRef]
34. Bakir, B.; Chiarella, A.M.; Pitarresi, J.R.; Rustgi, A.K. EMT, MET, Plasticity, and Tumor Metastasis. *Trends Cell Biol.* **2020**, *30*, 764–776. [CrossRef] [PubMed]
35. Wu, J.; Minikes, A.M.; Gao, M.; Bian, H.; Li, Y.; Stockwell, B.R.; Chen, Z.N.; Jiang, X. Intercellular interaction dictates cancer cell ferroptosis via NF2-YAP signalling. *Nature* **2019**, *572*, 402–406. [CrossRef]
36. Levitt, M.; Mosher, M.B.; DeConti, R.C.; Farber, L.R.; Skeel, R.T.; Marsh, J.C.; Mitchell, M.S.; Papac, R.J.; Thomas, E.D.; Bertino, J.R. Improved therapeutic index of methotrexate with "leucovorin rescue". *Cancer Res.* **1973**, *33*, 1729–1734.
37. Bailey, B.O. Acetaminophen hepatotoxicity and overdose. *Am. Fam. Physician* **1980**, *22*, 83–87. [PubMed]
38. Guo, J.; Xu, B.; Han, Q.; Zhou, H.; Xia, Y.; Gong, C.; Dai, X.; Li, Z.; Wu, G. Ferroptosis: A Novel Anti-tumor Action for Cisplatin. *Cancer Res. Treat.* **2018**, *50*, 445–460. [CrossRef]
39. Zhu, S.; Zhang, Q.; Sun, X.; Zeh, H.J., 3rd; Lotze, M.T.; Kang, R.; Tang, D. HSPA5 Regulates Ferroptotic Cell Death in Cancer Cells. *Cancer Res.* **2017**, *77*, 2064–2077. [CrossRef]
40. Roh, J.L.; Kim, E.H.; Jang, H.J.; Park, J.Y.; Shin, D. Induction of ferroptotic cell death for overcoming cisplatin resistance of head and neck cancer. *Cancer Lett.* **2016**, *381*, 96–103. [CrossRef]
41. Liu, D.S.; Read, M.; Cullinane, C.; Azar, W.J.; Fennell, C.M.; Montgomery, K.G.; Haupt, S.; Haupt, Y.; Wiman, K.G.; Duong, C.P.; et al. APR-246 potently inhibits tumour growth and overcomes chemoresistance in preclinical models of oesophageal adenocarcinoma. *Gut* **2015**, *64*, 1506–1516. [CrossRef] [PubMed]
42. Fransson, A.; Glaessgen, D.; Alfredsson, J.; Wiman, K.G.; Bajalica-Lagercrantz, S.; Mohell, N. Strong synergy with APR-246 and DNA-damaging drugs in primary cancer cells from patients with TP53 mutant High-Grade Serous ovarian cancer. *J. Ovarian Res.* **2016**, *9*, 27. [CrossRef] [PubMed]

43. Mohell, N.; Alfredsson, J.; Fransson, A.; Uustalu, M.; Bystrom, S.; Gullbo, J.; Hallberg, A.; Bykov, V.J.; Bjorklund, U.; Wiman, K.G. APR-246 overcomes resistance to cisplatin and doxorubicin in ovarian cancer cells. *Cell Death Dis.* **2015**, *6*, e1794. [CrossRef] [PubMed]

44. Ali, D.; Jonsson-Videsater, K.; Deneberg, S.; Bengtzen, S.; Nahi, H.; Paul, C.; Lehmann, S. APR-246 exhibits anti-leukemic activity and synergism with conventional chemotherapeutic drugs in acute myeloid leukemia cells. *Eur. J. Haematol.* **2011**, *86*, 206–215. [CrossRef]

45. Lachaier, E.; Louandre, C.; Godin, C.; Saidak, Z.; Baert, M.; Diouf, M.; Chauffert, B.; Galmiche, A. Sorafenib induces ferroptosis in human cancer cell lines originating from different solid tumors. *Anticancer Res.* **2014**, *34*, 6417–6422. [PubMed]

46. Gout, P.W.; Buckley, A.R.; Simms, C.R.; Bruchovsky, N. Sulfasalazine, a potent suppressor of lymphoma growth by inhibition of the x(c)- cystine transporter: A new action for an old drug. *Leukemia* **2001**, *15*, 1633–1640. [CrossRef] [PubMed]

47. Kim, W.; Lee, S.; Seo, D.; Kim, D.; Kim, K.; Kim, E.; Kang, J.; Seong, K.M.; Youn, H.; Youn, B. Cellular Stress Responses in Radiotherapy. *Cells* **2019**, *8*, 1105. [CrossRef] [PubMed]

48. Lang, X.; Green, M.D.; Wang, W.; Yu, J.; Choi, J.E.; Jiang, L.; Liao, P.; Zhou, J.; Zhang, Q.; Dow, A.; et al. Radiotherapy and Immunotherapy Promote Tumoral Lipid Oxidation and Ferroptosis via Synergistic Repression of SLC7A11. *Cancer Discov.* **2019**, *9*, 1673–1685. [CrossRef]

49. Lei, G.; Zhang, Y.; Koppula, P.; Liu, X.; Zhang, J.; Lin, S.H.; Ajani, J.A.; Xiao, Q.; Liao, Z.; Wang, H.; et al. The role of ferroptosis in ionizing radiation-induced cell death and tumor suppression. *Cell Res.* **2020**, *30*, 146–162. [CrossRef]

50. Soupene, E.; Kuypers, F.A. Mammalian long-chain acyl-CoA synthetases. *Exp. Biol. Med. (Maywood)* **2008**, *233*, 507–521. [CrossRef]

51. Dixon, S.J.; Winter, G.E.; Musavi, L.S.; Lee, E.D.; Snijder, B.; Rebsamen, M.; Superti-Furga, G.; Stockwell, B.R. Human Haploid Cell Genetics Reveals Roles for Lipid Metabolism Genes in Nonapoptotic Cell Death. *ACS Chem. Biol.* **2015**, *10*, 1604–1609. [CrossRef]

52. Wang, W.; Green, M.; Choi, J.E.; Gijon, M.; Kennedy, P.D.; Johnson, J.K.; Liao, P.; Lang, X.; Kryczek, I.; Sell, A.; et al. CD8(+) T cells regulate tumour ferroptosis during cancer immunotherapy. *Nature* **2019**, *569*, 270–274. [CrossRef]

53. Jiang, Z.; Lim, S.O.; Yan, M.; Hsu, J.L.; Yao, J.; Wei, Y.; Chang, S.S.; Yamaguchi, H.; Lee, H.H.; Ke, B.; et al. TYRO3 induces anti-PD-1/PD-L1 therapy resistance by limiting innate immunity and tumoral ferroptosis. *J. Clin. Investig.* **2021**, *131*. [CrossRef]

54. Ma, X.; Xiao, L.; Liu, L.; Ye, L.; Su, P.; Bi, E.; Wang, Q.; Yang, M.; Qian, J.; Yi, Q. CD36-mediated ferroptosis dampens intratumoral CD8(+) T cell effector function and impairs their antitumor ability. *Cell Metab.* **2021**, *33*, 1001–1012.e5. [CrossRef]

55. Ghosh, A.; Michel, J.; Dong, L.; Suek, N.; Zhong, H.; Budhu, S.; Henau, O.d.; Wolchok, J.; Merghoub, T. Abstract 4843: TP53-stabilization with APR-246 enhances antitumor effects of immune checkpoint blockade in preclinical models. *Cancer Res.* **2019**, *79*, 4843.

56. Warner, G.J.; Berry, M.J.; Moustafa, M.E.; Carlson, B.A.; Hatfield, D.L.; Faust, J.R. Inhibition of selenoprotein synthesis by selenocysteine tRNA[Ser]Sec lacking isopentenyladenosine. *J. Biol. Chem.* **2000**, *275*, 28110–28119. [CrossRef] [PubMed]

57. Bersuker, K.; Hendricks, J.M.; Li, Z.; Magtanong, L.; Ford, B.; Tang, P.H.; Roberts, M.A.; Tong, B.; Maimone, T.J.; Zoncu, R.; et al. The CoQ oxidoreductase FSP1 acts parallel to GPX4 to inhibit ferroptosis. *Nature* **2019**, *575*, 688–692. [CrossRef]

58. Doll, S.; Freitas, F.P.; Shah, R.; Aldrovandi, M.; da Silva, M.C.; Ingold, I.; Goya Grocin, A.; Xavier da Silva, T.N.; Panzilius, E.; Scheel, C.H.; et al. FSP1 is a glutathione-independent ferroptosis suppressor. *Nature* **2019**, *575*, 693–698. [CrossRef]

59. Garcia-Bermudez, J.; Baudrier, L.; Bayraktar, E.C.; Shen, Y.; La, K.; Guarecuco, R.; Yucel, B.; Fiore, D.; Tavora, B.; Freinkman, E.; et al. Squalene accumulation in cholesterol auxotrophic lymphomas prevents oxidative cell death. *Nature* **2019**, *567*, 118–122. [CrossRef] [PubMed]

60. LeBoeuf, S.E.; Wu, W.L.; Karakousi, T.R.; Karadal, B.; Jackson, S.R.; Davidson, S.M.; Wong, K.K.; Koralov, S.B.; Sayin, V.I.; Papagiannakopoulos, T. Activation of Oxidative Stress Response in Cancer Generates a Druggable Dependency on Exogenous Non-essential Amino Acids. *Cell Metab.* **2020**, *31*, 339–350.e4. [CrossRef] [PubMed]

61. Sayin, V.I.; LeBoeuf, S.E.; Singh, S.X.; Davidson, S.M.; Biancur, D.; Guzelhan, B.S.; Alvarez, S.W.; Wu, W.L.; Karakousi, T.R.; Zavitsanou, A.M.; et al. Activation of the NRF2 antioxidant program generates an imbalance in central carbon metabolism in cancer. *eLife* **2017**, *6*, e28083. [CrossRef] [PubMed]

62. Conlon, M.; Poltorack, C.D.; Forcina, G.C.; Armenta, D.A.; Mallais, M.; Perez, M.A.; Wells, A.; Kahanu, A.; Magtanong, L.; Watts, J.L.; et al. A compendium of kinetic modulatory profiles identifies ferroptosis regulators. *Nat. Chem. Biol.* **2021**, *17*, 665–674. [CrossRef] [PubMed]

 *antioxidants*

Article

PKCα Inhibition as a Strategy to Sensitize Neuroblastoma Stem Cells to Etoposide by Stimulating Ferroptosis

Lorenzo Monteleone [1,†], Andrea Speciale [2,†], Giulia Elda Valenti [1], Nicola Traverso [1], Silvia Ravera [3], Ombretta Garbarino [4], Riccardo Leardi [5], Emanuele Farinini [5], Antonella Roveri [6], Fulvio Ursini [6], Claudia Cantoni [1,7], Maria Adelaide Pronzato [1], Umberto Maria Marinari [1], Barbara Marengo [1,*,†] and Cinzia Domenicotti [1,*,†]

[1] General Pathology Section, Department of Experimental Medicine, University of Genoa, 16126 Genoa, Italy; lorenzo.monteleone@edu.unige.it (L.M.); giuliaelda.valenti@edu.unige.it (G.E.V.); nicola.traverso@unige.it (N.T.); claudia.cantoni@unige.it (C.C.); maidep@unige.it (M.A.P.); umm@unige.it (U.M.M.)

[2] Mutagenesis and Cancer Prevention Unit, IRCCS Ospedale Policlinico San Martino, 16132 Genoa, Italy; andrea.speciale@hsanmartino.it

[3] Human Anatomy Section, Department of Experimental Medicine, University of Genoa, 16126 Genoa, Italy; silvia.ravera@unige.it

[4] Department of Cardiovascular Medicine, IRCCS Humanitas Clinical and Research Center, Rozzano, 20089 Milan, Italy; ombretta.garbarino@humanitasresearch.it

[5] Department of Pharmacy, University of Genoa, 16126 Genoa, Italy; riclea@difar.unige.it (R.L.); farinini@difar.unige.it (E.F.)

[6] Department of Molecular Medicine, University of Padua, 35122 Padua, Italy; antonella.roveri@unipd.it (A.R.); fulvio.ursini@unipd.it (F.U.)

[7] Laboratory of Clinical and Experimental Immunology, Integrated Department of Services and Laboratories, IRCCS Istituto G. Gaslini, 35122 Genoa, Italy

[*] Correspondence: barbara.marengo@unige.it (B.M.); cinzia.domenicotti@unige.it (C.D.); Tel.: +39-010-3538831 (B.M.); +39-010-3538830 (C.D.)

[†] These authors contributed equally to this work.

Citation: Monteleone, L.; Speciale, A.; Valenti, G.E.; Traverso, N.; Ravera, S.; Garbarino, O.; Leardi, R.; Farinini, E.; Roveri, A.; Ursini, F.; et al. PKCα Inhibition as a Strategy to Sensitize Neuroblastoma Stem Cells to Etoposide by Stimulating Ferroptosis. *Antioxidants* **2021**, *10*, 691. https://doi.org/10.3390/antiox10050691

Academic Editor: José M. Matés

Received: 29 March 2021
Accepted: 24 April 2021
Published: 28 April 2021

Publisher's Note: MDPI stays neutral with regard to jurisdictional claims in published maps and institutional affiliations.

Abstract: Cancer stem cells (CSCs) are a limited cell population inside a tumor bulk characterized by high levels of glutathione (GSH), the most important antioxidant thiol of which cysteine is the limiting amino acid for GSH biosynthesis. In fact, CSCs over-express xCT, a cystine transporter stabilized on cell membrane through interaction with CD44, a stemness marker whose expression is modulated by protein kinase Cα (PKCα). Since many chemotherapeutic drugs, such as Etoposide, exert their cytotoxic action by increasing reactive oxygen species (ROS) production, the presence of high antioxidant defenses confers to CSCs a crucial role in chemoresistance. In this study, Etoposide-sensitive and -resistant neuroblastoma CSCs were chronically treated with Etoposide, given alone or in combination with Sulfasalazine (SSZ) or with an inhibitor of PKCα (C2-4), which target xCT directly or indirectly, respectively. Both combined approaches are able to sensitize CSCs to Etoposide by decreasing intracellular GSH levels, inducing a metabolic switch from OXPHOS to aerobic glycolysis, down-regulating glutathione-peroxidase-4 activity and stimulating lipid peroxidation, thus leading to ferroptosis. Our results suggest, for the first time, that PKCα inhibition inducing ferroptosis might be a useful strategy with which to fight CSC chemoresistance.

Keywords: cancer stem cells; chemoresistance; glutathione; lipid peroxidation; ZEB-1; GPX4; ferroptosis

1. Introduction

Neuroblastoma (NB), a pediatric cancer originating from the neural crest cells, accounts for about 6% of all cancers in children [1]. The development of the neural crest is correlated with the epithelial–mesenchymal transition (EMT) [2,3], a phenomenon that is strictly related to MYCN amplification [4], a negative prognostic factor that characterizes advanced tumor stages and aggressive progression of NB [5,6].

Several studies have reported that the genes involved in the EMT process are also implicated in regulating cancer stem cell (CSC) generation and propagation [7]. CSCs are a limited cell population found in different tumors, including NB [8], and are involved in various stages of the tumorigenic process and in cancer response to therapy [9–11]. In this regard, it has been demonstrated that CSCs, having physiologically low levels of reactive oxygen species (ROS) and high antioxidant defenses in comparison to cancer cells, are strongly involved in chemoresistance [12]. The maintenance of these low ROS levels is due to the ability of CSCs to synthesize high amounts of glutathione (GSH) by over-expressing xCT, a membrane transporter able to modulate the intracellular levels of cysteine, the limiting amino acid involved in GSH synthesis [13]. Moreover, it has also been found that CSCs express high levels of CD44, a well-known stemness marker [14] that is able to induce xCT membrane stabilization [15] and whose expression is modulated by protein kinase C α (PKCα) [16,17].

Based on these considerations, anti-tumor research should be focused on discovering innovative approaches able to limit GSH availability in cancer cells and, in particular, in CSCs. In fact, it has been shown that L-Buthionine-sulfoximine (BSO), a GSH-depleting agent, sensitizes NB cells to chemotherapeutic drugs [18–21] and CSCs to radiotherapy [13], by inducing ROS over-production. However, the clinical use of BSO is not actually feasible due to the secondary toxic effects of this compound, in addition to those resulting from the use of traditional radiation and drug therapies [22].

Therefore, in order to investigate new nontoxic therapeutic approaches, both Etoposide-sensitive and Etoposide-resistant NB-derived CSCs were long-term treated with Etoposide, alone or in combination with sulfasalazine (SSZ), or with an inhibitor of the PKCα-dependent pathway (C2-4, a peptide inhibitor), which can target xCT directly or indirectly, respectively. The herein results demonstrate that both combinations of Etoposide with SSZ or C2-4 efficiently counteract the propagation of Etoposide-resistant CSCs by preventing EMT transition and by triggering ferroptosis via down-regulation of GPX4 activity.

2. Materials and Methods

2.1. Chemicals

Etoposide was purchased from Calbiochem (Merk KGaA, Darmstadt, Germany) and xCT inhibitor sulfasalazine (SSZ) and PKCα peptide inhibitor (C2-4) from Cayman Chemical (Ann Arbor, MI, USA). The stock solutions of these compounds were prepared using dimethyl sulfoxide (DMSO, Sigma-Aldrich, St. Louis, MO, USA) as a solvent.

2.2. Cancer Stem Cell (Csc) Generation

To select neurospheres (3D cultures), floating parental cells derived from 2D cultures of HTLA-230 and HTLA-ER [23] cells were harvested, centrifuged (117 rcf $\times$ 8 min), seeded (16×10^4) and grown in DMEM-F12 Knock-out (Life Technologies, Carlsbad, CA, USA) containing 1% penicillin/streptomycin (Euroclone S.p.A., Pavia, Italy), 2% B27 (Life Technologies), 40 ng/mL basal growth factor for fibroblasts (bFGF) (R&D Systems, Inc., Minneapolis, MN, USA) and 20 ng/mL epidermal growth factor (EGF) (Life Technologies) [24]. It is important to outline that the expression of stem cell markers was evaluated (see Section 2.4) in order to verify the staminality and, only after this characterization, were the neurospheres renamed as CSCs. The HTLA-CSCs and ER-CSCs were split once a week to favor their propagation and, at each passage, the culture medium consisted of 50% fresh medium and 50% of the medium in which the cells were grown [24]. In order to analyze CSC propagation, at any split, the CSCs were collected, centrifuged, mechanically dissociated in PBS and then counted using a Burker chamber (Marienfeld, Germany).

Parental HTLA-230 and HTLA-ER cells were maintained in RPMI 1640 medium (Euroclone) containing 10% fetal bovine serum (FBS; Euroclone), 2 mM glutamine (Euroclone), 1% penicillin/streptomycin (Euroclone), 1% sodium pyruvate and 2% amino acid solution (Sigma-Aldrich, St. Louis, MO, USA). All cell lines were periodically tested with semi-quantitative PCR for the evaluation of any mycoplasma contamination (Mycoplasma Reagent Set; Euroclone).

2.3. Treatments

Both CSC populations were treated once a week with 1 mM BSO or three times a week with 0.1 µM C2-4, with 5 µM SSZ or with 1.25 µM Etoposide, administered once a week alone or in combination with C2-4 or SSZ. The treatments were protracted for 6 weeks and stock solutions (1 mM C2-4, 50 mM SSZ and 50 mM Etoposide) were prepared by dissolving each drug in DMSO. In order to demonstrate that the doses of DMSO used to dissolve the compounds did not affect the subsequent analyses, pilot studies were performed in parallel with the experiments.

2.4. Immunofluorescence and Cytofluorimetric Analysis

In order to investigate the expression of CD44, Oct3/4, CD44v9, xCT, and vimentin (used as a control of cell permeabilization), cells were analyzed by flow cytometry. In detail, both untreated CSC populations were collected, centrifuged (117 rcf × 8 min), and then re-suspended in the TrypLE ™ Express Enzyme 1X solution (Invitrogen, Carlsbad, CA, USA). After monitoring CSC disaggregation by microscope analysis, cells were diluted in PBS (137 mM NaCl, 2.6 mM KCl, 10 mM Na_2HPO_4, 1.8 mM KH_2PO_4, pH 7.4). For intracellular staining, prior to incubation with primary mAbs, cells were permeabilized and fixed for 30 min at 4 °C with the BD Cytofix/Cytoperm ™ kit (BD Biosciences, Franklin Lakes, NJ, USA). After this step, CSCs were re-suspended in 1% PBS/Saponin solution (Sigma-Aldrich, Milano, Italy) and then centrifuged (365 rcf × 10 min). Subsequently, CSCs were incubated for 30 min at 4 °C with primary antibodies: rabbit polyclonal anti-CD44 (Cosmo Bio, Tokio, Japan), rat monoclonal anti-Oct3/4 (EM92, PE-linked; eBioscience, San Diego, CA, USA), rat monoclonal anti-CD44v9 (RV3, Cosmo Bio), rabbit polyclonal anti-xCT (H-121, Santa Cruz Biotechnology, Dallas, TX, USA), and mouse monoclonal anti-vimentin (V9, Santa Cruz Biotechnology).

Labelled cells were then washed with 2% FBS solution in PBS and incubated for 30 min at 4 °C with the appropriate isotype-matched secondary antibodies: goat anti-rabbit-IgG-PE (Southern Biotech, Birmingham, AL, USA), goat anti-rat-IgG-FITC (Sigma-Aldrich), goat anti-mouse-IgG-PE (Southern Biotech). Finally, all samples were fixed with PBS/1% formaldehyde and analyzed with the FACScalibur flow cytometer (BD Biosciences). Data were analyzed by the CellQuest software (BD Biosciences).

2.5. HPLC Analysis of Intracellular Glutathione Levels

Intracellular levels of reduced glutathione (GSH) and oxidized glutathione (GSSG) were assessed by high performance liquid chromatography (HPLC) using the methods reported by Reed for total GSH [25] and Asensi for GSSG quantification [26].

Untreated and treated CSCs (1 × 10⁶ cells) were collected and centrifuged (117 rcf × 8 min). Then, the pellets were washed in PBS, precipitated with 10% perchloric acid (PCA) and the thiol groups were blocked with iodacetic acid (alkaline pH 8–9, 10 min, room temperature, in the dark). Subsequently, the analytes were converted to 2,4-dinitrophenyl derivatives by incubating the samples with 1% 1-Fluoro-2,4-dinitrobenzene (FDNB) at 4 °C in the dark overnight. The quantitative determination of the derivatized analytes was carried out by HPLC equipped with an NH2 Spherisorb column and a UV detector set at 365 nm with a flow rate of 1.25 mL/min. The mobile phase was maintained at 80% solution A (80% methanol in water) and 20% solution B (0.5 M sodium acetate in 64% methanol in water) for 5 min, followed by a 10-min-linear gradient to 1% A and 99% B. The chromatography was performed with gradient elution.

In order to evaluate GSSG, the samples were incubated with 20 mM N-ethyl-maleimide (NEM) in PCA and, after precipitation and alkalization of the sample (alkaline pH 8–9, 10 min, room temperature, in the dark), derivatization was carried out by adding 1% FDNB.

In order to allow the elution of GSSG, the mobile phase was kept at 99% solution B for a further 15 min.

The data obtained was normalized by the intracellular amount of protein and expressed as µEq/mg of protein.

2.6. ATP Synthesis

The ATP content was measured using the luciferin/luciferase ATP bioluminescence assay kit CLSII (Roche, Basel, Switzerland) on a Luminometer (Triathler, Bioscan, Washington, DC, USA) [23]. ATP standard solutions (Roche, Basel, Switzerland) in the concentration range of 10^{-10}–10^{-7} M were used for calibration. Assay was carried out at 37 °C over 2 min by measuring formed ATP from added ADP. Untreated and treated CSCs were incubated for 10 min with the assay buffer (10 mM Tris-HCl pH 7.4, 50 mM KCl, 1 mM EGTA, 2 mM EDTA, 5 mM KH_2PO_4, 2 mM $MgCl_2$, 0.6 mM Ouabain, 0.040 mg/mL Ampicillin, 0.2 mM di(adenosine-5′) penta-phosphate, 0.2 mM and 5 mM pyruvate plus 2.5 mM malate). Afterwards, ATP synthesis was induced by the addition of 0.3 mM ADP.

2.7. Oxygen Consumption Rate (OCR)

In order to measure the respiratory activity, 2×10^5 cells were analyzed by using an amperometric electrode for O2 placed in an isolated chamber and stirring maintained at 37 °C, as previously reported [23]. Untreated and treated CSCs were collected and centrifuged (117 rcf × 8 min). Then, the pellets were suspended in the assay buffer (137 mM NaCl, 5 mM KH_2PO_4, 5 mM KCl, 0.5 mM EDTA, 3 mM $MgCl_2$ and 25 mM Tris–HCl, pH 7.4), and permeabilized with 0.3% digitonin for 10 min. Then, the sample was transferred to the chamber; so as to measure the maximum respiration rate, 5 mM pyruvate plus 2.5 mM malate were added.

2.8. Glucose Consumption

Glucose consumption was assessed by measuring its concentration using a double beam spectrophotometer (UNICAM UV2, Analytical S.n.c., Langhirano, PR, Italy), by the hexokinase (HK) and glucose 6 phosphate dehydrogenase (G6PD) coupling system, following the reduction of NADP at 340 nm. [23]. The buffer assay contained 100 mM Tris HCl, pH 7.4, 2 mM ATP, 10 mM NADP, 2 mM $MgCl_2$, 2 IU of HK and 2 IU of G6PD. The reaction was started after the addition of 5 µL of CSC medium.

2.9. Lactate Release

The concentration of lactic acid released by CSCs in the culture medium was analyzed by spectrophotometric analysis as previously reported [23]. The assay buffer contained 100 mM Tris/HCl (pH 8), 5 mM NAD+ and 1 IU/mL of lactate dehydrogenase. Samples were analyzed before and after the addition of 4 µg of purified lactate dehydrogenase.

2.10. MDA Production

Malondialdehyde (MDA) levels were analyzed by thiobarbituric acid assay (TBARS) as previously reported [23,27]. Briefly, 50 µg of total proteins, derived from each sample, were dissolved in 300 µL of Milli-Q water and added to 600 µL of TBARS solution containing 15% trichloroacetic acid (TCA) and 26 mM thiobarbituric acid (TBA) in 0.25 N HCl. This mixture was incubated for 40 min at 100 °C and then centrifuged (20,800 rcf × 2 min). Subsequently, the supernatant was collected and analyzed on the spectrophotometer using a wavelength of 532 nm. In order to calculate MDA concentration, the absorbance of each sample was compared with that obtained from a standard curve evaluated with known concentrations of MDA (0.75, 1, and 2 µM).

2.11. ROS Production

Detection of ROS levels was evaluated by incubating CSCs with 5µM 2′-7′ dichlorofluorescein-diacetate (DCFH-DA; Sigma-Aldrich) as previously reported [23].

2.12. Western Blot Analyses

Immunoblots were carried out according to standard methods [28] using rabbit antibody anti-N-Cadherin (D4R1H), anti-ZEB-1 (D80D3), anti-Vimentin (D21H3), anti-β-Catenin (D10A8), anti-Claudin-1 (D5H1D) (Cell Signaling Technology Inc., Danvers, MA, USA Upstate, Lake Placid, NY, USA) and anti-GPX4 (Abcam, Cambridge, UK). Anti-rabbit secondary antibodies coupled with horseradish peroxidase (Cell signaling Technologies) were utilized.

2.13. GPX4 Activity

GPX4 activity was evaluated by standard methods [29]. In detail, cell pellets were re-suspended in 0.75 mL lysis buffer (0.1 M Tris-HCl, 0.25 M sucrose, protease inhibitors, pH 7.5) and then sonicated and used in the test (0.1–0.2 mL of sample per test). Samples were mixed with the assay buffer (0.1 M Tris-HCl pH 7.8, 5 mM EDTA, 5 mM GSH, 0.1% (v/v) Triton X-100, 0.16 mM NADPH and 0.6 IU/mL Glutathione Reductase (GR)) and incubated for 5 min at 25 °C. Then the baseline was recorded at 340 nm for about 1 min and finally the enzymatic activity was started by adding phosphatidylcholine hydroperoxide (0.020 mM). The quantification of the activity was done on the basis of the net speed with which the absorbance decreases after the addition of the substrate (net speed = speed after the addition of the substrate − baseline speed).

2.14. Principal Component Analyses (PCA)

PCA is a data display method for multivariate data. Given a data set in which each sample is described by n variables, PCA aims to find new directions, linear combinations of the original ones [30].

The first component (PC1) corresponds to the direction explaining the maximum variance, while PC2 is the direction, orthogonal to PC1, explaining the maximum variance not explained by PC1, and so on. The result of such a transformation is that a limited number of components is sufficient to explain the relevant part of the information.

The loadings are the coefficients of the linear combinations corresponding to the PCs. By plotting them in a loading plot it is possible to understand the relationships among the variables in the multivariate space.

On the other side, the score plot (the scores being the coordinates of the samples in the new space defined by the PCs) allows visualization of the location of the samples in the space described by the PCs, hence making possible to check similarities and differences among the samples. The elaborations and the plots were carried out by using CAT software [31].

2.15. Statistical Analysis

Results were expressed as mean ± SEM from at least four independent experiments. The statistical significance of parametric differences among the sets of experimental data was evaluated by one-way ANOVA and Dunnett's test for multiple comparisons.

3. Results and Discussion

3.1. Neurospheres, Isolated from Parental HTLA-230 and HTLA-ER Cells, Are Cancer Stem Cells (CSCs) and GSH Plays a Crucial Role in Their Generation and Propagation

In order to verify if the floating neurospheres, isolated from HTLA-230 and HTLA-ER cells [23], were formed by stem cells, the expression of CD44 and Oct-3/4, known markers of stemness, was analyzed. As shown in Figure 1a, isolated neurospheres expressed both CD44 and Oct-3/4, demonstrating that both cell populations have staminality characteristics and, consequently, have been referred to as HTLA- and ER-cancer stem cells (HTLA-CSCs and ER-CSCs).

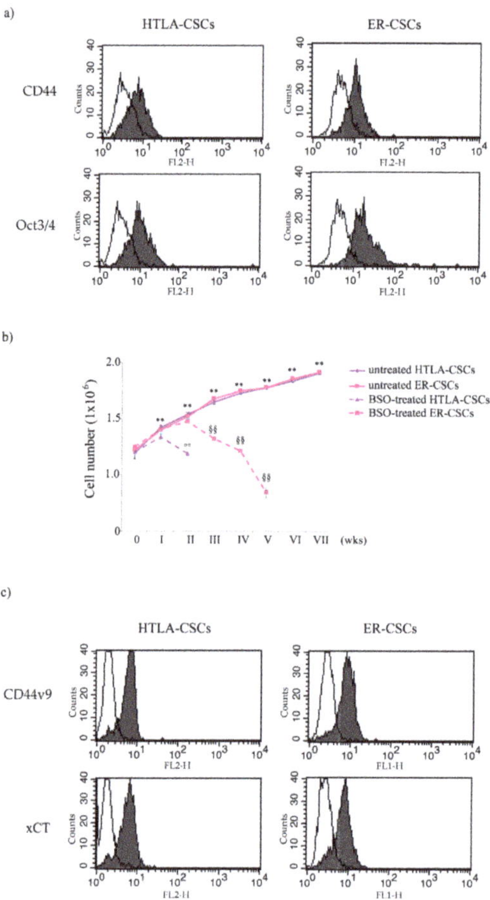

Figure 1. Flow cytometric analysis of CD44, Oct3/4 (**a**) and of CD44v9 and xCT (**c**) expression in untreated HTLA-CSCs and ER-CSCs. Cells were permeabilized, stained with mAbs to the indicated molecules, and analyzed by flow cytometry. Grey profiles indicate cells stained with the different mAbs, while white profiles correspond to isotype control. One representative experiment of five is shown. (**b**) Evaluation of cell number in HTLA-CSCs and ER-CSCs chronically treated with 1mM BSO. Both CSC populations were treated (once a week for six weeks) with 1 mM BSO and, at any split, disaggregated CSCs were counted using a Burker chamber as described in Materials and Methods. ** $p < 0.01$ vs. time 0; °° $p < 0.01$ vs. untreated HTLA-CSCs; §§ $p < 0.01$ vs. untreated ER-CSCs.

Since the presence of CSCs [12] and glutathione (GSH) play a crucial role in chemoresistance [23], both CSC populations were chronically treated (once a week for 6 weeks) with 1 mM BSO in order to investigate whether a relationship between GSH levels and stemness exists.

As shown in Figure 1b, BSO markedly counteracted the formation of HTLA-CSCs after two weeks of treatment, while a similar effect was observed in ER-CSCs after five weeks. These results demonstrate that GSH plays a crucial role in the formation and maintenance of CSCs and suggest that its depletion could be employed to increase sensitivity of CSCs to therapeutic approaches.

However, although BSO has been found to counteract CSC formation, both in vitro and in vivo [32–34], and has been included in NB clinical trials [35], its therapeutic use is limited by its short half-life and by its non-selective depleting effect on neoplastic cells [22].

In order to overcome this limitation, an effective strategy could be the indirect modulation of GSH levels by acting on GSH-related targets. In this context, it has been reported that CD44, a known stemness membrane marker [17,36], and, in particular, its variant 9 (CD44v9) contributes to the stabilization of xCT, an anti-port protein involved in the uptake of cystine, an amino acid that is essential for GSH biosynthesis. Considering that the expression of CD44 is modulated by protein kinase C (PKC)α [17,36], an opportunity would be to reduce the CD44 expression by inhibiting this kinase. As shown in Figure 1c, both CSC populations expressed CD44v9 and xCT, thus justifying the use of C2-4, a PKCα inhibitor, or of sulphasalazine (SSZ), a xCT inhibitor, in the reported studies.

3.2. Etoposide Prevents the Formation of HTLA-CSCs after 3 Weeks of Treatment While It Completely Counteracts the Formation of ER-CSCs when It Is Combined with SSZ or C2-4 for 6 Weeks

By analyzing HTLA-CSCs, it was observed that Etoposide reduced the propagation of the formed CSCs by 35% after only one week of exposure and totally prevented CSC formation after three weeks (Figure 2a). Moreover, similar effects were obtained in CSCs co-treated with either Etoposide and SSZ (Figure 2a, left panel) or Etoposide and C2-4 (Figure 2a, right panel). However, treatments with SSZ (Figure 2a, left panel) or C2-4 alone (Figure 2a, right panel) did not significantly alter this parameter during the analyzed time points.

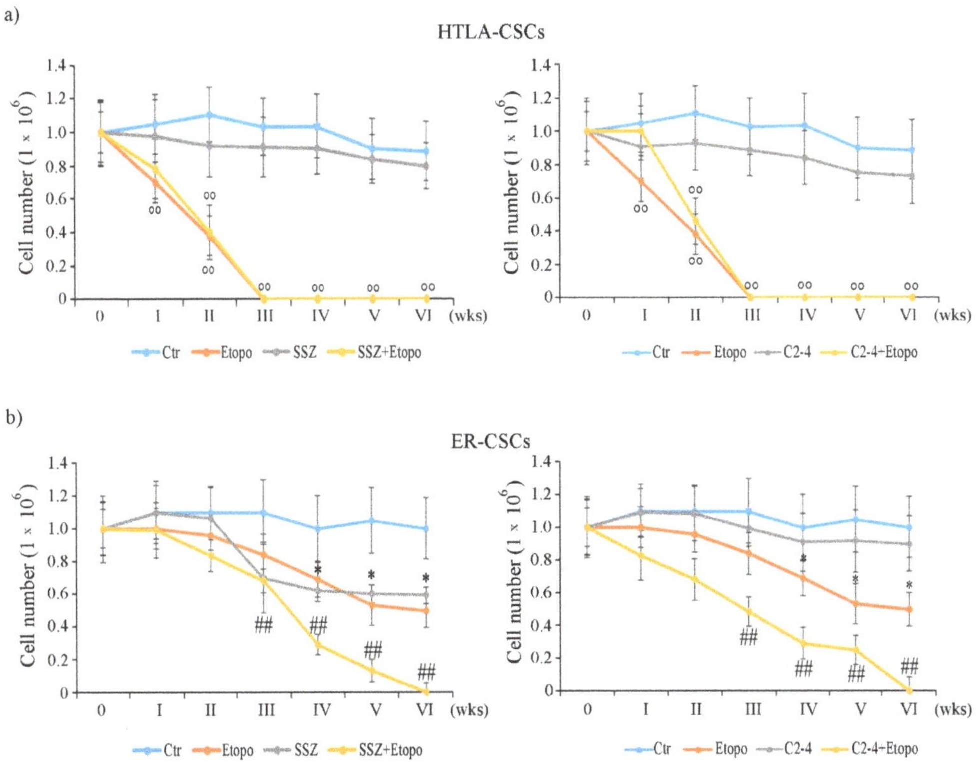

Figure 2. Evaluation of cell number in HTLA-CSCs (**a**) and ER-CSCs (**b**) treated three times a week with 0.1 µM C2-4, with 5 µM SSZ or with 1.25 µM Etoposide, administered once a week alone or in combination with C2-4 or SSZ. The treatments were protracted for six weeks. In order to evaluate CSC propagation, at any split, disaggregated CSCs were counted using a Burker chamber as described in Materials and Methods. °° $p < 0.01$ vs. untreated HTLA-CSCs (Ctr); * $p < 0.05$ vs. untreated ER-CSCs (Ctr); ## $p < 0.01$ vs. Etoposide-treated ER CSCs.

In contrast, the same treatments did not induce any significant alteration in the formation/propagation of ER-CSCs until two weeks of exposure (Figure 2b). In fact, Etoposide treatment reduced CSC propagation by 38% at four weeks, and this effect remained unchanged until six weeks of treatment (Figure 2b). By analyzing the results obtained in co-treated CSCs, it was observed that Etoposide combined with SSZ or Etoposide combined with C 2-4 reduced the ER-CSC propagation by 30% and 55%, respectively, after three weeks and totally prevent the formation of CSCs after six weeks (Figure 2b). Moreover, as shown in Figure 2b left panel, the results obtained in ER-CSCs, treated with SSZ alone for six weeks, were comparable to those found in CSCs exposed to only Etoposide, whereas the treatment with C2-4 did not affect CSC formation and propagation (Figure 2b, right panel).

Since the formation of HTLA-CSCs was totally inhibited after three weeks of treatments and that of ER-CSCs after six weeks, the following analyses were performed after two and five weeks respectively.

Together these results demonstrate that ER-CSCs maintain their resistance to Etoposide and show that ER-CSC generation is totally inhibited by co-treatments with SSZ or C2-4. Moreover, the data obtained confirm that the inhibition of xCT is a useful strategy for enhancing the effect of chemotherapy [37–42] and, although the use of SSZ is limited due to its low bioavailability, its employment has been recently considered to treat high risk NB [43]. In addition, the results obtained in C2-4-co-treated cells suggest that the inhibition of PKCα could be a novel way to sensitize CSCs to therapy by sustaining GSH depletion. In this regard, several clinical trials (NCT00042679; NCT00003236; NCT00017407) have been carried out using Aprinocarsen, an anti-sense oligonucleotide that targets PKCα mRNA, even if, unfortunately, the results obtained are not encouraging [44]. However, a large number of studies have reported the efficacy of PKCα inhibitors in sensitizing cancer cells to chemotherapeutic drugs [45–48].

3.3. Etoposide Induces a Marked Depletion of GSH in HTLA-CSCs after Two Weeks While the Same Effect Is Not Observed in ER-CSCs until Five Weeks of Treatment

Untreated HTLA-CSCs and ER-CSCs displayed comparable GSH levels (about 17.07 µEq/g and 17.37 µEq/g proteins, respectively). After two weeks of Etoposide exposure, an 80% reduction in GSH was observed in HTLA-CSCs with respect to untreated CSCs (Figure 3a). At the same time, both co-treatment conditions (C2-4 + Etoposide or SSZ + Etoposide), exerted similar effects to those detected in Etoposide-treated CSCs (Figure 3a), while C2-4 or SSZ alone did not significantly alter the GSH amount (Figure 3a). Notably, the same two-week treatments in ER-CSCs did not induce any significant changes in GSH levels (Figure 3b).

After five weeks, untreated HTLA-CSCs and ER-CSCs displayed a different GSH content of 4.7 µEq/g and 8.99 µE/g protein, respectively. Since the five-week-treatment with Etoposide, alone or in combination, totally counteracted HTLA-CSC formation (Figure 2a), it was not possible to analyze GSH. As shown in Figure 3c, Etoposide decreased GSH of ER-CSCs by 60% in comparison to untreated CSCs, and co-treatments, with C2-4 or SSZ, which did not modify the GSH-depleting effect of Etoposide.

The amount of GSSG, the oxidized form of GSH, was below the detection limits in both CSC populations after two weeks while it was depleted by 60–65% in ER-CSCs after the five-week-treatment with Etoposide, alone or in combination with C2-4 or SSZ (data not shown).

These results demonstrate that five-week treatment with Etoposide markedly reduces GSH levels in ER-CSCs and that this effect is only due to the action of Etoposide, given that a similar loss of GSH levels is observed in co-treated CSCs and neither C2-4 nor SSZ, alone or combined with Etoposide, modulate GSH levels.

This data demonstrates that the survival of Etoposide-resistant CSCs is not totally dependent on GSH since, even though Etoposide exerts a severe GSH-depleting action, it does not totally counteract CSC generation, suggesting that other factors might contribute to the maintenance of cancer stem cell survival. In this regard, it has been widely demonstrated that cancer cells usually depend on glycolysis for energy production [49,50] while chemoresistant cancer cells [23,51,52] and CSCs undergo a metabolic rewiring, stimulating OXPHOS with a major ATP production [53–56]. Therefore, based on these findings, the metabolic profile of untreated and treated CSCs was analyzed.

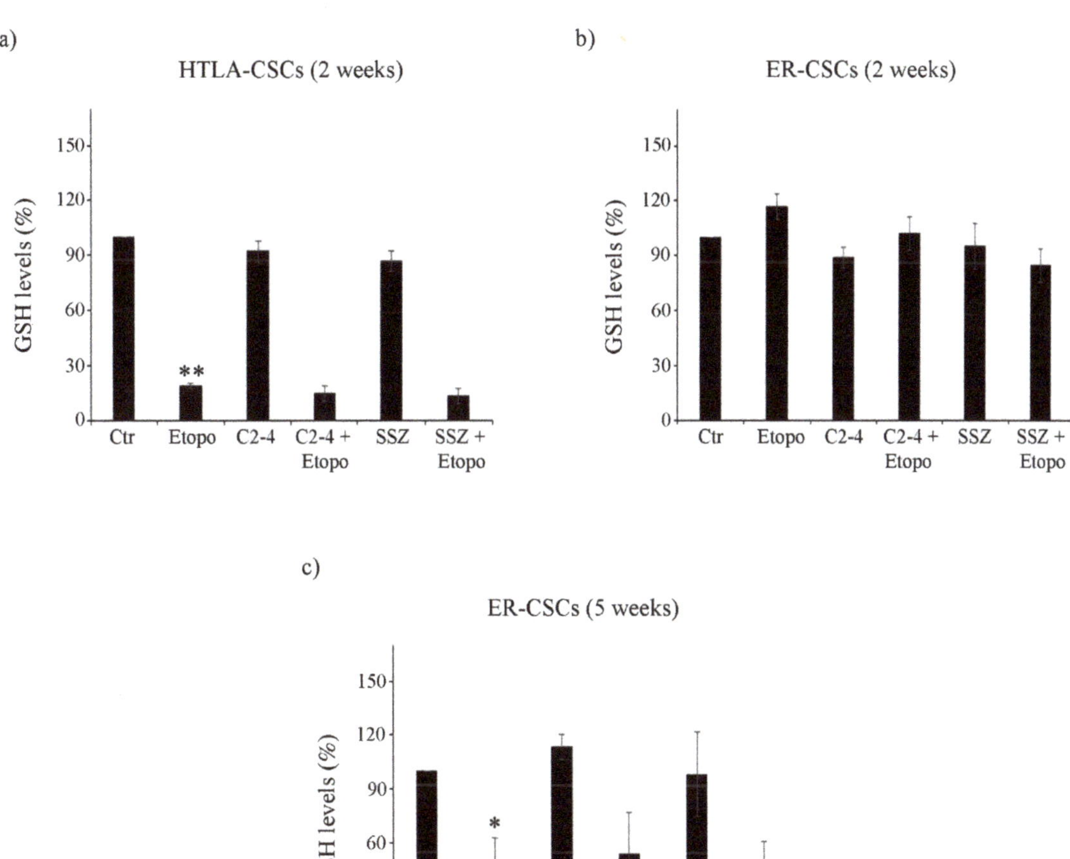

Figure 3. Glutathione (GSH) levels in both CSC populations treated for two (**a,b**) or five weeks (**c**). HTLA-CSCs and ER-CSCs were treated three times a week with 0.1 μM C2-4, with 5 μM SSZ or with 1.25 μM Etoposide, administered once a week alone or in combination with C2-4 or SSZ. GSH concentrations were determined by HPLC analysis and expressed as percentages of the control value. * $p < 0.05$ vs. untreated CSCs (Ctr); ** $p < 0.01$ vs. untreated CSCs (Ctr).

3.4. ER-CSCs Maintain an Efficient OXPHOS Metabolism Which Is Impaired Only by Co-Treatments

As shown in Figure 4a, ATP synthesis, detected in untreated HTLA-CSCs, after two weeks, was doubled in respect of that of ER-CSCs. In HTLA-CSCs, two weeks of Etoposide, alone or in combination with C2-4 or SSZ, reduced ATP production by 70% in respect of untreated CSCs (Figure 4a). Notably, in ER-CSCs, two weeks of Etoposide was ineffective and only co-treatment of Etoposide with SSZ reduced ATP synthesis by 25% compared to both untreated and Etoposide treated CSCs (Figure 4b).

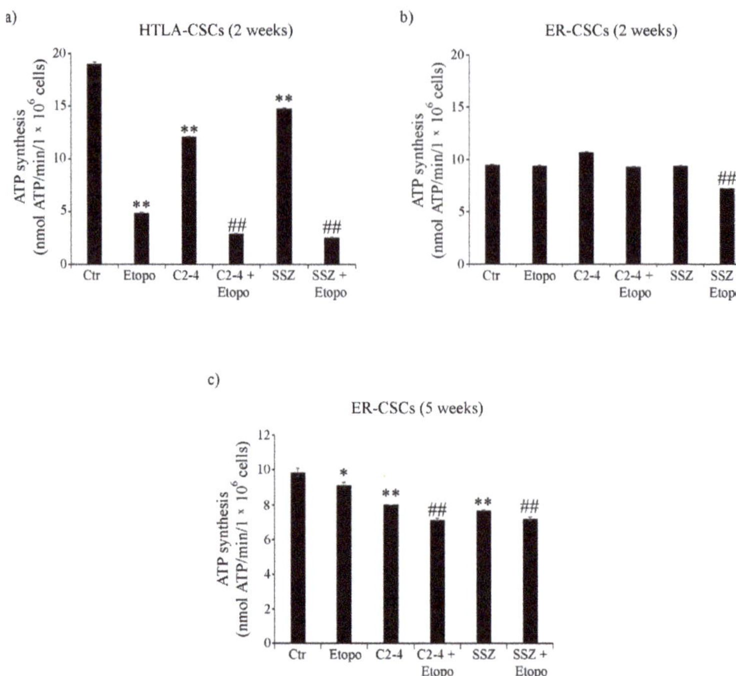

Figure 4. ATP synthesis in both CSC populations treated for two (**a,b**) or five weeks (**c**). HTLA-CSCs and ER-CSCs were treated three times a week with 0.1 μM C2-4, with 5 μM SSZ or with 1.25 μM Etoposide, administered once a week alone or in combination with C2-4 or SSZ. Results were reported as nmol ATP/min/10^6 cells. Bar graph summarizes quantitative data of means ± S.E.M. of three independent experiments. * $p < 0.05$ vs. untreated CSCs (Ctr); ** $p < 0.01$ vs. untreated CSCs (Ctr); ## $p < 0.01$ vs. Etoposide-treated CSCs.

After five weeks, untreated CSC populations had comparable levels of ATP (about 10 nmol/min/1 $\times$ 10^6 cells). Since five weeks of treatment with Etoposide, alone or in combination, totally counteracted HTLA-CSC formation (Figure 2a), the evaluation of ATP synthesis was not possible while in ER-CSCs, ATP decreased by 10% after Etoposide and by 35% by the co-treatments (Figure 4c).

Moreover, in HTLA-CSCs, two weeks of Etoposide treatment reduced the oxygen consumption rate (OCR) by 75% in respect of untreated CSCs (Figure 5a) and similar results were obtained in co-treated ones (Figure 5a). At the same time point, in ER-CSCs, the OCR was not modified either by Etoposide or co-treatments and it was increased by 15% and 35% only by C2-4 or SSZ respectively (Figure 5b). Since the five-week-treatment with Etoposide, alone or in combination, totally counteracted HTLA-CSC formation (Figure 2a), OCR analysis was not possible. Notably, in ER-CSCs, none of the treatments significantly altered the OCR (Figure 5c).

To verify the efficiency of OXPHOS, the P/O value was calculated. In detail, in both CSC populations, the P/O ratio was 2.3 ± 0.2, which is comparable to the physiological level of 2.5 as reported by Hinkle [57] and interestingly, only SSZ, administered alone or in combination with Etoposide for two or five weeks, led to a P/O value of 1.8 ± 0.15.

In order to complete the analysis of cell metabolism, glucose consumption and lactate release were evaluated. As shown in Figure 6a, untreated HTLA-CSCs consumed about 40% more glucose than untreated ER-CSCs. In HTLA-CSCs, two-week-Etoposide exposure increased glucose consumption by 60% in respect of untreated CSCs and the co-treatments did not modify the Etoposide-induced effect (Figure 6a). In ER-CSCs, treatments with either Etoposide, C2-4 or SSZ increased glucose consumption by 30% as compared to untreated CSCs, and the co-treatments with C2-4 or SSZ contributed to stimulating this parameter by a further 60% and 25%, respectively (Figure 6b).

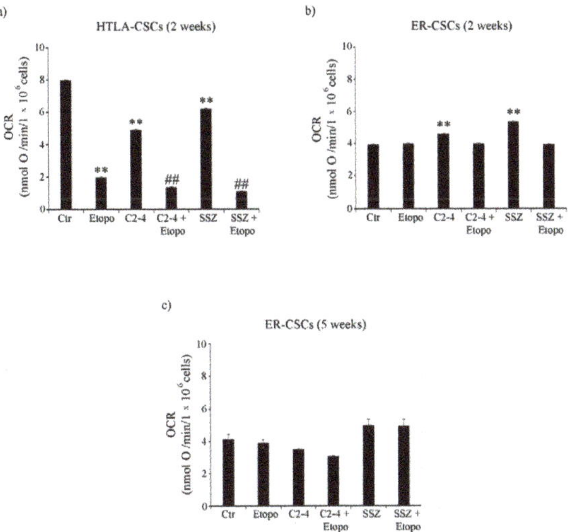

Figure 5. Oxygen consumption rate (OCR) in both CSC populations treated for two (**a,b**) or five weeks (**c**). HTLA-CSCs and ER-CSCs were treated three times a week with 0.1 μM C2-4, with 5 μM SSZ or with 1.25 μM Etoposide, administered once a week alone or in combination with C2-4 or SSZ. Results were reported as nmol O/min/10^6 cells. Bar graph summarizes quantitative data of means ± S.E.M. of three independent experiments. ** $p < 0.01$ vs. untreated CSCs (Ctr); ## $p < 0.01$ vs. Etoposide-treated CSCs.

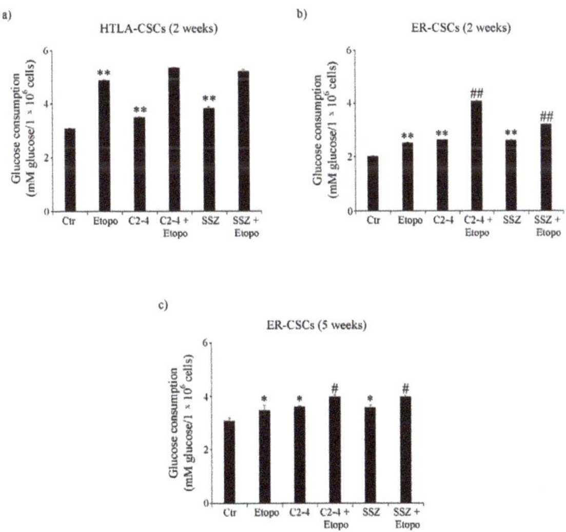

Figure 6. Glucose consumption in both CSC populations treated for two (**a,b**) or five weeks (**c**). HTLA-CSCs and ER-CSCs were treated three times a week with 0.1 μM C2-4, with 5 μM SSZ or with 1.25 μM Etoposide, administered once a week alone or in combination with C2-4 or SSZ. Results were reported as mM glucose/10^6 cells. Bar graph summarizes quantitative data of means ± S.E.M. of three independent experiments. * $p < 0.05$ vs. untreated CSCs (Ctr); ** $p < 0.01$ vs. untreated CSCs (Ctr); # $p < 0.05$ vs. Etoposide-treated CSCs; ## $p < 0.01$ vs. Etoposide-treated CSCs.

Since five-week-treatment with Etoposide, alone or in combination, totally counteracted HTLA-CSC formation (Figure 2a), glucose consumption analysis was not possible. Noteworthy, in ER-CSCs, the trend observed after five weeks was similar to that observed after two weeks and, also in this case, the co-treatments stimulated the Etoposide-induced effect on glucose consumption (Figure 6c).

As expected, in line with the highest level of glucose consumption observed in untreated HTLA-CSCs, an increase of 50% in the lactate release was detected in comparison with untreated ER-CSCs (Figure 7a,b). After two weeks, the Etoposide exposure of HTLA-CSCs stimulated the lactate release by 80% in respect of untreated CSCs, and the co-treatments did not modify the Etoposide-induced effect (Figure 7a). In ER-CSCs, a similar result was obtained only in the presence of C2-4 or SSZ, given alone or in combination with Etoposide, while Etoposide per se did not induce any effect (Figure 7b). Interestingly, in ER-CSCs, after five weeks, the co-treatments with C2-4 or SSZ stimulated lactate release by 30% in respect of both untreated and Etoposide-treated ER-CSCs (Figure 7c).

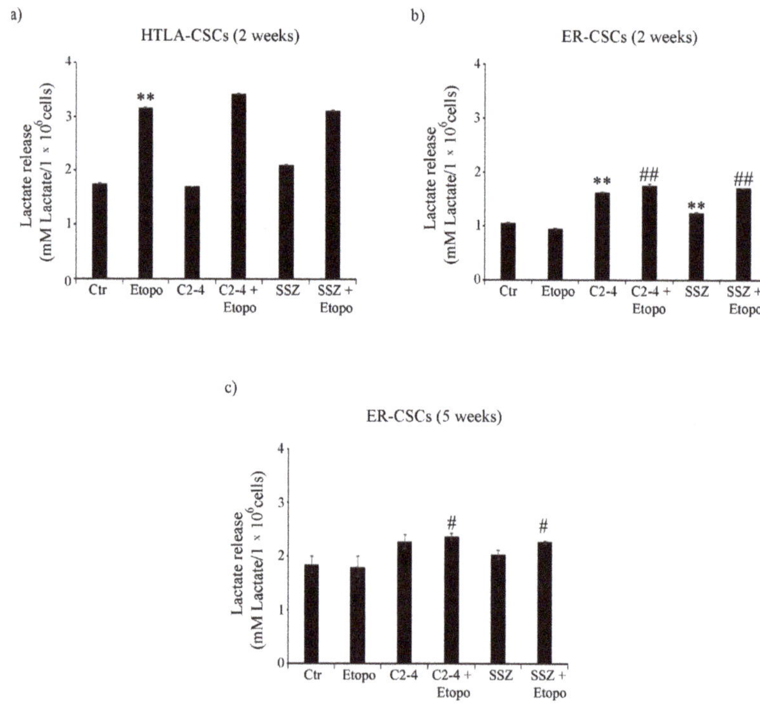

Figure 7. Lactate release in both CSC populations treated for two (**a,b**) or five weeks (**c**). HTLA-CSCs and ER-CSCs were treated three times a week with 0.1 μM C2-4, with 5 μM SSZ or with 1.25 μM Etoposide, administered once a week alone or in combination with C2-4 or SSZ. Results were reported as mM lactate/10^6 cells. Bar graph summarizes quantitative data of means ± S.E.M. of three independent experiments. ** $p < 0.01$ vs. untreated CSCs (Ctr); # $p < 0.05$ vs. Etoposide-treated CSCs; ## $p < 0.01$ vs. Etoposide-treated CSCs.

Taken together, the metabolic findings indicate that, although both untreated CSC populations have an efficient OXPHOS metabolism, HTLA-CSCs, following Etoposide exposure, activate glycolysis while ER-CSCs maintain their metabolic adaptation and are able to choose glycolytic metabolism after five-week-co-treatments with C2-4 or SSZ. This data, besides confirming the propensity of CSCs to obtain energy by stimulating OXPHOS [58], also demonstrates that this metabolic rewiring is reversible [59] and that, in having a crucial role in CSC survival, it could be targeted in order to eradicate chemoresistant CSC populations. In agreement with these findings, OXPHOS inhibitors (e.g., metformin) have been found to increase therapy sensitivity of cancer stem cells [60,61].

3.5. Co-Treatments Are Able to Induce Lipid Peroxidation in ER-CSCs

In order to better characterize cell damage, the production of malondialdehyde (MDA), a marker of membrane lipid peroxidation, was evaluated in both CSC populations. As observed in Figure 8a, two-week-treatments of HTLA-CSCs with Etoposide or SSZ alone increased MDA levels by 100% and the co-treatment with SSZ increased this parameter by a further 30% (Figure 8a). Notably, in ER-CSCs, only two-week-exposure to SSZ, alone or in combination with Etoposide, induced an 85% increase in MDA production (Figure 8b). After five weeks, in ER-CSCs, all treatments maintained MDA at similar levels to those observed after two weeks. In addition, a 35% increase in MDA levels was found in C2-4-co-treated ER-CSCs (Figure 8c).

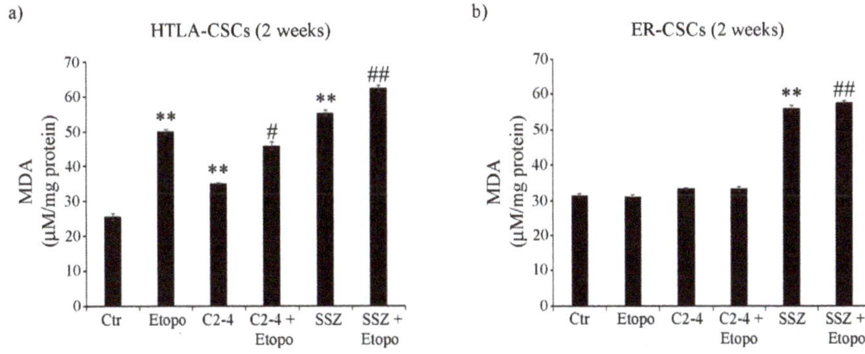

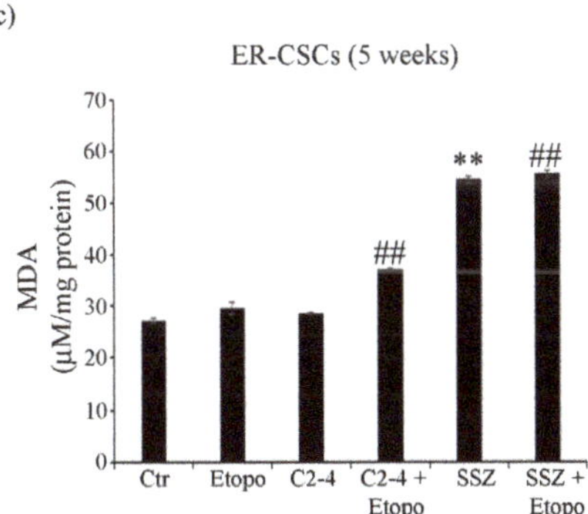

Figure 8. MDA production in both CSC populations treated for two (**a,b**) or five weeks (**c**). HTLA-CSCs and ER-CSCs were treated three times a week with 0.1 μM C2-4, with 5 μM SSZ or with 1.25 μM Etoposide, administered once a week alone or in combination with C2-4 or SSZ. Results were reported as μM/mg protein. Bar graph summarizes quantitative data of means ± S.E.M. of three independent experiments. ** $p < 0.01$ vs. untreated CSCs (Ctr); # $p < 0.05$ vs. Etoposide-treated CSCs; ## $p < 0.01$ vs. Etoposide-treated CSCs.

Considering that mitochondrial oxidative phosphorylation can generate ROS, this parameter was evaluated in both CSC populations.

As shown in Figure 9a, after two-week-exposure to Etoposide, alone or in combination with C2-4 or SSZ, no changes in ROS production were observed in HTLA-CSCs in respect of untreated CSCs. Instead, two-week treatment of ER-CSCs with Etoposide, alone or combined with C2-4 or SSZ, induced a 60% reduction of ROS levels in respect of untreated CSCs (Figure 9b). After five weeks of all treatments, no changes in ROS production were detected in ER-CSCs (Figure 9c).

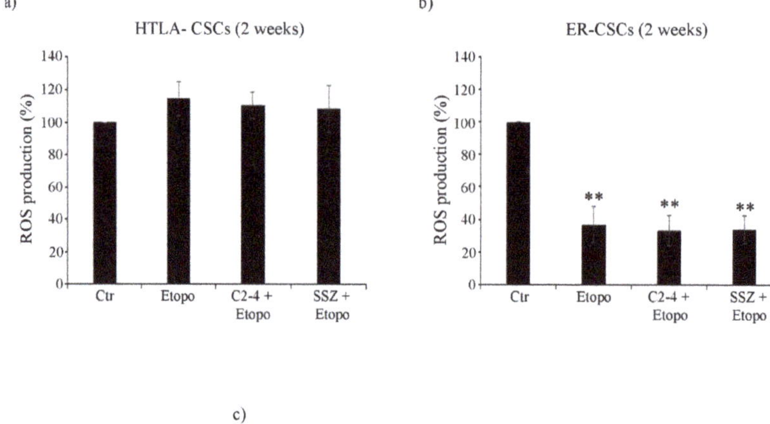

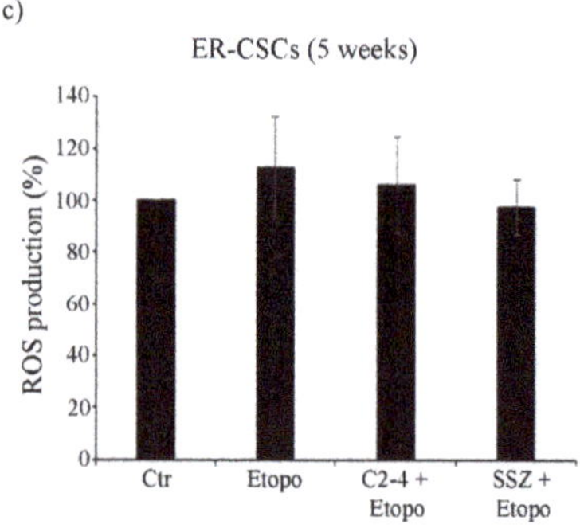

Figure 9. ROS production in both CSC populations treated for two (**a,b**) or five weeks (**c**). HTLA-CSCs and ER-CSCs were treated three times a week with 0.1 μM C2-4, with 5 μM SSZ or with 1.25 μM Etoposide, administered once a week alone or in combination with C2-4 or SSZ. Results were expressed as percentages of the control value. Bar graph summarizes quantitative data of means ± S.E.M. of three independent experiments. ** $p < 0.01$ vs. untreated CSCs (Ctr).

Altogether these results demonstrate that although Etoposide alone significantly decreased the level of GSH, GSSG amount was not enhanced and, in parallel, ROS and MDA production did not increase. In this context, we cannot exclude that other antioxidant systems might contribute to the maintenance of CSC oxidative status [23,62] and in support of these findings, it has been recently found that the ability of breast CSCs to maintain low ROS production confers cancer stemness and drug resistance [63]. Interestingly, the co-treatments which are able to activate glycolysis, reduce GSH levels without altering cell redox state and induce lipid peroxidation are also effective in counteracting ER-CSC propagation.

3.6. Co-Treatments Are Able to Induce Down-Regulation of GPX4 Activity and of ZEB-1 Expression

Considering that xCT inhibition by SSZ or erastin has been demonstrated to induce lipid peroxide production and ferroptosis, a form of non-apoptotic cell death consequent to a loss of GSH and of Glutathione peroxidase 4 (GPX4) activity [64,65], the role of this enzyme was investigated.

As shown in Figure 10, after five weeks, in untreated ER-CSCs, the activity of GPX4 was reduced by 20% in respect of that observed after two weeks. Two-week-exposure to Etoposide, alone or in combination with SSZ, reduced the activity of GPX4 by 30% compared to that analyzed in untreated CSCs. Instead, two-week co-treatments with C2-4 further decreased GPX4 activity by 38% in respect of Etoposide alone and by 55% in comparison to the control. Moreover, after five weeks, Etoposide alone or in combination with C2-4 decreased GPX4 activity by 40% in respect of untreated CSCs, while SSZ co-treatments further inhibited the activity by 60% in comparison to Etoposide-treated CSCs and by 75% in respect of the control (Figure 10).

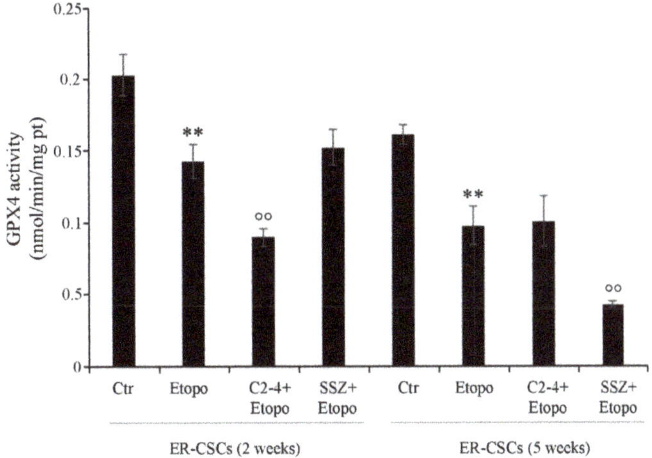

Figure 10. GPX4 activity in ER-CSC populations treated for two or five weeks. ER-CSCs were treated three times a week with 0.1 µM C2-4, with 5 µM SSZ or with 1.25 µM Etoposide, administered once a week alone or in combination with C2-4 or SSZ. Results were reported as nmol/min/mg protein. Bar graph summarizes quantitative data of means $\pm$ S.E.M. of three independent experiments. ** $p < 0.01$ vs. untreated ER-CSCs (Ctr); °° $p < 0.01$ vs. Etoposide-treated ER-CSCs.

Since GPX4 plays a crucial role in the suppression of lipid peroxidation by stimulating a GSH-dependent reduction of phospholipid hydroperoxides [64,66], GPX4 inhibitors could be employed as anticancer drugs capable of inducing ferroptosis [67]. Interestingly, it has been reported that chemoresistant cancer cells, that are preserved from ferroptosis through GPX4 activation, expressed high levels of ZEB-1 [68]. ZEB-1 is a protein involved in the epithelial to mesenchymal transition (EMT) process, which plays a fundamental role in supporting CSC generation [69]. In fact, ZEB-1 has been found to have a pleiotropic role in controlling lipid metabolism, growth, proliferation and death of CSCs [68,70].

Based on these considerations and since Etoposide, alone or in combination with C2-4 or SSZ, was able to totally prevent ER-CSC formation, the subsequent analyses were focused on the role played by EMT in drug resistance.

As shown in Figure 11, both untreated HTLA- and ER-CSCs expressed N-Cadherin and ZEB-1, two proteins favoring the EMT process. Two-week-treatments of HTLA-CSCs fully counteracted the expression of these two proteins while, interestingly, two-week-Etoposide treatment of ER-CSCs reduced the expression of N-Cadherin and ZEB-1 by 40% and 70%, respectively, in comparison to untreated CSCs (Figure 11a,b). After five weeks, the combination of Etoposide with C2-4 reduced the expression of both proteins by 50% and co-treatment with SSZ decreased ZEB-1 levels by 90% in respect of untreated CSCs (Figure 11a,b).

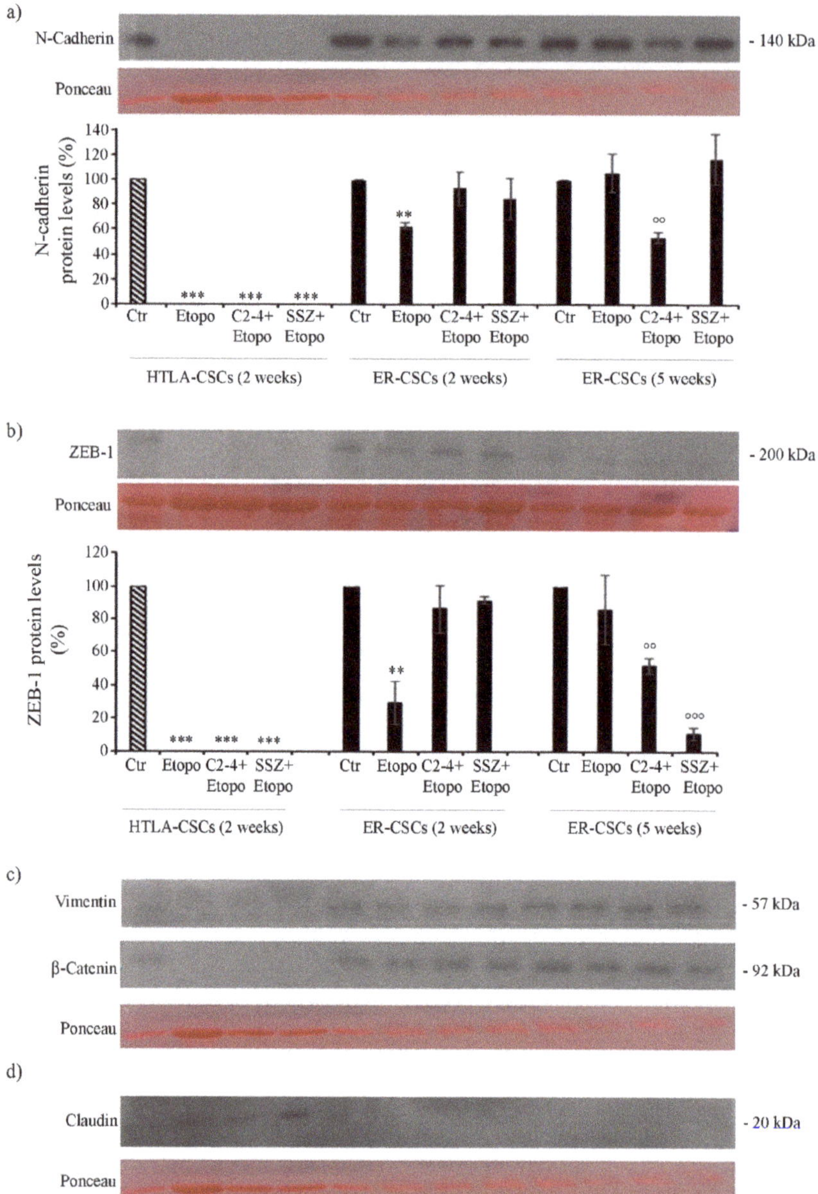

Figure 11. Protein levels of N-Cadherin (**a**), ZEB-1 (**b**), Vimentin (**c**), β-Catenin (**c**) and Claudin (**d**) in both CSC populations treated for two or five weeks. HTLA-CSCs and ER-CSCs were treated three times a week with 0.1 μM C2-4, with 5 μM SSZ or with 1.25 μM Etoposide, administered once a week alone or in combination with C2-4 or SSZ. Immunoblots shown are representative of three independent experiments. To ensure normalized protein content all filters were stained with Red Ponceau. Bar graph summarizes quantitative data of means ± S.E.M. of three independent experiments. ** $p < 0.01$ vs. untreated CSCs (Ctr); *** $p < 0.001$ vs. untreated CSCs (Ctr); °° $p < 0.01$ vs. Etoposide-treated CSCs; °°° $p < 0.001$ vs. Etoposide-treated CSCs.

Changes in the expression of N-Cadherin and ZEB-1 were not accompanied by significant alterations in the expression of Vimentin and β-catenin, another two EMT-related proteins (Figure 11c). In addition, as shown in Figure 11d, only treated HTLA-CSCs expressed Claudin, a protein inhibiting the EMT process.

Therefore, it is conceivable that co-treatments of Etoposide with SSZ/C2-4 are able to counteract the ability of ER-CSCs to generate spheres and to maintain the EMT phenotype. This data, although in vitro, supports the results obtained in early clinical trials demonstrating that the disruption of cancer stemness and/or the EMT process is a useful approach in fighting tumor recurrence [71,72].

Furthermore, in ER-CSCs, which become sensitive to the drug, GPX4 activity is reduced and is accompanied by ZEB-1 down-regulation leading to ferroptotic death [73,74].

The results obtained in co-treated CSCs confirm the ferroptotic-inducer action of SSZ [75] and suggest that C2-4, the PKCα inhibitor, is able to trigger ferroptosis of ER-CSCs. The relationship between PKCα and ferroptosis has been previously suggested in dopaminergic cells [76] but, to our knowledge, this is the first time it has been demonstrated that the anti-PKCα peptide can sensitize CSCs to chemotherapy by inhibiting GPX4 and inducing ferroptosis. PKCα activation has been implicated in promoting cancer progression [18,77] and in the formation and survival of cancer stem cells [78] Moreover, it has been demonstrated that the pharmacologic inhibition of PKCα can target breast CSCs [78] and override ZEB1-induced chemoresistance in hepatocarcinoma [79].

3.7. Chemoresistance of ER-CSCs Is Characterized by An Efficient OXPHOS Metabolism, High GSH Levels, ZEB-1 Up-Regulation and GPX4 Activation

In order to identify which variables are more determinant in maintaining cancer cell survival and, consequently, responsible for chemoresistance, Principal Component Analysis (PCA) was carried out by collecting all the analyzed metabolic, biochemical and morphological variables. The first two components explain 45.2% and 19.5% of the variance (64.6% in total).

The loading plot (Figure 12a) shows the correlation between the following variables: P/O ratio, ATP production, GSH and ZEB-1 expression and GPX4 activity. They are all characterized by negative loadings on PC1 (grouped in the green rectangle) and are all related since they allow the propagation of ER-CSCs. On the other side of the plot (positive loadings on PC1) another group of correlated variables can be detected (red rectangle): lactate release, glucose consumption and MDA production. They reduce ER-CSC propagation, and are responsible for their susceptibility to the co-treatments. The two groups of correlated variables are negatively correlated, since they are opposite compared to the origin (red cross in the plot). A third group of correlated variables can be detected, made by N-cad, OCR and Vim. These variables, having high loadings on PC2, are uncorrelated with the other ones. As expected, the score plot (Figure 12b) shows that the two-week and five-week controls are close to each other, having negative PC1 scores. Therefore, they are characterized by the high values of the variables grouped into the green rectangle and the low values of the variables grouped into the red rectangle in the loading plot.

The effect of the three different treatments and of the different times, together with the metabolic modifications involved, can be estimated by taking into account the distance and the direction of the different points in the score plot, compared to the controls.

First of all, it can be seen that for all of them the effect after five weeks is larger than the effect after two weeks.

It can also be seen that the treatment with Etoposide by itself is much less efficient than the other two treatments in which Etoposide is combined with another drug. This treatment has almost no effect after two weeks, while after five weeks it leads to a more positive score on PC1, this meaning a reduction of the "green" variables and an increase of the "red" ones.

a)

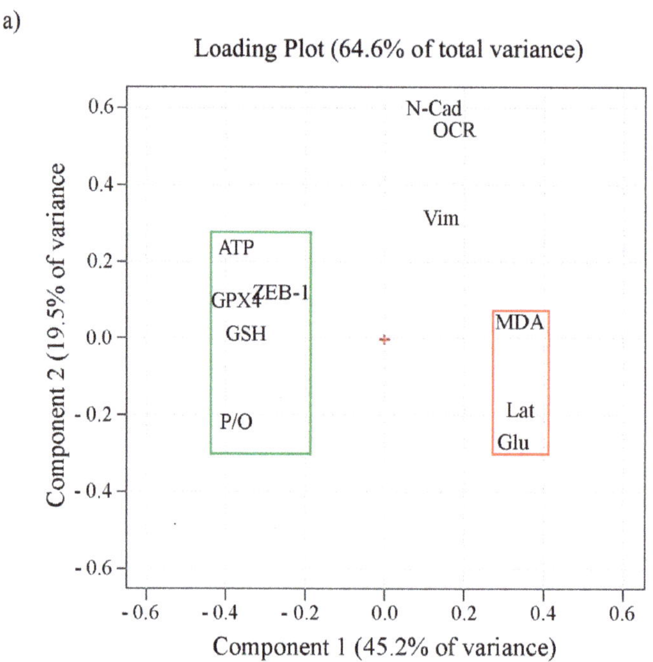

b)

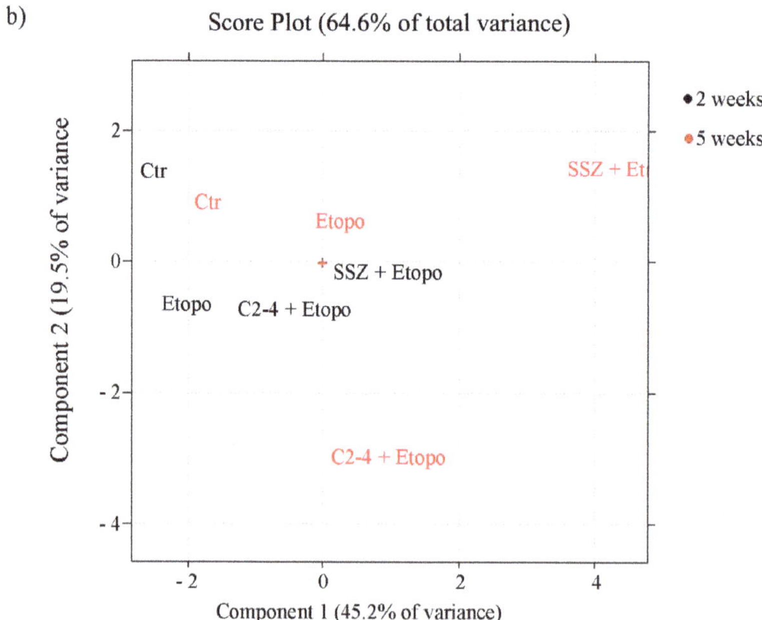

Figure 12. Loading (**a**) and score (**b**) plots. (**a**) The variables which sustain chemoresistance (green rectangle: ATP, GSH, P/O, ZEB-1, GPX4) have negative loadings on PC1 values, while the variables that are able to induce chemosensitivity (red rectangle: MDA production, lactate release and glucose consumption) have positive loadings on PC1; variables N-CAD, OCR and Vim have instead high loadings on PC2 (**b**) In the score plot, the black-colored samples were treated for two weeks, while the red-colored ones were treated for five weeks.

The co-treatment with SSZ has a much greater effect, already relevant after two weeks (corresponding to the effect after five weeks of Etoposide by itself). Since the direction is the same as the one already observed with Etoposide, it can be concluded that SSZ just amplifies the effect of Etoposide.

Instead, when looking at the co-treatment with C2-4 it can be seen that a strong effect is also present already at two weeks, but the trajectory followed is quite different. In this case an increase along PC1 together with a decrease along PC2 is observed, corresponding to a decrease of variables N-Cad, OCR and Vim. It can therefore be concluded that C2-4 not only amplifies the effect of Etoposide, but also produces different metabolic variations.

Collectively, these results further underline that CSC chemoresistance can be the result of a complex cellular adaptation that involves several metabolic and biochemical pathways and, therefore, it is necessary "to attack the enemy" simultaneously from several fronts in order to bypass this adaptation.

4. Conclusions

Even if the mechanism underlying CSC chemoresistance has not yet been fully clarified, the results herein reported suggest that drug refractoriness is maintained as long as CSCs are able to keep a fine balance between correlated molecular actors. In fact, the maintenance of an oxidative metabolism, high levels of GSH and the expression of ZEB-1 concur to allow the survival of resistant CSCs. Therefore, an approach able to down-regulate these molecular targets and to induce ferroptosis may be the winning strategy to counteract drug resistance. In this context, the results herein reported suggest that PKCα might be a new pharmacological target able to fight chemoresistance of cancer stem cells by preventing drug-induced metabolic adaptation and triggering ferroptotic death.

Author Contributions: Conceptualization, C.D. and B.M.; methodology, B.M., N.T., L.M., A.S., A.R. and F.U.; formal analysis, C.D., B.M., R.L., E.F.; software, R.L. and E.F.; validation, C.D., B.M.; investigation, B.M., L.M., A.S., N.T., C.C., S.R., O.G., G.E.V. and A.R.; data curation, C.D. and B.M.; writing-original draft, B.M.; writing-review and editing, C.D., B.M., N.T., M.A.P. and U.M.M.; supervision, C.D. and B.M. All authors have read and agreed to the published version of the manuscript.

Funding: This research was supported by University of Genoa.

Institutional Review Board Statement: Not applicable.

Informed Consent Statement: Not applicable.

Data Availability Statement: Not applicable.

Acknowledgments: We would like to thank Giuseppe Catalano (DIMES-University of Genoa) for his technical assistance and Suzanne Patten for her language editing.

Conflicts of Interest: The authors declare no conflict of interest.

References

1. L'Abbate, A.; Macchia, G.; D'Addabbo, P.; Lonoce, A.; Tolomeo, D.; Trombetta, D.; Kok, K.; Bartenhagen, C.; Whelan, C.W.; Palumbo, O.; et al. Genomic organization and evolution of double minutes/homogeneously staining regions with MYC amplification in human cancer. *Nucleic Acids Res.* **2014**, *42*, 9131–9145. [CrossRef]
2. Strobl-Mazzulla, P.H.; Bronner, M.E. Epithelial to mesenchymal transition: New and old insights from the classical neural crest model. *Semin. Cancer Biol.* **2012**, *22*, 411–416. [CrossRef]
3. Denecker, G.; Vandamme, N.; Akay, O.; Koludrovic, D.; Taminau, J.; Lemeire, K.; Gheldof, A.; De Craene, B.; Van Gele, M.; Brochez, L.; et al. Identification of a ZEB2-MITF-ZEB1 transcriptional network that controls melanogenesis and melanoma progression. *Cell Death Differ.* **2014**, *21*, 1250–1261. [CrossRef] [PubMed]
4. Maris, J.M.; Hogarty, M.D.; Bagatell, R.; Cohn, S.L. Neuroblastoma. *Lancet* **2007**, *369*, 2106–2120. [CrossRef]
5. Brodeur, G.M.; Seeger, R.C.; Schwab, M.; Varmus, H.E.; Bishop, J.M. Amplification of N-myc in untreated human neuroblastomas correlates with advanced disease stage. *Science* **1984**, *224*, 1121–1124. [CrossRef] [PubMed]
6. Seeger, R.C.; Brodeur, G.M.; Sather, H.; Dalton, A.; Siegel, S.E.; Wong, K.Y.; Hammond, D. Association of multiple copies of the N-myc oncogene with rapid progression of neuroblastomas. *N. Engl. J. Med.* **1985**, *313*, 1111–1116. [CrossRef]

7. Gupta, P.B.; Fillmore, C.M.; Jiang, G.; Shapira, S.D.; Tao, K.; Kuperwasser, C.; Lander, E.S. Stochastic state transitions give rise to phenotypic equilibrium in populations of cancer cells. *Cell* **2011**, *146*, 633–644. [CrossRef] [PubMed]
8. Walton, J.D.; Kattan, D.R.; Thomas, S.K.; Spengler, B.A.; Guo, H.F.; Biedler, J.L.; Cheung, N.K.; Ross, R.A. Characteristics of stem cells from human neuroblastoma cell lines and in tumors. *Neoplasia* **2004**, *6*, 838–845. [CrossRef]
9. Visvader, J.E.; Lindeman, G.J. Cancer stem cells in solid tumours: Accumulating evidence and unresolved questions. *Nat. Rev. Cancer* **2008**, *8*, 755–768. [CrossRef]
10. Clevers, H. The cancer stem cell: Premises, promises and challenges. *Nat. Med.* **2011**, *17*, 313–319. [CrossRef] [PubMed]
11. Kreso, A.; Dick, J.E. Evolution of the cancer stem cell model. *Cell Stem Cell* **2014**, *14*, 275–291. [CrossRef]
12. Nagano, O.; Okazaki, S.; Saya, H. Redox regulation in stem-like cancer cells by CD44 variant isoforms. *Oncogene* **2013**, *32*, 5191–5198. [CrossRef] [PubMed]
13. Diehn, M.; Cho, R.W.; Lobo, N.A.; Kalisky, T.; Dorie, M.J.; Kulp, A.N.; Qian, D.; Lam, J.S.; Ailles, L.E.; Wong, M.; et al. Association of reactive oxygen species levels and radioresistance in cancer stem cells. *Nature* **2009**, *458*, 780–783. [CrossRef]
14. Yan, Y.; Zuo, X.; Wei, D. Concise Review: Emerging Role of CD44 in Cancer Stem Cells: A Promising Biomarker and Therapeutic Target. *Stem Cells Transl. Med.* **2015**, *4*, 1033–1043. [CrossRef]
15. Ishimoto, T.; Nagano, O.; Yae, T.; Tamada, M.; Motohara, T.; Oshima, H.; Oshima, M.; Ikeda, T.; Asaba, R.; Yagi, H.; et al. CD44 variant regulates redox status in cancer cells by stabilizing the xCT subunit of system xc(-) and thereby promotes tumor growth. *Cancer Cell* **2011**, *19*, 387–400. [CrossRef] [PubMed]
16. Nagano, O.; Murakami, D.; Hartmann, D.; De Strooper, B.; Saftig, P.; Iwatsubo, T.; Nakajima, M.; Shinohara, M.; Saya, H. Cell-matrix interaction via CD44 is independently regulated by different metalloproteinases activated in response to extracellular Ca(2+) influx and PKC activation. *J. Cell Biol.* **2004**, *165*, 893–902. [CrossRef] [PubMed]
17. Nagano, O.; Saya, H. Mechanism and biological significance of CD44 cleavage. *Cancer Sci.* **2004**, *95*, 930–935. [CrossRef]
18. Domenicotti, C.; Marengo, B.; Verzola, D.; Garibotto, G.; Traverso, N.; Patriarca, S.; Maloberti, G.; Cottalasso, D.; Poli, G.; Passalacqua, M.; et al. Role of PKC-delta activity in glutathione-depleted neuroblastoma cells. *Free Radic Biol. Med.* **2003**, *35*, 504–516. [CrossRef]
19. Marengo, B.; Raffaghello, L.; Pistoia, V.; Cottalasso, D.; Pronzato, M.A.; Marinari, U.M.; Domenicotti, C. Reactive oxygen species: Biological stimuli of neuroblastoma cell response. *Cancer Lett.* **2005**, *228*, 111–116. [CrossRef]
20. Marengo, B.; De Ciucis, C.; Verzola, D.; Pistoia, V.; Raffaghello, L.; Patriarca, S.; Balbis, E.; Traverso, N.; Cottalasso, D.; Pronzato, M.A.; et al. Mechanisms of BSO (L-buthionine-S,R-sulfoximine)-induced cytotoxic effects in neuroblastoma. *Free Radic Biol. Med.* **2008**, *44*, 474–482. [CrossRef]
21. Marengo, B.; De Ciucis, C.; Ricciarelli, R.; Passalacqua, M.; Nitti, M.; Zingg, J.M.; Marinari, U.M.; Pronzato, M.A.; Domenicotti, C. PKCδ sensitizes neuroblastoma cells to L-buthionine-sulfoximine and etoposide inducing reactive oxygen species overproduction and DNA damage. *PLoS ONE* **2011**, *6*, e14661. [CrossRef]
22. Hamilton, D.; Wu, J.H.; Batist, G. Structure-based identification of novel human gamma-glutamylcysteine synthetase inhibitors. *Mol. Pharmacol.* **2007**, *71*, 1140–1147. [CrossRef]
23. Colla, R.; Izzotti, A.; De Ciucis, C.; Fenoglio, D.; Ravera, S.; Speciale, A.; Ricciarelli, R.; Furfaro, A.L.; Pulliero, A.; Passalacqua, M.; et al. Glutathione-mediated antioxidant response and aerobic metabolism: Two crucial factors involved in determining the multi-drug resistance of high-risk neuroblastoma. *Oncotarget* **2016**, *7*, 70715–70737. [CrossRef]
24. Toma, J.G.; McKenzie, I.A.; Bagli, D.; Miller, F.D. Isolation and characterization of multipotent skin-derived precursors from human skin. *Stem Cells* **2005**, *23*, 727–737. [CrossRef] [PubMed]
25. Fariss, M.W.; Reed, D.J. High-performance liquid chromatography of thiols and disulfides: Dinitrophenol derivatives. *Methods Enzymol.* **1987**, *143*, 101–109. [CrossRef]
26. Asensi, M.; Sastre, J.; Pallardo, F.V.; Lloret, A.; Lehner, M.; Garcia-de-la Asuncion, J.; Viña, J. Ratio of reduced to oxidized glutathione as indicator of oxidative stress status and DNA damage. *Methods Enzymol.* **1999**, *299*, 267–276. [CrossRef]
27. Ravera, S.; Bartolucci, M.; Cuccarolo, P.; Litamè, E.; Illarcio, M.; Calzia, D.; Degan, P.; Morelli, A.; Panfoli, I. Oxidative stress in myelin sheath: The other face of the extramitochondrial oxidative phosphorylation ability. *Free Radic Res.* **2015**, *49*, 1156–1164. [CrossRef]
28. Marengo, B.; De Ciucis, C.G.; Ricciarelli, R.; Furfaro, A.L.; Colla, R.; Canepa, E.; Traverso, N.; Marinari, U.M.; Pronzato, M.A.; Domenicotti, C. p38MAPK inhibition: A new combined approach to reduce neuroblastoma resistance under etoposide treatment. *Cell Death Dis.* **2013**, *4*, e589. [CrossRef] [PubMed]
29. Roveri, A.; Maiorino, M.; Ursini, F. Enzymatic and immunological measurements of soluble and membrane-bound phospholipid-hydroperoxide glutathione peroxidase. *Methods Enzymol.* **1994**, *233*, 202–212. [CrossRef] [PubMed]
30. Leardi, R. Chemometric methods in food authentication. In *Modern Techniques for Food Authentication*, 2nd ed.; Sun, D.W., Ed.; Academic Press: Cambridge, MA, USA; Elsevier: Amsterdam, The Netherlands, 2018; Chapter 17; pp. 687–729.
31. Leardi, R.; Melzi, C.; Polotti, G. CAT (Chemometric Agile Software). Available online: http://gruppochemiometria.it/index.php/software (accessed on 27 April 2021).
32. Miran, T.; Vogg, A.T.J.; Drude, N.; Mottaghy, F.M.; Morgenroth, A. Modulation of glutathione promotes apoptosis in triple-negative breast cancer cells. *FASEB J.* **2018**, *32*, 2803–2813. [CrossRef] [PubMed]

33. Boivin, A.; Hanot, M.; Malesys, C.; Maalouf, M.; Rousson, R.; Rodriguez-Lafrasse, C.; Ardail, D. Transient alteration of cellular redox buffering before irradiation triggers apoptosis in head and neck carcinoma stem and non-stem cells. *PLoS ONE* **2011**, *6*, e14558. [CrossRef] [PubMed]

34. Siemann, D.W.; Beyers, K.L. In vivo therapeutic potential of combination thiol depletion and alkylating chemotherapy. *Br. J. Cancer* **1993**, *68*, 1071–1079. [CrossRef] [PubMed]

35. Villablanca, J.G.; Volchenboum, S.L.; Cho, H.; Kang, M.; Cohn, S.L.; Anderson, C.P.; Marachelian, A.; Groshen, S.; Tsao-Wei, D.; Matthay, K.K.; et al. A Phase I New Approaches to Neuroblastoma Therapy Study of Buthionine Sulfoximine and Melphalan With Autologous Stem Cells for Recurrent/Refractory High-Risk Neuroblastoma. *Pediatr. Blood Cancer* **2016**, *63*, 1349–1356. [CrossRef]

36. Okamoto, I.; Kawano, Y.; Matsumoto, M.; Suga, M.; Kaibuchi, K.; Ando, M.; Saya, H. Regulated CD44 cleavage under the control of protein kinase C, calcium influx, and the Rho family of small G proteins. *J. Biol. Chem.* **1999**, *274*, 25525–25534. [CrossRef]

37. Zheng, Z.; Luo, G.; Shi, X.; Long, Y.; Shen, W.; Li, Z.; Zhang, X. The Xc- inhibitor sulfasalazine improves the anti-cancer effect of pharmacological vitamin C in prostate cancer cells via a glutathione-dependent mechanism. *Cell Oncol.* **2020**, *43*, 95–106. [CrossRef] [PubMed]

38. Wei, C.W.; Yu, Y.L.; Lu, J.Y.; Hung, Y.T.; Liu, H.C.; Yiang, G.T. Anti-Cancer Effects of Sulfasalazine and Vitamin E Succinate in MDA-MB 231 Triple-Negative Breast Cancer Cells. *Int. J. Med. Sci.* **2019**, *16*, 494–500. [CrossRef] [PubMed]

39. Ogihara, K.; Kikuchi, E.; Okazaki, S.; Hagiwara, M.; Takeda, T.; Matsumoto, K.; Kosaka, T.; Mikami, S.; Saya, H.; Oya, M. Sulfasalazine could modulate the CD44v9-xCT system and enhance cisplatin-induced cytotoxic effects in metastatic bladder cancer. *Cancer Sci.* **2019**, *110*, 1431–1441. [CrossRef]

40. Miyoshi, S.; Tsugawa, H.; Matsuzaki, J.; Hirata, K.; Mori, H.; Saya, H.; Kanai, T.; Suzuki, H. Inhibiting xCT Improves 5-Fluorouracil Resistance of Gastric Cancer Induced by CD44 Variant 9 Expression. *Anticancer Res.* **2018**, *38*, 6163–6170. [CrossRef]

41. Wada, F.; Koga, H.; Akiba, J.; Niizeki, T.; Iwamoto, H.; Ikezono, Y.; Nakamura, T.; Abe, M.; Masuda, A.; Sakaue, T.; et al. High expression of CD44v9 and xCT in chemoresistant hepatocellular carcinoma: Potential targets by sulfasalazine. *Cancer Sci.* **2018**, *109*, 2801–2810. [CrossRef] [PubMed]

42. Balza, E.; Castellani, P.; Moreno, P.S.; Piccioli, P.; Medraño-Fernandez, I.; Semino, C.; Rubartelli, A. Restoring microenvironmental redox and pH homeostasis inhibits neoplastic cell growth and migration: Therapeutic efficacy of esomeprazole plus sulfasalazine on 3-MCA-induced sarcoma. *Oncotarget* **2017**, *8*, 67482–67496. [CrossRef]

43. Mooney, M.R.; Geerts, D.; Kort, E.J.; Bachmann, A.S. Anti-tumor effect of sulfasalazine in neuroblastoma. *Biochem. Pharmacol.* **2019**, *162*, 237–249. [CrossRef] [PubMed]

44. Mochly-Rosen, D.; Das, K.; Grimes, K.V. Protein kinase C, an elusive therapeutic target? *Nat. Rev. Drug Discov.* **2012**, *11*, 937–957. [CrossRef] [PubMed]

45. Lei, J.; Li, Q.; Gao, Y.; Zhao, L.; Liu, Y. Increased PKCα activity by Rack1 overexpression is responsible for chemotherapy resistance in T-cell acute lymphoblastic leukemia-derived cell line. *Sci. Rep.* **2016**, *6*, 33717. [CrossRef] [PubMed]

46. Kim, C.W.; Asai, D.; Kang, J.; Kishimura, A.; Mori, T.; Katayama, Y. Reversal of efflux of an anticancer drug in human drug-resistant breast cancer cells by inhibition of protein kinase Cα (PKCα) activity. *Tumour. Biol.* **2016**, *37*, 1901–1908. [CrossRef] [PubMed]

47. Abera, M.B.; Kazanietz, M.G. Protein kinase Cα mediates erlotinib resistance in lung cancer cells. *Mol. Pharmacol.* **2015**, *87*, 832–841. [CrossRef] [PubMed]

48. Ruvolo, P.P.; Zhou, L.; Watt, J.C.; Ruvolo, V.R.; Burks, J.K.; Jiffar, T.; Kornblau, S.; Konopleva, M.; Andreeff, M. Targeting PKC-mediated signal transduction pathways using enzastaurin to promote apoptosis in acute myeloid leukemia-derived cell lines and blast cells. *J. Cell Biochem.* **2011**, *112*, 1696–1707. [CrossRef]

49. Warburg, O.; Wind, F.; Negelein, E. The metabolism of tumors in the body. *J. Gen. Physiol.* **1927**, *8*, 519–530. [CrossRef]

50. Koppenol, W.H.; Bounds, P.L.; Dang, C.V. Otto Warburg's contributions to current concepts of cancer metabolism. *Nat. Rev. Cancer* **2011**, *11*, 325–337. [CrossRef]

51. Vazquez, F.; Lim, J.H.; Chim, H.; Bhalla, K.; Girnun, G.; Pierce, K.; Clish, C.B.; Granter, S.R.; Widlund, H.R.; Spiegelman, B.M.; et al. PGC1α expression defines a subset of human melanoma tumors with increased mitochondrial capacity and resistance to oxidative stress. *Cancer Cell* **2013**, *23*, 287–301. [CrossRef]

52. Viale, A.; Pettazzoni, P.; Lyssiotis, C.A.; Ying, H.; Sánchez, N.; Marchesini, M.; Carugo, A.; Green, T.; Seth, S.; Giuliani, V.; et al. Oncogene ablation-resistant pancreatic cancer cells depend on mitochondrial function. *Nature* **2014**, *514*, 28–32. [CrossRef]

53. Shin, M.K.; Cheong, J.H. Mitochondria-centric bioenergetic characteristics in cancer stem-like cells. *Arch. Pharm. Res.* **2019**, *42*, 113–127. [CrossRef]

54. Lagadinou, E.D.; Sach, A.; Callahan, K.; Rossi, R.M.; Neering, S.; Minhajuddin, M.; Ashton, J.M.; Pei, S.; Grose, V.; O'Dwyer, K.M.; et al. BCL-2 inhibition targets oxidative phosphorylation and selectively eradicates quiescent human leukemia stem cells. *Cell Stem Cell* **2013**, *12*, 329–341. [CrossRef]

55. Molina, J.R.; Sun, Y.; Protopopova, M.; Gera, S.; Bandi, M.; Bristow, C.; McAfoos, T.; Morlacchi, P.; Ackroyd, J.; Agip, A.A.; et al. An inhibitor of oxidative phosphorylation exploits cancer vulnerability. *Nat. Med.* **2018**, *24*, 1036–1046. [CrossRef] [PubMed]

56. Pollyea, D.A.; Stevens, B.M.; Jones, C.L.; Winters, A.; Pei, S.; Minhajuddin, M.; D'Alessandro, A.; Culp-Hill, R.; Riemondy, K.A.; Gillen, A.E.; et al. Venetoclax with azacitidine disrupts energy metabolism and targets leukemia stem cells in patients with acute myeloid leukemia. *Nat. Med.* **2018**, *24*, 1859–1866. [CrossRef] [PubMed]

57. Hinkle, P.C. P/O ratios of mitochondrial oxidative phosphorylation. *Biochim. Biophys. Acta* **2005**, *1706*, 1–11. [CrossRef]

58. Sato, M.; Kawana, K.; Adachi, K.; Fujimoto, A.; Yoshida, M.; Nakamura, H.; Nishida, H.; Inoue, T.; Taguchi, A.; Takahashi, J.; et al. Spheroid cancer stem cells display reprogrammed metabolism and obtain energy by actively running the tricarboxylic acid (TCA) cycle. *Oncotarget* **2016**, *7*, 33297–33305. [CrossRef] [PubMed]

59. Denise, C.; Paoli, P.; Calvani, M.; Taddei, M.L.; Giannoni, E.; Kopetz, S.; Kazmi, S.M.; Pia, M.M.; Pettazzoni, P.; Sacco, E.; et al. 5-fluorouracil resistant colon cancer cells are addicted to OXPHOS to survive and enhance stem-like traits. *Oncotarget* **2015**, *6*, 41706–41721. [CrossRef] [PubMed]

60. Saini, N.; Yang, X. Metformin as an anti-cancer agent: Actions and mechanisms targeting cancer stem cells. *Acta Biochim. Biophys. Sin.* **2018**, *50*, 133–143. [CrossRef]

61. Samuel, S.M.; Varghese, E.; Koklesová, L.; Líšková, A.; Kubatka, P.; Büsselberg, D. Counteracting Chemoresistance with Metformin in Breast Cancers: Targeting Cancer Stem Cells. *Cancers* **2020**, *12*, 2482. [CrossRef]

62. Trachootham, D.; Alexandre, J.; Huang, P. Targeting cancer cells by ROS-mediated mechanisms: A radical therapeutic approach? *Nat. Rev. Drug Discov.* **2009**, *8*, 579–591. [CrossRef]

63. Čipak Gašparović, A.; Milković, L.; Dandachi, N.; Stanzer, S.; Pezdirc, I.; Vrančić, J.; Šitić, S.; Suppan, C.; Balic, M. Chronic Oxidative Stress Promotes Molecular Changes Associated with Epithelial Mesenchymal Transition, NRF2, and Breast Cancer Stem Cell Phenotype. *Antioxidants* **2019**, *8*, 633. [CrossRef]

64. Yang, W.S.; SriRamaratnam, R.; Welsch, M.E.; Shimada, K.; Skouta, R.; Viswanathan, V.S.; Cheah, J.H.; Clemons, P.A.; Shamji, A.F.; Clish, C.B.; et al. Regulation of ferroptotic cancer cell death by GPX4. *Cell* **2014**, *156*, 317–331. [CrossRef] [PubMed]

65. Maiorino, M.; Conrad, M.; Ursini, F. GPx4, Lipid Peroxidation, and Cell Death: Discoveries, Rediscoveries, and Open Issues. *Antioxid. Redox Signal.* **2018**, *29*, 61–74. [CrossRef] [PubMed]

66. Angeli, J.P.F.; Schneider, M.; Proneth, B.; Tyurina, Y.Y.; Tyurin, V.A.; Hammond, V.J.; Herbach, N.; Aichler, M.; Walch, A.; Eggenhofer, E.; et al. Inactivation of the ferroptosis regulator Gpx4 triggers acute renal failure in mice. *Nat. Cell Biol.* **2014**, *16*, 1180–1191. [CrossRef] [PubMed]

67. Seibt, T.M.; Proneth, B.; Conrad, M. Role of GPX4 in ferroptosis and its pharmacological implication. *Free Radic Biol. Med.* **2019**, *133*, 144–152. [CrossRef] [PubMed]

68. Viswanathan, V.S.; Ryan, M.J.; Dhruv, H.D.; Gill, S.; Eichhoff, O.M.; Seashore-Ludlow, B.; Kaffenberger, S.D.; Eaton, J.K.; Shimada, K.; Aguirre, A.J.; et al. Dependency of a therapy-resistant state of cancer cells on a lipid peroxidase pathway. *Nature* **2017**, *547*, 453–457. [CrossRef]

69. Celià-Terrassa, T.; Jolly, M.K. Cancer Stem Cells and Epithelial-to-Mesenchymal Transition in Cancer Metastasis. *Cold Spring Harb. Perspect. Med.* **2020**, *10*, a036905. [CrossRef]

70. Caramel, J.; Ligier, M.; Puisieux, A. Pleiotropic Roles for ZEB1 in Cancer. *Cancer Res.* **2018**, *78*, 30–35. [CrossRef]

71. Rodriguez-Aznar, E.; Wiesmüller, L.; Sainz, B., Jr.; Hermann, P.C. EMT and Stemness-Key Players in Pancreatic Cancer Stem Cells. *Cancers* **2019**, *11*, 1136. [CrossRef]

72. Shibue, T.; Weinberg, R.A. EMT, CSCs, and drug resistance: The mechanistic link and clinical implications. *Nat. Rev. Clin. Oncol.* **2017**, *14*, 611–629. [CrossRef]

73. Angeli, J.P.F.; Shah, R.; Pratt, D.A.; Conrad, M. Ferroptosis Inhibition: Mechanisms and Opportunities. *Trends Pharmacol. Sci.* **2017**, *38*, 489–498. [CrossRef]

74. Angeli, J.P.F.; Krysko, D.V.; Conrad, M. Ferroptosis at the crossroads of cancer-acquired drug resistance and immune evasion. *Nat. Rev. Cancer* **2019**, *19*, 405–414. [CrossRef] [PubMed]

75. Su, Y.; Zhao, B.; Zhou, L.; Zhang, Z.; Shen, Y.; Lv, H.; AlQudsy, L.H.H.; Shang, P. Ferroptosis, a novel pharmacological mechanism of anti-cancer drugs. *Cancer Lett.* **2020**, *483*, 127–136. [CrossRef]

76. Do Van, B.; Gouel, F.; Jonneaux, A.; Timmerman, K.; Gelé, P.; Pétrault, M.; Bastide, M.; Laloux, C.; Moreau, C.; Bordet, R.; et al. Ferroptosis, a newly characterized form of cell death in Parkinson's disease that is regulated by PKC. *Neurobiol. Dis.* **2016**, *94*, 169–178. [CrossRef] [PubMed]

77. Griner, E.M.; Kazanietz, M.G. Protein kinase C and other diacylglycerol effectors in cancer. *Nat. Rev. Cancer* **2007**, *7*, 281–294. [CrossRef]

78. Tam, W.L.; Lu, H.; Buikhuisen, J.; Soh, B.S.; Lim, E.; Reinhardt, F.; Wu, Z.J.; Krall, J.A.; Bierie, B.; Guo, W.; et al. Protein kinase C α is a central signaling node and therapeutic target for breast cancer stem cells. *Cancer Cell* **2013**, *24*, 347–364. [CrossRef] [PubMed]

79. Sreekumar, R.; Emaduddin, M.; Al-Saihati, H.; Moutasim, K.; Chan, J.; Spampinato, M.; Bhome, R.; Yuen, H.M.; Mescoli, C.; Vitale, A.; et al. Protein kinase C inhibitors override ZEB1-induced chemoresistance in HCC. *Cell Death Dis.* **2019**, *10*, 703. [CrossRef]

MDPI

Review

Anti-Tumor Activity of *Hypericum perforatum* L. and Hyperforin through Modulation of Inflammatory Signaling, ROS Generation and Proton Dynamics

Marta Menegazzi [1,*], **Pellegrino Masiello** [2] and **Michela Novelli** [2]

[1] Department of Neuroscience, Biomedicine and Movement Sciences, Biochemistry Section, School of Medicine, University of Verona, Strada Le Grazie 8, I-37134 Verona, Italy

[2] Department of Translational Research and New Technologies in Medicine and Surgery, School of Medicine, University of Pisa, Via Roma 55, I-56126 Pisa, Italy; pellegrino.masiello@med.unipi.it (P.M.); michela.novelli@med.unipi.it (M.N.)

[*] Correspondence: marta.menegazzi@univr.it; Tel.: +39-045-802-7168

Abstract: In this paper we review the mechanisms of the antitumor effects of *Hypericum perforatum* L. (St. John's wort, SJW) and its main active component hyperforin (HPF). SJW extract is commonly employed as antidepressant due to its ability to inhibit monoamine neurotransmitters re-uptake. Moreover, further biological properties make this vegetal extract very suitable for both prevention and treatment of several diseases, including cancer. Regular use of SJW reduces colorectal cancer risk in humans and prevents genotoxic effects of carcinogens in animal models. In established cancer, SJW and HPF can still exert therapeutic effects by their ability to downregulate inflammatory mediators and inhibit pro-survival kinases, angiogenic factors and extracellular matrix proteases, thereby counteracting tumor growth and spread. Remarkably, the mechanisms of action of SJW and HPF include their ability to decrease ROS production and restore pH imbalance in tumor cells. The SJW component HPF, due to its high lipophilicity and mild acidity, accumulates in membranes and acts as a protonophore that hinders inner mitochondrial membrane hyperpolarization, inhibiting mitochondrial ROS generation and consequently tumor cell proliferation. At the plasma membrane level, HPF prevents cytosol alkalization and extracellular acidification by allowing protons to re-enter the cells. These effects can revert or at least attenuate cancer cell phenotype, contributing to hamper proliferation, neo-angiogenesis and metastatic dissemination. Furthermore, several studies report that in tumor cells SJW and HPF, mainly at high concentrations, induce the mitochondrial apoptosis pathway, likely by collapsing the mitochondrial membrane potential. Based on these mechanisms, we highlight the SJW/HPF remarkable potentiality in cancer prevention and treatment.

Keywords: *Hypericum perforatum*; hyperforin; reactive oxygen species; pH regulation; tumor prevention; tumor therapy; apoptosis; cancerogenesis; inflammatory signaling; natural compounds

Citation: Menegazzi, M.; Masiello, P.; Novelli, M. Anti-Tumor Activity of *Hypericum perforatum* L. and Hyperforin through Modulation of Inflammatory Signaling, ROS Generation and Proton Dynamics. *Antioxidants* **2021**, *10*, 18. https://dx.doi.org/10.3390/antiox10010018

Received: 27 November 2020
Accepted: 21 December 2020
Published: 28 December 2020

1. Introduction

Despite the enormous and enduring effort undertaken by biomedical scientists in the investigation of tumor pathogenesis and therapy, cancer remains the second leading cause of death worldwide (World Health Organization, 2019 https://www.who.int/health-topics/cancer#tab=tab_1). Among the numerous antitumor drugs approved in the last fifty years, a large percentage originates from natural products or their derivatives [1,2]. Actually, natural antitumor agents show a broad spectrum of mechanisms to inhibit cancer development, through reduction of proliferation rate of malignant cells, induction of apoptosis, blockade of invasiveness and neo-angiogenesis [2]. Furthermore, they generally display lower side effects than other antitumor drugs [3].

In this review, among various natural compounds endowed with antitumor properties, we want to aim attention at *Hypericum perforatum* L., also known as St. John's wort, and its principal constituent hyperforin, by reviewing research advancements about their key molecular mechanisms and highlighting their remarkable potential in cancer prevention and treatment.

2. Signaling Pathways in Cancer and Molecular Targets Susceptible to Therapeutic Intervention by Phytochemicals

2.1. Oncogenesis

Oncogenesis involves dysregulation of proto-oncogenes or tumor suppressor genes, which upon mutation can modify key cellular processes linked to cell proliferation and its control [4]. Despite significant advances in cancer diagnostics, the detection of a tumor growth in an early stage remains difficult due to the not infrequent lack or paucity of symptoms for long time. Thus, prevention of the early events in carcinogenesis appears crucial for protection against tumor development. A primary preventive role can actually be played by the use of dietary phytochemical supplements which would help to shield genotoxic insults, reverse the promotion stage of multistep carcinogenesis, and also halt or retard the progression of transformed cells [5]. Phytochemicals can veritably prevent these processes by activating antioxidant and detoxification pathways and exerting an anti-inflammatory action [6,7].

There is nowadays a wide consensus that inflammation plays a relevant role in carcinogenetic events. Many experimental results document a pro-tumor activity of inflammatory mediators and epidemiological studies reveal that chronic inflammation predisposes to various types of cancer [8]. In the attempt to replace lost cells or repair damaged tissues, cytokines, chemokines, angiogenic factors and extracellular matrix-degrading enzymes might in some cases increase the risk of cell transformation and most often sustain high proliferation rate of transformed cells [8,9]. Among the factors involved in carcinogenesis-related inflammation, cytokines like tumor necrosis factor (TNF)-α, interleukin (IL)-1β, interferon (IFN)-γ, and IL-6, appear to play a pivotal role. Cytokine signaling results in the activation of transcription factors, in particular nuclear factor kappa-light-chain-enhancer of activated B cells (NF-κB) and signal transducer and activator of transcription (STAT)-1/3, which are points of convergence for several pathways promoting malignancy, as evidenced in many studies [10–13]. In particular, NF-κB is a coordinator of innate immunity and inflammation and has emerged as an important endogenous tumor promoter [14,15]. NF-κB also enhances expression of IL-6, that further elicits cell proliferation and inhibits apoptosis via STAT-3 signaling pathway [16,17].

Interestingly, a number of phytochemicals impair both inflammation and oncogenesis by inhibiting activation of transcription factors [6]. For instance, ginger extract protects against ethionine-induced rat hepato-carcinogenesis restraining TNF-α production and NF-κB activation [18]; astaxanthin exerts anti-inflammatory effects by hindering NF-κB signaling pathway and prevents hamster buccal pouch carcinogenesis [19]; ursolic acid and resveratrol inhibit skin tumor development by hampering pro-inflammatory cytokine expression and NF-κB/STAT-3 activation [20].

2.2. Tumor Growth, Angiogenesis, Invasiveness, and Metastasis

In the complex network regulating tumor development, the role of mitogen activated protein kinases (MAPK) activity appears crucial for tumor cell survival. Growth factors, such as epidermal growth factor (EGF), platelet derived growth factor (PDGF), and insulin growth factor, act through extracellular-regulated kinase (ERK)1/2 and protein kinase B (Akt) signaling pathways to foster cell proliferation [21]. Phosphatidylinositol-3 kinase (PI3K)/Akt/mammalian target of rapamycin (mTOR) pathway represents another critical signaling axis which supports tumor growth [22], also through stimulation of protein synthesis and angiogenesis [23,24]. These kinases have connections to each other and to downstream transcription factors [25]. For instance, kinase-induced NF-κB activation favors cell survival in most tumor types, also by enhancing the expression of anti-apoptotic genes [15,25,26]. Promotion of apoptosis in cancer cells is indeed considered a powerful and reliable therapeutic tool. Notably, in both cancer cell lines and experimental tumor models, several phytochemicals have been characterized as pro-apoptotic agents effectively curtailing tumor growth [27–30].

Furthermore, natural products counteract tumor progression by hindering malignant cells ability to invade and seed in other body sites [27,28,31]. Invasiveness and metastatic spread are complex phenomena involving interactions between malignant and inflammatory micro-environmental cells [32–34]. Actually, tumor-associated macrophages potently induce angiogenesis by producing plentiful vasoactive factors, such as IL-8, vascular endothelial growth factor (VEGF), basic fibroblast growth factor (bFGF), and PDGF [33,35]. Surrounding fibroblasts and macrophages share with tumor cells the ability to release proteolytic enzymes, like metalloproteinases (MMPs), elastase and trypsin, which degrade extracellular matrix and cause abnormal blood vessel permeability [36–38]. Moreover, proliferation and migration of lymphatic endothelial cells, leading to neogenesis or remodeling of lymphatic vessels, can favor cancer metastasis [39].

Scientific evidence suggests that the daily consumption of adequate amounts of bioactive phytochemicals improves cancer prognosis, as recently reviewed [40]. For instance, herbal extracts from *Astragalus membranaceus*, *Angelica gigas*, and *Trichosanthes kirilowii maximowicz* were reported to suppress lung metastasis in vivo through inhibition of IL-6/STAT-3 signaling pathway [40,41]. The ethanol extract of *Rhizoma amorphophalli* significantly decreased proliferation and migration of breast cancer cells in the lung [42]. Daucosterol linoleate downregulated the expression of VEGF, MMP-2 and MMP-9 in breast and lung cancers [43], as *Manuka* honey did in human colon cancer cell lines [44]. Thus, the treatment with various natural product extracts represents a promising approach to limit both progression and spread of tumors.

3. Reactive Oxygen Species as a Double-Edged Sword in the Fight against Cancer

Free radical oxygen species (ROS) are highly reactive byproducts of cell metabolism. ROS level in normal cells is tightly regulated by redox homeostasis based on several antioxidant detoxifying agents [45]. However, if ROS amount exceeds a certain threshold, serious damages occur in cells, like DNA deletions, insertions, single- and double-strand breaks that, if not repaired, might lead to either tumor transformation or cell death [46].

Mitochondria are considered the major production site of ROS, generated through the electron transport chain (ETC). The free energy made available by the electron transfer drives a proton-motive force (pmf) across the inner mitochondrial membrane. As protons are positively charged, pmf determines a charge gradient which constitutes the inner mitochondrial membrane potential ($\Delta\psi_m$). When the $\Delta\psi_m$ is high, like in cancer cells, the electron transfer is slowed down, as protons have to be pumped against an enhanced electrochemical force, so that an electron leakage and, consequently, high ROS production are favored [47] (Figure 1). By lowering the $\Delta\psi_m$ toward that of normal cells, the electron flow would accelerate with the result that (i) electrons spend less time in ETC, thus decreasing their chance of reducing oxygen to superoxide radical; (ii) the increase in oxygen consumption due to the faster electron transfer limits free oxygen availability for superoxide generation [47]. In this regard, tumor cells have a more negative mitochondrial membrane potential inside the matrix ($\Delta\psi_m \sim -210$ mV), than normal cells ($\Delta\psi_m \sim -140$ mV) [28,48] (Figure 1).

Quickly after superoxide generation, other reactive species can be produced, including hydroxyl and thiol radicals, hydrogen peroxide (H_2O_2) and peroxy-nitrites, the latter derived from the interaction between superoxide and nitric oxide produced by inducible nitric oxide synthase (iNOS) [48,49]. Overexpression of iNOS is common in different tumor types [49,50] and in surrounding microenvironmental cells, due to cytokine-elicited activation of NF-κB and STAT-1 signaling [51,52].

Besides their direct production in mitochondria, superoxide radicals are enzymatically generated by NADPH oxidases (NOXs) located in plasmatic and organelle membranes [53], and can also arise as byproducts from COX-2 and 5-lipoxygenase (5-LO) activities. Anyway, superoxide radical can be easily converted into H_2O_2, that exhibits higher stability and target selectivity than all other reactive species [53,54]. Thus, H_2O_2 can operate as a bona fide second messenger for the intracellular transduction signaling of hormones, growth factors, cytokines and other inflammatory mediators, by oxidizing the thiol groups of cysteine residues of target proteins, including tyrosine phosphatases (PTPs) [49,54–56]. A generalized inhibition of PTPs can actually shift the dynamic equilibrium between PTPs and protein kinases, favoring kinase activation cascade [56].

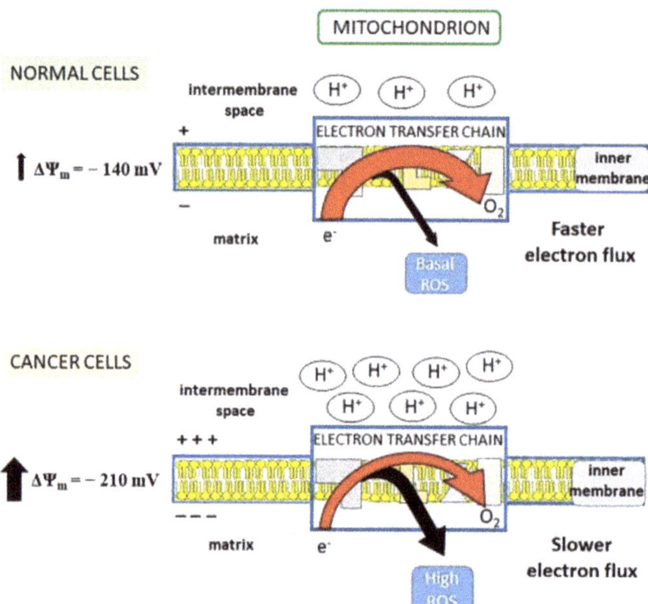

Figure 1. Electron flux in electron transport chain (ETC) and mitochondrial (mt)ROS production. Normal cells: the physiological electron flux in ETC determines robust proton pumping from the matrix to the intermembrane space and maintains stable negative mitochondrial membrane potential ($\Delta\psi_m = -140$ mV). In this situation, electron transfer to O_2 is fast and the electron leakage for ROS generation is small (basal ROS production). Cancer cells: the electron transfer to O_2 is slow because the associated proton pump is impaired by the fact that protons must be pumped against a stronger electrochemical force ($\Delta\psi_m = -210$ mV). Thus, a larger number of electrons leaks out ETC, increasing ROS production.

A tight crosstalk between ROS, NF-κB and inflammatory mediators is documented. Pro-inflammatory cytokines, through binding to their receptors and the subsequent ROS/H_2O_2 signal transduction, drive the phosphorylation and activation of IκB kinases (IKKs) and downstream NF-κB [57–59]. NF-κB signaling is able to prevent ROS-elicited apoptosis by activating gene transduction for antioxidant enzymes, thereby further sustaining a pro-survival response [57,60]. Thus, despite enhanced ROS, malignant cells escape death by maintaining ROS levels below the lethal threshold through the production of large amounts of ROS scavengers. At the same time, cancer cells can utilize ROS signaling to enhance growth by promoting the activity of pro-survival kinases (i.e., PI3K, Akt, ERK1/2, mTOR) [48,61]. Additionally, high ROS levels can trigger angiogenesis through stabilization of hypoxia inducing factor (HIF)-1, that drives the expression of hypoxia-responsive genes leading to upregulation of MMP-1/2/9 and favoring tumor cell infiltration in other tissues [62].

There is increasing consensus that natural products, even at low concentrations, are able to decrease carcinogen-elicited DNA damage, thereby exerting tumor preventive effects through the regulation of ROS-related processes [45], although the mechanisms involved are still unclear. Low concentrations of natural agents are probably ineffective in vivo as ROS scavengers, because both normal and even more cancer cells possess an ample reserve of antioxidants, such as vitamin C and glutathione (GSH) [28]. It is more likely that natural products act at low concentration by interfering with specific steps of signaling pathways [28,63]. Other mechanisms may involve modulation of expression and/or activity of oxidative stress-related enzymes [64,65], as well as limitation of ROS generation through a partial dissipation of the inner mitochondrial membrane potential [28,45,66]. Conversely, it has also been suggested that, in the context of established cancer, several natural bioactive compounds, especially at high concentrations, can exert pro-oxidant rather than anti-oxidant activity, further enhancing oxidative stress and driving tumor cells to death [67,68]. This hypothesis originates from in vitro evidences that catechins and other natural products could actually induce ROS-dependent apoptosis in various types of tumor cells [69–72]. Anyhow, to explain the large variability of the experimental data reported in literature, a number of limitations of the in vitro studies on the anti-tumor effects of natural products should be mentioned: (i) the high concentrations of natural compounds often used in such studies raise the question whether they are really relevant to the in vivo situation [28]; (ii) the majority of the in vitro experiments has been performed with cell lines cultured in a medium often deficient in antioxidants, that can lead to overstate the beneficial effects of added antioxidant products [73]; (iii) some polyphenols are instable in standard culture media, in which they can generate oxidation products capable of depleting cellular glutathione and eventually resulting in pro-oxidant effects [73]. For these reasons, in vivo experiments are expected to make us better understand at the molecular level the real effects of natural products on the reactive oxygen radicals. Indeed, although it has been well documented that polyphenols and other natural compounds are able to inhibit tumorigenesis and tumor growth in several animal models, a putative ROS-dependent pro-oxidant mechanism in such in vivo inhibition has not been elucidated yet [67].

4. Dysregulation of Proton and Ion Content in Cancer Cells as a Possible Target for Phytochemical Therapy

Another hallmark of neoplastic cells is the aberrant regulation of ion flow affecting the intra- and extra-cellular pH homeostasis. Many channels able to modify proton concentration are present in the plasma membrane. The most important is the Na^+/H^+ exchanger (NHE) that uses the natural inward Na^+ gradient to move Na^+ into the cytosol and H^+ in the extracellular space [74,75]. Remarkably, the expression of the ubiquitous NHE1 antiporter increases by two-three-fold in many different tumor types [76]. The key role of NHE1 in carcinogenesis has been well elucidated by Reshkin et al., who provided evidence that the earlier event in tumor development was the cytoplasmic alkalization consequent to the transformation-dependent activation of the NHE1 channel, driven by an increased affinity of NHE1 allosteric proton regulatory site [77]. In addition to protons, NHE1 activity can be amplified by osmotic or ischemic stress, as well as by growth factor and MAPK signaling [74,78,79].

Besides NHE1, pH can be regulated by the Na^+/HCO_3^- cotransporter, that is overexpressed in hypoxic tumors [74,80]. In addition, other plasma membrane channels are effectors of pH dynamics, such as the voltage-gated proton channel, the vacuolar type H^+-ATPase (V-ATPase), the H^+/K^+-ATPase and the $H^+/$monocarboxylate transporters, the latter extruding from the cell lactate or other monocarboxylate metabolites together with a proton [81,82]. Importantly, both pharmacological and phytochemical inhibitors of these channels were found to revert or at least attenuate the malignant phenotype of cancer cells, suggesting that a crucial role is played by pH and ion concentration changes in tumor development, as reviewed in [74,78,81,83,84].

Aside the pH effectors above described, a number of pH sensors (i.e., proteins whose activities or ligand binding affinities are regulated by pH) have been recognized [83]. Actually, changes in proton concentration (i.e., pH dynamics) are reflected by the amount of the protonated form of histidine residues which are able to buffer pH variations [84]. In this way, pH dysregulation can simultaneously modify charge and function of many protein targets, facilitating gain- or loss-of-function of mutated proteins as it occurs in the R337H substitution of the tumor suppressor p53 in cancer [84].

It should be reminded that in normal cells proton concentration is slightly higher in cytosol (pHi ~7.2) than in extracellular space (pHe ~7.4) [83]. Conversely, in tumor cells pHi increases to ~7.6 and pHe decreases to ~6.8, thereby reversing the pH gradient (ΔpH) direction. Such pH dysregulation, that occurs early in cancer development, is further exacerbated during tumor progression [83].

The alterations in proton concentration imply serious consequences in the cellular functions. The acidification of extracellular tumor microenvironment contributes to enhance invasiveness and metastatic spread, suppress immune responses and induce chemotherapy resistance, whereas the cytosolic alkalization favors cell proliferation, protects from apoptosis, and supports tumor neo-vascularization by promoting VEGF expression [74,81,83–85]. It has been observed that, if the dysregulation of proton concentration is reversed by inhibiting NHE1 transporter activity, the resulting intracellular acidification contributes to cell death via apoptosis [85–87]. Reciprocally, the observation that upon exposure to apoptotic stimuli, an intracellular acidification occurs in mammalian cells undergoing apoptosis [87] confirms the strong relationship between acidic pHi and apoptosis pathway [88]. As counteracting pHi increase and Na^+ overload reverts cancer cell phenotype, it is worth to highlight that the use of proton and ion transporters inhibitors such as ameloride, cariporide and other synthetic inhibitors, can be a new promising therapeutic approach against tumor development [78,83]. In this regard, also some natural products can affect pH-regulating ions channels. For instance, ginsenoside induces apoptosis in hepatocellular carcinoma by decreasing NHE1 expression and activity, via EGF/ERK/HIF-1 signaling pathway [89]. Ellagic acid markedly down-regulates ROS formation and NHE1 expression leading to decreased NHE1 activity, pHi, glucose uptake and lactate release in endometrial cancer cells [90]. Corncob extract decreases proliferation and viability in glioma cells by reducing ROS amount, Bcl-2, and lactate monocarboxylate transporter 1 expression levels [91]. Moreover, the increase in plasmatic and mitochondrial membrane proton conductance could represent an alternative mechanism by which natural compounds can restore pH homeostasis without affecting a specific ion channel, but still allowing protons re-entry into the cytosol or the mitochondrial matrix. For instance, in the mitochondria, any compound acting to decrease differences in pH across the membrane, lowers $\Delta\psi_m$ and can also cause uncoupling of the electron flow with the oxidative phosphorylation [92]. In general, lipophilic, weakly acidic compounds can work as protonophores [92]. Due to their lipophilicity, they enter and remain in the membrane in which they can shuttle from one membrane-water interface to the other. For their weakly acidic nature, they can assume either the neutral state or the anionic form. In the latter form, stabilization occurs by delocalization of negative charge by resonance, so that their lipophilicity is not compromised. A protonophore in anionic form (e.g., a phenolate) is attracted by the positive side of the membrane-water interface close to the intermembrane space (where the pH is 6.8). There, it can bound a proton and move to the other membrane-water interface close to the matrix, where it releases the proton, due to the negative charge on this side of the membrane and the high pH (7.6) of the matrix. Then, it comes back for another cycle [28,92] (Figure 2).

It is known that a slight increase of mitochondrial membrane proton conductance, without affecting ATP production (mild uncoupling), can inhibit ROS formation by reducing $\Delta\psi_m$ [93] (through the mechanism described in Figure 1). Such protonophore activity has been established for the acylphloroglucinol family of natural products, such as clusianone, isolated from the roots of *Clusia congestiflora*, which is a mitochondrial uncoupler and a well-known cytotoxic anti-cancer agent in hepatocarcinoma [94], the bis-geranylacylphloroglucinol moronone, contained in the *Moronobea coccinea* extract, that displays cytotoxic activity in breast cancer cells by dissipating the mitochondrial proton gradient [95], and hyperforin, from *Hypericum perforatum*, which is able to increase proton conductance of both plasma and organelle membranes [96], as it will be detailed below.

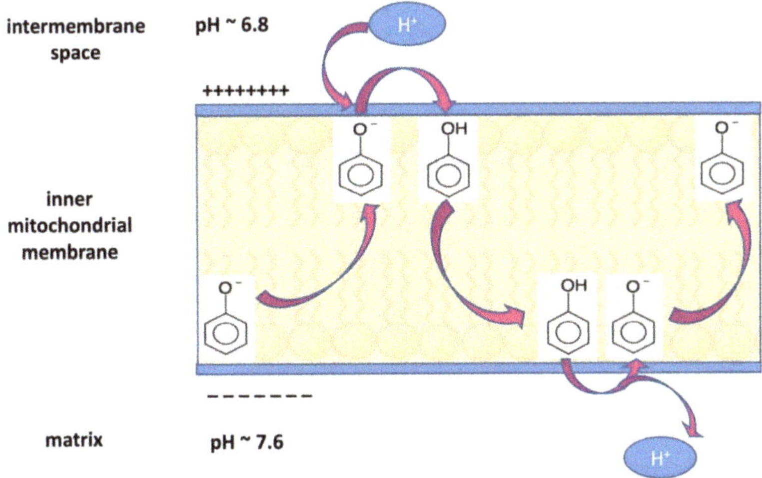

Figure 2. Mechanism of protonophore activity in the inner mitochondrial membrane. For its high lipophilicity, a protonophore like the phenolate anion depicted in the Figure (or hyperforin, as pointed out in the text) is going to be located inside the membrane and can move from one membrane-water interface to the other. As regards the inner mitochondrial membrane, a protonophore molecule captures a proton at the interface with the positive charged intermembrane space, where the proton concentration is high (pH 6.8). In the protonated form, it moves to the interface with the matrix, where the low proton concentration (pH 7.6) and the presence of negative charges favor proton extrusion, and it assumes an anionic form. This form does not interfere with its lipophilicity because it is stabilized by resonance, thus allowing its presence in the membrane and recycling. In this way, the protonophore can dissipate the transmembrane potential ($\Delta\psi_m$).

5. *Hypericum perforatum* or St. John's Wort (SJW) and Hyperforin (HPF)

SJW is an herbaceous plant containing many bioactive molecules [97]. For long time SJW has been used in the traditional medicine for its anti-inflammatory and lenitive properties and recently it has been mostly employed for the treatment of anxiety and depression to replace conventional antidepressant drugs [97,98], with which it shares the inhibition of the uptake of monoamine neurotransmitters as a mechanism of action [99]. Several randomized controlled clinical trials have been performed and their results indicate that SJW treatment of mild/moderate depression is effective and well-tolerated with low occurrence of adverse effects [98,100,101]. Besides such anti-depressant properties, other biological activities of SJW make this vegetal extract very suitable for both prevention and treatment of a number of diseases in which the inflammatory events play a relevant role. The pharmacological properties of *Hypericum perforatum* are due to a number of bioactive molecules, among which the most important and characteristic of this species are hyperforin (HPF) and hypericin, usually present in the total hydro-alcoholic SJW extract within a range of concentrations of 1–5% and 0.1–0.3%, respectively. Additional active compounds, such as hyperoside, rutin, quercetin, several catechins and other polyphenols are often contained in the total extract of SJW, although with a large variability of concentrations, mainly depending on seasonal fluctuations and the geographic origin of plant [102–104].

Among the major SJW components, we will not consider here the antitumor effect of hypericin, whose mechanism of action is based on its phototoxic properties, currently exploited in the so-called photodynamic therapy applied in several tumor types, especially of cutaneous origin [105,106]. We will actually focus on HPF, that is the primary active principle responsible for both the antidepressant and the anti-inflammatory properties of SJW [52,99,107–109].

HPF is a bio-active acylphloroglucinol abundantly present in the apical flowers of *Hypericum perforatum* plant. Notably, pharmacokinetic data, obtained in healthy volunteers, indicate that oral administration of anti-depressive therapeutic doses of SJW hydro-alcoholic extract (3×300 mg daily, containing 15 mg HPF) results in a maximal HPF circulating concentration of 0.28 μM, at 3.5 h after a single dose, and a steady-state concentration of 0.18 μM [110]. Similar results were obtained in a successive study [111], confirming the bioavailability of HPF and the achievement of suitable blood levels after ingestion of SJW extract. Altogether, the available data show that HPF contained in SJW extract is effectively absorbed and can protect cells against pro-inflammatory cytokine-induced damage. Moreover, it should be stressed that HPF is able to induce long-lasting changes in the cell signaling and transcriptional pathways even after being withdrawn from the cell culture medium following a pre-incubation period [112,113]. While HPF is stabilized by the presence of antioxidant compounds in the whole SJW extract, purified HPF is relatively unstable in presence of light and oxygen [114]. Therefore, various hyperforin salt preparations [110] and stabilized hyperforin analogs, such as DCHA-hyperforin [115,116] or aristoforin [117] were tested and proved to retain HPF biological activity.

5.1. Protective Effects of SJW and HPF against Noxious Stimuli

SJW extract and HPF exert powerful anti-inflammatory effects by blocking the activation of several signaling pathways triggered by injuring stimuli and by slowing down the production of inflammatory mediators. SJW and HPF have been shown to suppress the activities of COX-1 and 5-LO in vitro and in vivo [107,108]. Strong inhibition of prostaglandin E_2 (PGE_2) has been confirmed using either SJW extract in a lipopolysaccharides (LPS)-stimulated RAW 264.7 mouse macrophage model [118], or HPF in an in vitro assay of LPS-stimulated human whole blood [119]. In our laboratory, we observed that, in pancreatic β cells as well as in rat and human pancreatic islets, SJW and HPF are able to markedly inhibit both IFN-γ-elicited STAT-1 and TNF-α/IL-1β-triggered NF-κB activities, leading to prevention of iNOS gene expression and protection against β-cell damage [52,120,121]. Furthermore, in the same cells supplemented with SJW or HPF, we showed a significant concentration-dependent reduction of the phosphorylation level of several cytokines-elicited kinases, such as IKK, Akt, p38, c-Jun N-terminal kinase (JNK), ERK1/2 [109]. HPF also influences pro-inflammatory and immunological responses of microglia that are involved in the progression of neuropathological disorders. Actually, Kraus et al. reported that HPF significantly suppresses NF-κB activity and strongly inhibits iNOS expression in N11 and BV2 mouse microglia and RAW 264.7 macrophages treated with LPS [112]. Additionally, 1 μM HPF triggers differentiation in primary cultures of human keratinocytes and in derived HaCaT cell lines and inhibits their proliferation [122]. Pretreatment with SJW extract preserved PC12 cell line from H_2O_2-induced ROS generation and damage in a concentration-dependent manner (1–100 μg/mL) [123]. Feisst and Werz [124] found that HPF at very low concentration (IC_{50}~0.3 μM) interferes with the induction of oxidative burst in polymorphonuclear leucocytes (PMN) by rather suppressing N-formyl-methionyl-leucil-phenylalanine (fMLP)-induced ROS production, than acting as a ROS scavenger. The authors reported that, after fMLP-elicited PMN stimulation, HPF is able to hinder ROS generation, by targeting an unknown component within the signaling cascade leading to Ca^{2+} mobilization [124].

Moreover, SJW and HPF exert potent anti-inflammatory effects in several animal model of acute and chronic inflammation by downregulating the expression or activity of inflammatory mediators and lowering ROS production. HPF attenuated microglial activation, p65 NFκB phosphorylation, and suppressed TNF-α expression in rat piriform cortex following status epilepticus [125]. HPF protected neuronal cells against aluminum maltolate-induced neurotoxicity by inhibiting ROS formation and enhancing superoxide dismutase and glutathione peroxidase activities [126]. In our laboratory, we have shown in various animal models, that SJW (30 mg/kg) prevents or attenuates tissue damage [127–131]. Acute lung inflammation induced by carrageenan is mediated by ROS production and activation of the redox-sensitive transcription factors NF-κB and STAT-3. SJW protect mice from induced pleurisy by reduction of PMNs infiltration in lung tissues and subsequent lipid peroxidation, by decrease of TNF-α and IL-1β production and nitro-tyrosine formation, as well as by inhibition of NF-κB and STAT-3 activity [127]. This finding is consistent with data reported in various other studies dealing with rodent models of carrageenan-induced paw edema, which was markedly reduced upon oral administration of SJW extract [132–134]. Zymosan administration in mice also resulted in excessive ROS formation by activated PMNs and lipid peroxidation in plasma, intestine, and lung [128]. In zymosan-injected mice pre-treated with SJW, we observed an increase in the intracellular amounts of both GSH and anti-oxidant enzymes associated with the reduction of iNOS expression as well as of ROS and nitro-tyrosine levels as compared to untreated controls [128]. Furthermore, SJW extract was able to mitigate cerulean-induced pancreatitis in mice by inhibiting edema, neutrophil infiltration and levels of intercellular adhesion molecule (ICAM)-1 and nitrosylated proteins in the injured pancreas, finally reducing the mortality of cerulean-treated animals [129]. In addition, SJW displayed a pronounced protective role against gastro-enteric inflammatory injury, as observed in rat gastric mucosa damaged by indomethacin administration [135] or in colonic mucosa after induction of experimental inflammatory bowel disease [136]. SJW was also able to protect rat intestinal epithelial architecture caused by irinotecan-elicited inflammation, likely dependent on the fact that SJW pretreatment limited the cytokine increase induced by irinotecan administration in the intestine and liver [137].

In summary, HPF-containing SJW extract or HPF can attenuate inflammatory response and subsequent tissue injury in several cell types and animal models by modulating a number of potentially harmful processes triggered by inflammatory signaling and ROS generation.

5.2. Molecular Mechanisms of HPF Related to Its Protonophore Activity

It has been suggested [96] that the molecular mechanism(s) underlying HPF effects toward various intracellular targets could be due to the fact that this phloroglucinol behaves as a protonophore, that is, as explained above, capable of increasing proton conductance of biological membranes. Interestingly, studies on the structure-activity relationship indicate that the protonophore activity of a compound, in most cases, is a linear function of the partition coefficient in the octanol and water (ow) system (Log K_{ow}) and also depends on its acidic dissociation constant (pK_a), an index of acidity strength [92]. For instance, the partition coefficient of several polyphenols ranges from 1 to 3 and their pK_a ranges from 4 to 9 [28]. Definitely, HPF is much more lipophilic than the majority of polyphenols as it displays a Log K_{ow} > 10, meanwhile its pK_a value of 6.3 allows effective buffering of pH changes in a neutral environment [138]. Both values represent ideal chemical features for a compound to act as a protonophore. Indeed, the ability of HPF to increase proton conductance of plasmatic and organelle membranes was firstly disclosed by Roz and Rehavi [139] to explain its inhibitory action on neurotransmitter reuptake by a non-competitive mechanism. Taking into account that the proton gradient, induced by V-ATPase, is the major driving force for vesicular monoamine uptake and storage, these authors showed that HPF (at 0.4–1 μM), alike the synthetic protonophore carbonyl cyanide 4-(trifluoromethoxy)phenyl-hydrazone (FCCP), is able to dissipate the pH gradient across the synaptic vesicle membrane [139]. Moreover, Froestl et al. [140] have linked the HPF protonophore activity to its aptitude to enhance amyloid precursor protein extrusion from the cells, process that is highly depending on intracellular pH. In this model, HPF, by affecting proton content, can indeed protect neurons against development of Alzheimer's disease [140,141]. Sell et al. [96] confirmed that HPF, due to its physico-chemical properties, is easily incorporated into the membrane lipid bilayer, where it can function as a protonophore, triggering proton crossing regardless the involvement of any channel protein. Finally, in experiments carried out in isolated brain mitochondria, Tu et al. observed that HPF, likewise the protonophore and uncoupler FCCP, cause a concentration-dependent drop in the mitochondrial membrane potential [142]. Thus, HPF is also able to affect mitochondrial function, as well as mitochondrial (mt)ROS production [142].

6. Antitumor Activity of SJW Extract and Its Component HPF

6.1. Protective Effects of SJW and HPF against Carcinogenesis

The above described capability of SJW extract and HPF to defend normal cells and tissues from injurious stimuli could be the rationale for their preventive anti-tumor activity. As already mentioned, HPF suppresses 5-LO and COX-1 activity and PGE_2 production both in human whole blood at low concentrations range (0.03–2 μM) and in a number of in vivo models [107,108,118,119]. Actually, inhibition of eicosanoid synthesis provides a molecular basis for not only the anti-inflammatory but also the anti-carcinogenic properties of HPF [119]. Indeed, excessive PGE_2 formation has been found associated with tumorigenesis of colon, breast, prostate, and lung carcinoma [119]. Furthermore, overexpression of 5-LO protein was observed in cancer cell lines and tumor specimens [143].

Mainly due to their inhibitory action on inflammation-generated byproducts, SJW and HPF can also help preventing genotoxic effects. Actually, the anti-genotoxic ability of HPF has been verified by measuring the amount of bacterial gene mutations (Ames' test), the occurrence of DNA strand breaks in human lymphocytes (comet assay) and the induction of chromosome aberrations in a mammalian cell line [144]. The results showed that HPF (at a concentration of about 1 μM) displays indeed anti-mutagenic activity as well as DNA protective effect against zeocin-induced single- and double-strand breaks and reduction of chromosome aberrations in human liver cells exposed to benzo[a]pyrene (B[a]P) or cisplatin. This suggests that HPF could de facto contribute to reduce human environmental risk [144]. The SJW- and HPF-dependent prevention of ROS generation, observed in different in vitro and in vivo inflammatory models [123,124,126–128,145,146], may account for their protective effect against genotoxic insults. In particular, the HPF ability to decrease ROS production could be crucial to prevent the genotoxic action of zeocin, that was reported to cause DNA damage in two breast cancer cell lines mainly by increasing intracellular ROS level [147,148]. Furthermore, the aforementioned protonophore action of HPF, by allowing protons to re-enter the cytosol and thus hinder intracellular alkalization, might explain the HPF protective effect on B[a]P-induced carcinogenesis, reported by Imreova et al. [144]. In fact, intracellular alkalization was detected in rat hepatic epithelial F258 cells quickly upon exposure to B[a]P, resulting in alteration of pH dynamics in different cell compartments including mitochondria and affecting mitochondrial function [149].

A protective action against oncogenesis has also been shown by using SJW total extract. In a mouse experimental model of colorectal carcinogenesis induced by azoxymethane, Manna et al. [150] reported the preventive potential of a dietary supplementation with SJW extract. Azoxymethane induces alkylation of DNA leading to mutations and tumorigenesis [150]. Diet enrichment with 5% SJW decreases the incidence of colorectal polyps and large tumors in azoxymethane-treated mice, finally improving overall animals' survival. In this model, at molecular level, it has been observed that NF-κB and ERK1/2 pathways are hampered by SJW, with a reduction in the mRNA expression level of NF-κB-elicited genes, such as TNF-α, IL-1β, MMP-7 and -9 and iNOS [150]. These results clearly indicate that preventive administration of SJW can decrease both the occurrence of colorectal tumors and the associated induction of inflammatory signaling [150]. Interestingly, the same mouse model of colorectal carcinogenesis has recently been showed to depend on NOX/ROS activity, since a low dose of diphenyleneiodonium (DPI), a NOX inhibitor, prevents the formation of azoxymethane-induced adenomatous polyps and inhibits the intestinal inflammatory response [151]. Mechanistically, DPI decreases ROS production in the colon, resulting in inhibition of TNF-α, IL-6 and monocyte chemoattractant protein-1 (MCP-1) production, as well as ERK1/2, STAT-3 and NF-κB signaling, finally exerting a strong anti-inflammatory effect [151]. Thus, it is likely that DPI and SJW exert protective effects against azoxymethane-elicited colorectal carcinogenesis with a similar mechanism, i.e., by interfering with ROS-mediated inflammatory signaling.

It is meaningful to remark that the presence of HPF in SJW extract can affect the pharmacokinetics of many co-administered drugs by inducing a number of liver cytochrome P450 (CYP) isozymes, such as CYP3A4, CYP2C19, CYP2D2 [152]. It has been shown indeed that HPF is a potent ligand for the pregnane X receptor (PXR), which is the principal transcriptional regulator of CYP3A enzyme expression [152]. Notably, several CYP isozymes are implicated in the detoxification of xenobiotics, or otherwise in the biotransformation of pro-carcinogens requiring metabolic activation to exert their genotoxic effects [153,154]. Nevertheless, at variance of most in vitro studies showing carcinogen activation by CYPs, in vivo experiments performed after gene deletion of P450 isozymes have often revealed that these metabolizing enzymes have a major role in detoxification rather than activation of carcinogens [154]. For instance, a recent study of Beyerle et al. [155] reported downregulation of xenobiotic-metabolizing enzymes, including CYP3A4, in human colorectal cancer tissues with respect to normal mucosa samples. Thus, the activation of CYP isozymes involved in carcinogen detoxification might be another mechanism by which HPF can function as a tumor preventive phytochemical agent.

It has to be highlighted that the protective effects of SJW against colon carcinogenesis can occur also in humans. Remarkably, a very large epidemiologic study in USA (including 77,000 participants monitored for 5 years) explored whether the regular employment of various herbal supplements would lessen the risk for lung or colorectal cancers. The results showed that continuative use of SJW was associated with a 65% decrease in risk for colorectal cancer, although no protective effect was observed with regard to lung cancer [156].

It is worth mentioning that SJW could prevent tumor onset also by virtue of its antimicrobial and antiviral activities. For instance, quite low concentrations of an alcoholic SJW extract have been shown to exert powerful protection against *Helicobacter pylori*, a pathogen that is known to increase the risk of some forms of stomach cancer [157]. In a more recent study, the antimicrobial activity of SJW against *H. pylori* was confirmed by the observation that low concentrations of SJW extract were able to rapidly kill a high percentage of most strains of this bacterium [158]. SJW is also able to inhibit the growth of viruses implicated in the development of some cancers, such as hepatitis B virus (HBV) whose long-term infection could indeed result in development of cirrhosis and hepatocellular carcinoma. SJW ethanol extract has been reported to have strong inhibitory effects on HBV in vitro by decreasing the expression of HBV DNA and the secretion of HBV antigens [159].

6.2. Effects of SJW and HPF on Cell Proliferation and Apoptosis

When tumor has already grown, SJW and HPF can still exert therapeutic effects by down-modulating survival signaling and/or inducing apoptosis. Remarkably, in rats injected subcutaneously with MT-450 breast cancer cells and administered HPF daily at the site of the cell transfer, in vivo tumor growth was inhibited in the absence of any side effect [160].

Ten human and six rat cancer cell lines, derived from melanoma, breast and ovary carcinoma, prostate and pancreas tumors, glioblastoma, sarcoma and T cell leukemia, were found to be very sensitive to the antitumor effects of hyperforin-DCHA (in the range 5–20 μM) [160]. HPF-DCHA slowed down tumor cell proliferation and, at the highest concentration used in this study (20 μM), induced apoptosis as well as caspase-9 and -3 activities [160]. Apoptosis was triggered by a loss of mitochondrial transmembrane potential ($\Delta\psi_m$) very early after HPF exposure, accompanied by a rapid release of cytochrome (Cyt) c, as assessed in a mitochondria-enriched cell fraction [160]. Recently, Hsu et al. [161] reported that in two glioblastoma cell lines, HPF prompts apoptosis, increase of cytosolic [Ca^{2+}], loss of $\Delta\psi_m$, suppression of EGFR/ERK/NF-κB signaling, and decrease of anti-apoptotic proteins expression. Wiechmann et al. [162] also showed that HPF can directly impair viability of mitochondria isolated from HL-60 cells by affecting mitochondrial proton-motive force. Again, in hepatocellular carcinoma cell line, HPF significantly inhibits cell viability and cyclin D1 expression and induces loss of $\Delta\psi_m$ and downregulation of the anti-apoptotic proteins fetal liver LKB1 interacting protein c (c-FLIP), X linked inhibitor of apoptosis protein (XIAP) and myeloid leukemia cell differentiation protein (Mcl-1), finally triggering apoptosis [163].

By acting as a protonophore in the plasma membrane, HPF can counteract cytosolic alkalization, thus allowing apoptosis. Nevertheless, for the achievement of the apoptotic goal, the major target of HPF appears to be the inner mitochondrial membrane, whose hyperpolarization in cancer cells protects them from apoptosis [28]. In fact, HPF-induced severe dissipation of $\Delta\psi_m$ provokes pore formation and Cyt c release, sensitizing cancer cells to death [28]. Moreover, the fall in mitochondrial membrane potential and the consequent lowering of mtROS hinders both mitogen signals triggered by growth factors and activation of survival kinases, including ERK1/2 and Akt [48]. As reported above, mitochondrial membranes incorporate lipophilic HPF [96,164], so that, while in normal condition the inner membrane is impervious to protons, in the presence of HPF, protons can shuttle into the matrix following their concentration and electric gradients (as described in Figure 2), with consequent breakdown of $\Delta\psi_m$. Other natural active compounds, such as curcumin, epigallocatechin gallate, honokiol, myricetin, urolithin A, moronone, nemorosone, may also exert, at least partially, antitumor action by dissipation of $\Delta\psi_m$ [28,95,165], as do the synthetic uncoupling molecules [166]. In this regard, the protonophore carbonyl cyanide m-chlorophenyl hydrazone (CCCP) directly interferes with mitochondrial function and induces apoptosis [167,168]. It has also been reported that uncoupling agents such as 2,4-dinitro-phenol and others, acting as protonophores, can prevent inflammatory response even at low concentrations, by inhibiting TNF-α-dependent activation of NF-κB signaling [93,169]. Hence, NF-κB is confirmed to be a crucial target of SJW and HPF antitumor action also through mechanisms involving an uncoupling/protonophore activity that HPF shares with classical uncouplers. Actually, it has been documented that SJW extract promotes apoptosis and decreases NF-κB protein level in MCF-7 cells [170] and that SJW oil inhibits NF-κB activation in human chronic myelogenous leukemia K562 cells [171]. HPF decreases the expression level of anti-apoptotic proteins and induces apoptotic cell death through blockade of NF-κB activity in non-small cell lung cancer [147], bladder cancer [172] and U-87 and GBM-8401 glioblastoma [161] cell lines. In order to clarify the molecular mechanism leading to NF-κB inhibition by HPF, it should be reminded that this transcription factor can be activated by inflammatory mediators through ROS as second messengers [57,59]. Thus, HPF, by blocking NOX/ROS-dependent signal transduction elicited by cytokines and growth factors, and/or mtROS production by decreasing $\Delta\psi_m$, could affect NF-κB activation and its downstream signaling pathway.

Many pro-apoptotic proteins are encoded by NF-κB-responsive genes. Notably, several investigations document a pro-apoptotic effect of SJW and HPF depending on a complex modulation of pro- and anti-apoptotic members of the B cell lymphoma 2 (Bcl2) family of proteins. For instance, in MCF-7 human breast cancer cells, SJW extract induces apoptosis by increasing the pro-apoptotic proteins Bcl 2 associated X (Bax) and Bcl 2 antagonist of cell death (Bad) and decreasing the anti-apoptotic proteins Bcl-2, B cell lymphoma extra-large (Bcl-xL), and phosphorylated form of Bad (pBad), finally leading to caspase 7 activation and cell death [173]. In human myeloid tumor cells, Merhi et al. [174] show that HPF inhibits Akt kinase activity, determining dephosphorylation of Bad, an Akt substrate, and activation of its pro-apoptotic function. The authors conclude that HPF, as a negative regulator of Akt, could represent an interesting novel approach for treatment of AML and other malignancies. In human K562 T cell leukemia and U937 lymphoma cell lines, Hostanska et al. confirm the anti-carcinogenic property of HPF that triggers a caspase-dependent apoptotic cell death [175]. The HPF antitumor activity has been also tested in primary cells of hematological malignancies, i.e., in leukemic cells from B-cell chronic lymphocytic leukemia (B-CLL) patients [176]. HPF induces apoptosis in the B-CLL cells by eliciting a drop in mitochondrial transmembrane potential and activation of caspase 3. In this study, the anti-apoptotic factors Bcl-2 and Mcl-1 as well as the protein expressions of iNOS and p27^{kip1b} are downregulated [176]. In the B-CCL cells, Zaher et al. [177] show a strong correlation between the upregulation of Noxa protein expression and HPF-elicited cell apoptosis, that is in fact partially inhibited by RNA-interfering Noxa silencing. Notably, the pro-apoptotic activator Noxa, by binding to Mcl-1 and neutralizing its anti-apoptotic action, is involved in B-CCL cells death [178]. HPF is also able to elicit growth arrest and caspase-dependent apoptosis in acute myeloid leukemia (AML) cell lines and in primary AML cells, in which the apoptotic mechanisms are again traced back to mitochondrial membrane potential loss, Noxa overexpression and Mcl-1 downregulation [174].

As the ability of HPF to regulate the balance between pro- and anti-apoptotic proteins has been so frequently associated with mitochondrial membrane potential loss, due to its protonophore activity, it is actually surprising that in many experimental models this mechanism has not yet been extensively investigated. However, all studies which measured both apoptosis activity and mitochondrial membrane potential found a negative correlation between these two phenomena [160,162,163,174,176], suggesting a close link between HPF mitochondrial protonophore action and programmed cell death susceptibility in cancer cells.

It is worth to highlight that the HPF pro-apoptotic action is selectively observed in cancer cells, whereas healthy cells are much less sensitive to HPF cytotoxicity. It is known indeed that the same concentration of HPF that induces clear-cut damage in B-CLL cells is totally unable to compromise viability of human B lymphocytes from healthy donors [176]. The reason why normal cells are more resistant to HPF cytotoxic action could be related to their intrinsic differences in metabolism, pH and ionic concentrations in the cytosol and organelle compartments with respect to malignant cells. For instance, in normal cells the negative cytosolic side of the plasma membrane facilitates proton influx, although this influx is attenuated by the mild pH gradient in the opposite direction, resulting in only a faint proton entry in the presence of a protonophore such as HPF (Figure 3). Instead, in malignancy, the negative plasma membrane charges together with a reversed pH gradient, due to the highest activity of proton channels that extrude proton from the cells, determine a substantial driving force for robust proton influx. Thereby, when the proton conductance is empowered by the presence of a protonophore agent, a significant and persistent cytosolic proton influx occurs in cancer cells. Alike other protonophore, HPF elicits proton influx in tumor cells restoring cytosolic acidity and allowing apoptosis. At the same time, extracellular tumor microenvironment would be impoverished in proton concentration (i.e., pH_e tends to neutrality), thus impeding extracellular matrix digestion, tumor cells migration and invasiveness (Figure 3).

This hypothesis must be experimentally verified in the case of malignant cells, but is consistent with the findings of Sell et al. [96] in both plasmatic membrane of normal cells (primary microglia and chromaffin cells) and synthetic lipid bilayer (devoid of any channel protein), in which HPF was shown to mediate a proton conductance due to its intrinsic protonophore activity and resulting in proton entry into the cell and cytosol acidification. Notably, the extent of proton influx, in the presence of HPF, is directly depending on the proton gradient between the two sides of the membrane [96]. These authors also demonstrated, by measurement of capacitance changes in the presence of various HPF concentrations, that this lipophilic compound accumulates in the membrane, which might explain the effects of even low doses of HPF [96] and the persistence of the protonophore activity also when SJW or HPF is no more available in the cell culture medium [113].

At the present state of our knowledge, the hypothesis that HPF can change the proton flux in function of the proton gradient could explain why HPF is not cytotoxic for normal cells, whereas it can induce apoptosis in cancer cells. In cancer cells, both at the plasma membrane level, in which ΔpH favors high proton influx, and at the hyperpolarized inner mitochondria membrane, the presence of HPF could counteract malignant phenotype, or at higher concentration, promote cell death by collapsing inner mitochondrial membrane.

In general, other effects of natural products could be related to their uncoupling activity. As an example, curcumin, that can act as an uncoupler, is able to increase AMP/ATP ratio, thereby activating AMP-activated protein kinase (AMPK) [179]. In this way, curcumin can inhibit mTOR that decreases Ser727 phosphorylation of STAT-3, blocking cell proliferation [180]. On the other hand, activated AMPK stabilizes the tumor suppressor p53 favoring cellular sensitization to mitochondrial apoptosis [181]. Similarly, in HL60 leukemic cells, HPF and myrtucommulone, another phloroglucinol derivative, directly interact with the mitochondrial membrane, decreasing $\Delta\psi_m$, lowering the ATP level and activating AMPK [162]. Moreover, SJW treatment, in the concentration range of 10–50 µg/mL, enhances pAMPK and decreases pAkt, suppressing growth of breast cancer MCF7 cells. SJW is also able to repress protein synthesis by decreasing the phosphorylation levels of both mTOR and its downstream substrate 4E-BP1 [173].

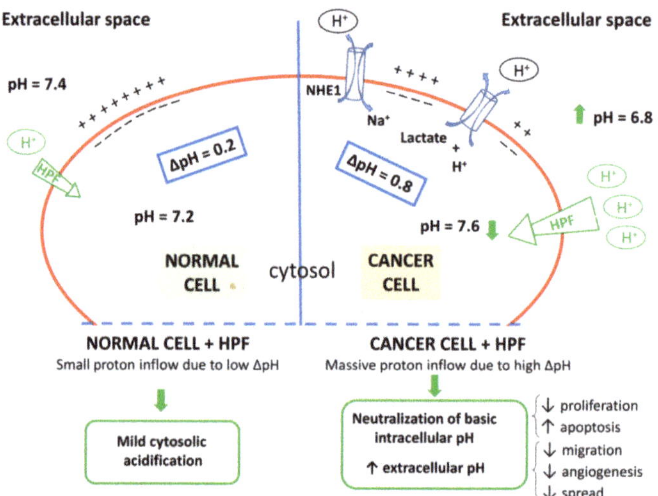

Figure 3. Proton dynamics in normal and cancer cells. Left: normal cells display mild difference in proton concentration between extracellular space and cytosol (ΔpH~0.2). Right: Differently from normal cells, cancer cells show high activity of a number of channels that regulate pH (NHE; Hv; V-H$^+$/ATPase; Na$^+$/HCO$_3$$^-$ cotransporter; H$^+$/K$^+$-ATPase and H$^+$/lactate cotransporter, two of them being represented as examples), which results in increase of proton concentration in the extracellular space combined with cytosol alkalization (ΔpH~0.8). In the presence of HPF, due to the protonophore activity of this compound, proton conductance increases. In both cell types exposed to HPF, the negative charges of the cytosolic side of plasma membrane foster a proton influx, which is small in normal cells and large in cancer cells displaying higher ΔpH. Thus, cancer cells exposed to HPF can import a larger amount of protons which counteracts intracellular alkalization and extracellular acidification, both hallmarks of malignancy. Green arrows and outlines refer to the putative effects of HPF.

6.3. SJW and HPF Affect Neo-Angiogenesis, Tumor Cell Invasion and Metastasis.

The capability of SJW and in particularly HPF to inhibit angiogenesis, cell invasion and metastasis has been largely investigated.

Martinez-Poveda et al. [182,183] showed that HPF can inhibit angiogenesis both in bovine aortic endothelial (BAE) cells in vitro and in chorioallantoid membrane in vivo. HPF also inhibits MMP-2 and urokinase secretion from BAE cells and restrains their invasive capability in a Matrigel layer. The authors claim that HPF could be considered a promising anti-angiogenic natural product interfering with key events in angiogenesis at concentrations (10 µM) that do not cause endothelial cell death [183]. The anti-angiogenic activity of HPF has been confirmed by many other studies. Lorusso et al. [184], using hyperforin-DCHA in human umbilical vascular endothelial cells (HUVEC), showed that HPF induces cytostatic but not cytotoxic effects, significantly reducing HUVEC cell migration triggered by chemoattractant stimuli. Again, the authors showed that in the presence of 1–3 µM HPF, the chemokine-elicited migration of neutrophils and monocytes as well as the TNF-α-stimulated nuclear translocation of NF-κB were markedly inhibited in HUVEC cells [184]. In an in vivo mouse model of conspicuous angiogenesis induced by subcutaneous injection of Matrigel implants, HPF potently prevents neo-vascularization as well as expression of pro-angiogenic IL-8 and MCP-1 chemokines [184]. In addition, the growth of Kaposi's sarcoma (a highly angiogenic tumor) in HPF-treated mice is markedly diminished both in size and vascularization with respect to untreated controls. The authors conclude that HPF can block many crucial events in tumor angiogenesis, most likely by inhibition of inflammatory signaling, thus opening new perspectives in the treatment of solid cancers [184].

Likewise, Rothley et al. [185] reported that HPF and its more stable analog aristoforin can block cell cycle and proliferation of either arterial or lymphatic endothelial cells at 5 μM concentration, whereas they induce apoptosis at higher concentrations (>10 μM). The authors analyzed the effect of these compounds in a rat model of tumor-induced lympho-angiogenesis, obtained by subcutaneous injection of MT-450 malignant cells. When HPF or aristoforin is administered daily in the peri-tumor area for two weeks, the lymphatic capillary outgrowth is significantly reduced, suggesting that these bio-active compounds can limit tumor-induced lympho-angiogenesis also in vivo [185].

Angiogenesis involves a tight regulation of multiple signaling pathways, among which VEGF is the most prominent effector, although PDGF and bFGF also play a role [186]. It is well-known that ROS signaling is linked to angiogenesis and involves induction of several kinase pathways as MAPKs, PI3K, Akt, and activation of transcription factors, including HIF-1, NF-κB, STAT3 [187]. For instance, ROS determine HIF1 stabilization, which in turn increases the transcription of many angiogenic genes, including VEGF [187]. Additionally, the promoters of the pro-angiogenic chemokines, such as IL-8 and MCP-1, contain binding sites for NF-κB which is regulated in a redox dependent manner [188]. Notably, NOX-derived ROS are required for the angiogenic response induced by various growth factors, with also the contribution of mtROS. It must be recalled that extracellular acidification supports angiogenesis, since it regulates VEGF expression in a both transcriptional and post-transcriptional manner, thus suggesting that an acidic tumor microenvironment can contributes to cancer progression [189]. Moreover, the G protein-coupled receptor (GPR)-4 expressed on endothelial membrane acts as a sensor of extracellular protons and stimulates intracellular signaling. Indeed, GPR4-deficient mice show strongly reduced responses to VEGF-driven angiogenesis, with a reduction of tumor growth that is correlated with impaired vessel structure and lower VEGF receptor 2 level [190]. Therefore, it can be argued that a change in proton concentration driven by the protonophore activity of HPF could affect angiogenic signals.

Interestingly, the contribution of enzymes such as metalloproteinase to tumor invasiveness and metastasis can be potentiated by their proton concentration dependency. Actually, the expression and secretion of MMP-9 increases at lower pH and higher pHi [84]. Additionally, acid-activated proteases, such as cathepsin B, cleave latent MMPs into active enzymes. These examples of pH-dependent proteins reveal potential therapeutic targets and strategies based on using changes in pH to control tumor metastatic process [84], as recently confirmed by Robey et al. [191], who were able to inhibit experimental and spontaneous metastases by increasing systemic buffering capacity and tumor pHe through oral bicarbonate administration to mice.

Anyway, HPF, besides showing anti-angiogenic activity, inhibits the MMP-9 production in B-CLL [192] and prevents the formation of microtubules in bone marrow endothelial cells, so that its therapeutic use in B-CLL patients has been suggested [192]. Furthermore, non-cytotoxic HPF concentrations can hinder cell invasiveness by downregulating the activities of MMP-2 and -9, elastase and cathepsin G, highly expressed in inflammatory and tumor cells [116]. Such downregulation is most likely dependent on HPF-induced inhibition of the constitutively activated ERK1/2 in malignant cells, responsible for the enhancement of MMPs expression [193]. In this way, HPF can counteract the capability of tumor cells to digest components of extracellular matrix, thereby impairing their spreading. Indeed, HPF has been shown to significantly reduce the number of lung metastatic foci in mice receiving colon carcinoma or melanoma cells, suggesting that this natural compound can effectively prevent cancer growth and metastatic spread in vivo [116].

Finally, it should be mentioned that an observational study suggests that SJW can exert anti-tumor effects also in humans. A clinical report described the beneficial effect of SJW in two cases of colon cancer and one of duodenal adenocarcinoma. None of the cases received neo-adjuvant chemo-radiotherapy or additional treatment, but only SJW oil (one teaspoon in the morning) before surgery. In each patient, an intense lympho-plasmocytic reaction occurred, resulting in a fibrosis surrounding the tumor, that created a defensive shield against the penetration of the tumor cells into underlying tissue [194]

7. Conclusions

HPF, the main active component of SJW, that has been shown to be the major responsible for the antidepressant effect of this plant extract, is provided of additional advantageous properties, including anti-inflammatory, antiangiogenic and antitumor activities.

The beneficial effects of HPF-containing SJW extract against cancer are summarized as follows (Figure 4).

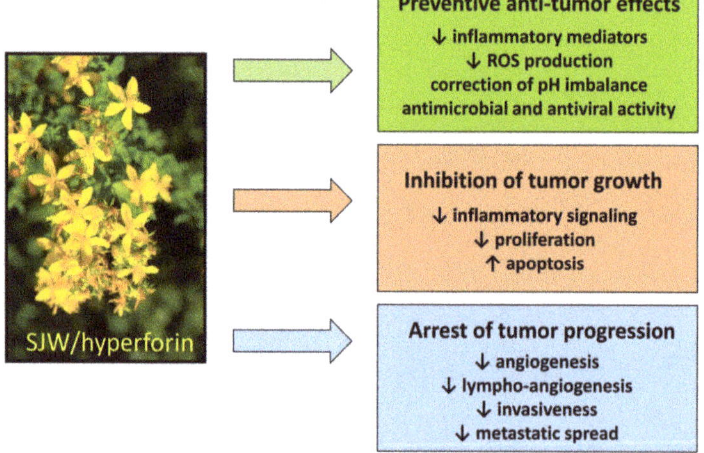

Figure 4. Antitumor effects of SJW and HPF. SJW and HPF display many antitumor effects with regard to both prevention and therapy of neoplastic diseases by acting on many different intracellular targets.

1. Tumor prevention. Regular use of SJW reduces cancer risk, by preventing the genotoxic effect of carcinogens [144]. This protective action is essentially based on the ability of HPF to slowdown inflammatory mediators and regulate ROS production and/or pH imbalance, resulting in counteraction of malignant phenotype. It should be emphasized that preventive dietary supplementation of SJW has been found to reduce the risk of colorectal cancer also in humans [156].
2. Effects on tumor growth and spread. The anti-inflammatory and anti-angiogenic effects of SJW/HPF, presumably due to inhibition of cytokine and chemokine production and hindrance of their downstream signaling, can protect from tumor expansion. Additionally, the protonophore property of HPF can avoid the acidification of the tumor extracellular milieu, impairing neo-angiogenesis and metalloproteinases activity and thereby tumor invasiveness and metastatic spread [116]. The HPF ability to drop off mitochondrial membrane hyperpolarization and consequently mtROS generation inhibits cell proliferation and favors apoptosis induction. Indeed, the pro-apoptotic effect of SJW/HPF is well documented in many malignant cell lines or in animal tumor models and appears determined by the imbalance between pro- and anti-apoptotic protein expression, even though the concentrations of HPF capable

of inducing apoptosis by such mechanism are quite high (at least 5–10 μM) and difficult to be achieved in clinical treatment. Nevertheless, it is well known that proliferating cancer cells display higher $\Delta\psi_m$ than normal cells, and that positive correlation exists between the malignancy grade of cell clones and their mitochondrial potential [195]. Consequently, malignant cells are more sensitive than normal cells to even small changes of the mitochondrial electro-chemical gradient. Thus, low doses of SJW/HPF, likely insufficient to induce cancer cell death, can nevertheless block ROS-elicited tumor growth and spread by hindering activity of pro-survival protein kinases and angiogenesis.

3. Bioavailability and broad spectrum of action. Furthermore, SJW/HPF, for their large bioavailability, persistence of protective benefits and substantial absence of adverse effects, are natural products of biological relevance for tumor prevention and treatment. As cancer therapy requires a multifactorial strategy, SJW/HPF treatment can actually meet this requirement, due to their pleiotropic effects against many different molecular targets along signaling pathways crucial for tumor growth and progression. Additionally, SJW/HPF, acting against different types of tumor cells, can be a broad-spectrum anti-tumor compound and should be tested in association with current chemotherapy drugs to achieve additive effects.

4. Limitations. The therapeutic use of HPF-containing SJW extract as anti-depressant has confirmed its very good tolerability and the paucity of adverse effects. However, a major concern of SJW/HPF treatment regards the possible occurrence of drug-drug interaction, due to HPF high affinity binding to PXR, resulting in increased expression levels of cytochrome P450 isoenzymes [141]. Actually, by activating CYP3A4, HPF can enhance drug metabolism and excretion, thereby reducing the effectiveness of a number of chemotherapeutic agents. Thus, the association of SJW/HPF to a chemotherapeutic drug should be thoughtfully decided and monitored to ensure that the drug efficacy is maintained. Very interestingly, several phloroglucinol derivatives devoid of PXR binding activity, have recently been characterized [196]. These HPF derivatives have already been proven to maintain antidepressant activity, although antitumor properties have not yet been tested and should be investigated in a near future.

On the basis of the studies presented in this review, SJW extract and its component hyperforin, already safely employed in antidepressant therapy, hold considerable promise to exert relevant beneficial effects as anti-tumor phytochemicals. These natural compounds have a remarkable potential for both prophylactic and therapeutic use against tumor development by mechanisms that involve modulation of ROS production and action along different steps of carcinogenesis as well as regulation of proton dynamics in cancer cells. Although further studies are undoubtedly required for a full clarification of mechanisms and effects, ample experimental evidence in vitro and in vivo already point out that SJW and HPF can play a remarkable role as nutraceutical supplement or in association with chemotherapy in the challenging fight against cancer.

Author Contributions: All authors have equally contributed to bibliographic research and selection, manuscript writing, review and editing, and figures preparation. All authors have read and agreed to the published version of the manuscript.

Funding: This research received no external funding.

References

1. Newman, D.J.; Cragg, G.M. Natural Products as Sources of New Drugs from 1981 to 2014. *J. Nat. Prod.* **2016**, *79*, 629–661. [CrossRef] [PubMed]
2. Choudhari, A.S.; Mandave, P.C.; Deshpande, M.; Ranjekar, P.; Prakash, O. Phytochemicals in Cancer Treatment: From Preclinical Studies to Clinical Practice. *Front. Pharm.* **2019**, *10*, 1614. [CrossRef] [PubMed]

3. Deng, L.-J.; Qi, M.; Li, N.; Lei, Y.-H.; Zhang, D.-M.; Chen, J.-X. Natural products and their derivatives: Promising modulators of tumor immunotherapy. *J. Leukoc. Biol.* **2020**. [CrossRef] [PubMed]

4. Russo, G.L. Ins and outs of dietary phytochemicals in cancer chemoprevention. *Biochem. Pharm.* **2007**, *74*, 533–544. [CrossRef] [PubMed]

5. Lee, J.H.; Khor, T.O.; Shu, L.; Su, Z.-Y.; Fuentes, F.; Kong, A.-N.T. Dietary phytochemicals and cancer prevention: Nrf2 signaling, epigenetics, and cell death mechanisms in blocking cancer initiation and progression. *Pharmacol. Ther.* **2013**, *137*, 153–171. [CrossRef] [PubMed]

6. Surh, Y.-J. Cancer chemoprevention with dietary phytochemicals. *Nat. Rev. Cancer* **2003**, *3*, 768–780. [CrossRef] [PubMed]

7. Li, W.; Guo, Y.; Zhang, C.; Wu, R.; Yang, A.Y.; Gaspar, J.; Kong, A.-N.T. Dietary Phytochemicals and Cancer Chemoprevention: A Perspective on Oxidative Stress, Inflammation, and Epigenetics. *Chem. Res. Toxicol.* **2016**, *29*, 2071–2095. [CrossRef] [PubMed]

8. Mantovani, A.; Allavena, P.; Sica, A.; Balkwill, F. Cancer-related inflammation. *Nature* **2008**, *454*, 436–444. [CrossRef] [PubMed]

9. Borrello, M.G.; Alberti, L.; Fischer, A.; Degl'innocenti, D.; Ferrario, C.; Gariboldi, M.; Marchesi, F.; Allavena, P.; Greco, A.; Collini, P.; et al. Induction of a proinflammatory program in normal human thyrocytes by the RET/PTC1 oncogene. *Proc. Natl. Acad. Sci. USA* **2005**, *102*, 14825–14830. [CrossRef] [PubMed]

10. Landskron, G.; De la Fuente, M.; Thuwajit, P.; Thuwajit, C.; Hermoso, M.A. Chronic Inflammation and Cytokines in the Tumor Microenvironment. *J. Immunol. Res.* **2014**, *2014*, 1–19. [CrossRef] [PubMed]

11. Moore, R.J.; Owens, D.M.; Stamp, G.; Arnott, C.; Burke, F.; East, N.; Holdsworth, H.; Turner, L.; Rollins, B.; Pasparakis, M.; et al. Mice deficient in tumor necrosis factor-alpha are resistant to skin carcinogenesis. *Nat. Med.* **1999**, *5*, 828–831. [CrossRef] [PubMed]

12. Szlosarek, P.; Charles, K.A.; Balkwill, F.R. Tumour necrosis factor-alpha as a tumour promoter. *Eur. J. Cancer* **2006**, *42*, 745–750. [CrossRef] [PubMed]

13. Karki, R.; Kanneganti, T.-D. Diverging inflammasome signals in tumorigenesis and potential targeting. *Nat. Rev. Cancer* **2019**, *19*, 197–214. [CrossRef] [PubMed]

14. Labbozzetta, M.; Notarbartolo, M.; Poma, P. Can NF-κB Be Considered a Valid Drug Target in Neoplastic Diseases? Our Point of View. *Int. J. Mol. Sci.* **2020**, *21*, 70. [CrossRef] [PubMed]

15. Karin, M. Nuclear factor-kappaB in cancer development and progression. *Nature* **2006**, *441*, 431–436. [CrossRef] [PubMed]

16. Bromberg, J.F.; Wrzeszczynska, M.H.; Devgan, G.; Zhao, Y.; Pestell, R.G.; Albanese, C.; Darnell, J.E. Stat3 as an oncogene. *Cell* **1999**, *98*, 295–303. [CrossRef]

17. Zhang, Q.; Raje, V.; Yakovlev, V.A.; Yacoub, A.; Szczepanek, K.; Meier, J.; Derecka, M.; Chen, Q.; Hu, Y.; Sisler, J.; et al. Mitochondrial localized Stat3 promotes breast cancer growth via phosphorylation of serine 727. *J. Biol. Chem.* **2013**, *288*, 31280–31288. [CrossRef] [PubMed]

18. Habib, S.H.M.; Makpol, S.; Abdul Hamid, N.A.; Das, S.; Ngah, W.Z.W.; Yusof, Y.A.M. Ginger extract (*Zingiber officinale*) has anti-cancer and anti-inflammatory effects on ethionine-induced hepatoma rats. *Clinics* **2008**, *63*, 807–813. [CrossRef] [PubMed]

19. Valenti, M.T.; Perduca, M.; Romanelli, M.G.; Mottes, M.; Dalle Carbonare, L. A potential role for astaxanthin in the treatment of bone diseases (Review). *Mol. Med. Rep.* **2020**, *22*, 1695–1701. [CrossRef] [PubMed]

20. Cho, J.; Rho, O.; Junco, J.; Carbajal, S.; Siegel, D.; Slaga, T.J.; DiGiovanni, J. Effect of Combined Treatment with Ursolic Acid and Resveratrol on Skin Tumor Promotion by 12-O-Tetradecanoylphorbol-13-Acetate. *Cancer Prev. Res.* **2015**, *8*, 817–825. [CrossRef] [PubMed]

21. Cristea, S.; Sage, J. Is the Canonical RAF/MEK/ERK Signaling Pathway a Therapeutic Target in SCLC? *J. Thorac. Oncol.* **2016**, *11*, 1233–1241. [CrossRef] [PubMed]

22. Mirza-Aghazadeh-Attari, M.; Ekrami, E.M.; Aghdas, S.A.M.; Mihanfar, A.; Hallaj, S.; Yousefi, B.; Safa, A.; Majidinia, M. Targeting PI3K/Akt/mTOR signaling pathway by polyphenols: Implication for cancer therapy. *Life Sci.* **2020**, *255*, 117481. [CrossRef] [PubMed]

23. Revathidevi, S.; Munirajan, A.K. Akt in cancer: Mediator and more. *Semin. Cancer Biol.* **2019**, *59*, 80–91. [CrossRef] [PubMed]

24. Rad, E.; Murray, J.T.; Tee, A.R. Oncogenic Signalling through Mechanistic Target of Rapamycin (mTOR): A Driver of Metabolic Transformation and Cancer Progression. *Cancers* **2018**, *10*, 5. [CrossRef] [PubMed]

25. Wang, T.; Hu, Y.-C.; Dong, S.; Fan, M.; Tamae, D.; Ozeki, M.; Gao, Q.; Gius, D.; Li, J.J. Co-activation of ERK, NF-kappaB, and GADD45beta in response to ionizing radiation. *J. Biol. Chem.* **2005**, *280*, 12593–12601. [CrossRef]

26. Kucharczak, J.; Simmons, M.J.; Fan, Y.; Gélinas, C. To be, or not to be: NF-kappaB is the answer–role of Rel/NF-kappaB in the regulation of apoptosis. *Oncogene* **2003**, *22*, 8961–8982. [CrossRef]

27. Dutta, S.; Mahalanobish, S.; Saha, S.; Ghosh, S.; Sil, P.C. Natural products: An upcoming therapeutic approach to cancer. *Food Chem. Toxicol.* **2019**, *128*, 240–255. [CrossRef]

28. Stevens, J.F.; Revel, J.S.; Maier, C.S. Mitochondria-Centric Review of Polyphenol Bioactivity in Cancer Models. *Antioxid. Redox Signal.* **2018**, *29*, 1589–1611. [CrossRef]

29. Kumar, M.; Kaur, V.; Kumar, S.; Kaur, S. Phytoconstituents as apoptosis inducing agents: Strategy to combat cancer. *Cytotechnology* **2016**, *68*, 531–563. [CrossRef]

30. Fontana, F.; Raimondi, M.; Marzagalli, M.; Di Domizio, A.; Limonta, P. The emerging role of paraptosis in tumor cell biology: Perspectives for cancer prevention and therapy with natural compounds. *Biochim. Biophys. Acta Rev. Cancer* **2020**, *1873*, 188338. [CrossRef]

31. Chaffer, C.L.; Weinberg, R.A. A perspective on cancer cell metastasis. *Science* **2011**, *331*, 1559–1564. [CrossRef] [PubMed]

32. Erez, N.; Truitt, M.; Olson, P.; Arron, S.T.; Hanahan, D. Cancer-Associated Fibroblasts Are Activated in Incipient Neoplasia to Orchestrate Tumor-Promoting Inflammation in an NF-kappaB-Dependent Manner. *Cancer Cell* **2010**, *17*, 135–147. [CrossRef] [PubMed]

33. Zubair, H.; Khan, M.A.; Anand, S.; Srivastava, S.K.; Singh, S.; Singh, A.P. Modulation of the tumor microenvironment by natural agents: Implications for cancer prevention and therapy. *Semin. Cancer Biol.* **2020**. [CrossRef] [PubMed]

34. Fu, L.-Q.; Du, W.-L.; Cai, M.-H.; Yao, J.-Y.; Zhao, Y.-Y.; Mou, X.-Z. The roles of tumor-associated macrophages in tumor angiogenesis and metastasis. *Cell. Immunol.* **2020**, *353*, 104119. [CrossRef] [PubMed]

35. Aras, S.; Zaidi, M.R. TAMeless traitors: Macrophages in cancer progression and metastasis. *Br. J. Cancer* **2017**, *117*, 1583–1591. [CrossRef]

36. Viallard, C.; Larrivée, B. Tumor angiogenesis and vascular normalization: Alternative therapeutic targets. *Angiogenesis* **2017**, *20*, 409–426. [CrossRef]

37. Baeriswyl, V.; Christofori, G. The angiogenic switch in carcinogenesis. *Semin. Cancer Biol.* **2009**, *19*, 329–337. [CrossRef]

38. Ribatti, D.; Crivellato, E. Mast cells, angiogenesis, and tumour growth. *Biochim. Biophys. Acta (Bba) Mol. Basis Dis.* **2012**, *1822*, 2–8. [CrossRef]

39. Stacker, S.A.; Williams, S.P.; Karnezis, T.; Shayan, R.; Fox, S.B.; Achen, M.G. Lymphangiogenesis and lymphatic vessel remodelling in cancer. *Nat. Rev. Cancer* **2014**, *14*, 159–172. [CrossRef]

40. Kapinova, A.; Kubatka, P.; Liskova, A.; Baranenko, D.; Kruzliak, P.; Matta, M.; Büsselberg, D.; Malicherova, B.; Zulli, A.; Kwon, T.K.; et al. Controlling metastatic cancer: The role of phytochemicals in cell signaling. *J. Cancer Res. Clin. Oncol.* **2019**, *145*, 1087–1109. [CrossRef]

41. Choi, Y.K.; Cho, S.-G.; Woo, S.-M.; Yun, Y.J.; Park, S.; Shin, Y.C.; Ko, S.-G. Herbal extract SH003 suppresses tumor growth and metastasis of MDA-MB-231 breast cancer cells by inhibiting STAT3-IL-6 signaling. *Mediat. Inflamm.* **2014**, *2014*, 492173. [CrossRef] [PubMed]

42. Wu, C.; Qiu, S.; Liu, P.; Ge, Y.; Gao, X. Rhizoma Amorphophalli inhibits TNBC cell proliferation, migration, invasion and metastasis through the PI3K/Akt/mTOR pathway. *J. Ethnopharmacol.* **2018**, *211*, 89–100. [CrossRef] [PubMed]

43. Han, B.; Jiang, P.; Liu, W.; Xu, H.; Li, Y.; Li, Z.; Ma, H.; Yu, Y.; Li, X.; Ye, X. Role of Daucosterol Linoleate on Breast Cancer: Studies on Apoptosis and Metastasis. *J. Agric. Food Chem.* **2018**, *66*, 6031–6041. [CrossRef] [PubMed]

44. Afrin, S.; Giampieri, F.; Gasparrini, M.; Forbes-Hernández, T.Y.; Cianciosi, D.; Reboredo-Rodriguez, P.; Manna, P.P.; Zhang, J.; Quiles, J.L.; Battino, M. The inhibitory effect of Manuka honey on human colon cancer HCT-116 and LoVo cell growth. Part 2: Induction of oxidative stress, alteration of mitochondrial respiration and glycolysis, and suppression of metastatic ability. *Food Funct.* **2018**, *9*, 2158–2170. [CrossRef] [PubMed]

45. NavaneethaKrishnan, S.; Rosales, J.L.; Lee, K.-Y. ROS-Mediated Cancer Cell Killing through Dietary Phytochemicals. *Oxid. Med. Cell. Longev.* **2019**, *2019*, 9051542. [CrossRef] [PubMed]

46. Yu, Y.; Cui, Y.; Niedernhofer, L.J.; Wang, Y. Occurrence, Biological Consequences, and Human Health Relevance of Oxidative Stress-Induced DNA Damage. *Chem. Res. Toxicol.* **2016**, *29*, 2008–2039. [CrossRef] [PubMed]

47. Berry, B.J.; Trewin, A.J.; Amitrano, A.M.; Kim, M.; Wojtovich, A.P. Use the Protonmotive Force: Mitochondrial Uncoupling and Reactive Oxygen Species. *J. Mol. Biol.* **2018**, *430*, 3873–3891. [CrossRef] [PubMed]

48. Kumari, S.; Badana, A.K.; Malla, R. Reactive Oxygen Species: A Key Constituent in Cancer Survival. *Biomark Insights* **2018**, *13*, 1177271918755391. [CrossRef] [PubMed]

49. Beckman, J.S.; Koppenol, W.H. Nitric oxide, superoxide, and peroxynitrite: The good, the bad, and ugly. *Am. J. Physiol.* **1996**, *271*, C1424–C1437. [CrossRef] [PubMed]

50. Murakami, A. Chemoprevention with phytochemicals targeting inducible nitric oxide synthase. *Forum Nutr.* **2009**, *61*, 193–203. [CrossRef]

51. Tedeschi, E.; Menegazzi, M.; Margotto, D.; Suzuki, H.; Förstermann, U.; Kleinert, H. Anti-inflammatory actions of St. John's wort: Inhibition of human inducible nitric-oxide synthase expression by down-regulating signal transducer and activator of transcription-1alpha (STAT-1alpha) activation. *J. Pharm. Exp.* **2003**, *307*, 254–261. [CrossRef] [PubMed]

52. Menegazzi, M.; Novelli, M.; Beffy, P.; D'Aleo, V.; Tedeschi, E.; Lupi, R.; Zoratti, E.; Marchetti, P.; Suzuki, H.; Masiello, P. Protective effects of St. John's wort extract and its component hyperforin against cytokine-induced cytotoxicity in a pancreatic beta-cell line. *Int. J. Biochem. Cell Biol.* **2008**, *40*, 1509–1521. [CrossRef] [PubMed]

53. Nordzieke, D.E.; Medraño-Fernandez, I. The Plasma Membrane: A Platform for Intra- and Intercellular Redox Signaling. *Antioxidants* **2018**, *7*, 168. [CrossRef] [PubMed]

54. Sies, H.; Jones, D.P. Reactive oxygen species (ROS) as pleiotropic physiological signalling agents. *Nat. Rev. Mol. Cell Biol.* **2020**, *21*, 363–383. [CrossRef] [PubMed]

55. Roy, K.; Wu, Y.; Meitzler, J.L.; Juhasz, A.; Liu, H.; Jiang, G.; Lu, J.; Antony, S.; Doroshow, J.H. NADPH oxidases and cancer. *Clin. Sci.* **2015**, *128*, 863–875. [CrossRef]

56. Landry, W.D.; Cotter, T.G. ROS signalling, NADPH oxidases and cancer. *Biochem. Soc. Trans.* **2014**, *42*, 934–938. [CrossRef]

57. Bubici, C.; Papa, S.; Dean, K.; Franzoso, G. Mutual cross-talk between reactive oxygen species and nuclear factor-kappa B: Molecular basis and biological significance. *Oncogene* **2006**, *25*, 6731–6748. [CrossRef]

58. Renard, P.; Zachary, M.D.; Bougelet, C.; Mirault, M.E.; Haegeman, G.; Remacle, J.; Raes, M. Effects of antioxidant enzyme modulations on interleukin-1-induced nuclear factor kappa B activation. *Biochem. Pharm.* **1997**, *53*, 149–160. [CrossRef]

59. Fouani, L.; Kovacevic, Z.; Richardson, D.R. Targeting Oncogenic Nuclear Factor Kappa B Signaling with Redox-Active Agents for Cancer Treatment. *Antioxid. Redox Signal.* **2019**, *30*, 1096–1123. [CrossRef]

60. Glasauer, A.; Chandel, N.S. Targeting antioxidants for cancer therapy. *Biochem. Pharm.* **2014**, *92*, 90–101. [CrossRef]

61. Lim, J.K.M.; Leprivier, G. The impact of oncogenic RAS on redox balance and implications for cancer development. *Cell Death Dis.* **2019**, *10*, 955. [CrossRef] [PubMed]

62. Kirtonia, A.; Sethi, G.; Garg, M. The multifaceted role of reactive oxygen species in tumorigenesis. *Cell. Mol. Life Sci.* **2020**. [CrossRef] [PubMed]

63. Owczarek, K.; Hrabec, E.; Fichna, J.; Sosnowska, D.; Koziołkiewicz, M.; Szymański, J.; Lewandowska, U. Flavanols from Japanese quince (*Chaenomeles japonica*) fruit suppress expression of cyclooxygenase-2, metalloproteinase-9, and nuclear factor-kappaB in human colon cancer cells. *Acta Biochim. Pol.* **2017**, *64*, 567–576. [CrossRef] [PubMed]

64. Moskaug, J.Ø.; Carlsen, H.; Myhrstad, M.C.W.; Blomhoff, R. Polyphenols and glutathione synthesis regulation. *Am. J. Clin. Nutr.* **2005**, *81*, 277S–283S. [CrossRef] [PubMed]

65. Stefanson, A.L.; Bakovic, M. Dietary regulation of Keap1/Nrf2/ARE pathway: Focus on plant-derived compounds and trace minerals. *Nutrients* **2014**, *6*, 3777–3801. [CrossRef] [PubMed]

66. Sandoval-Acuña, C.; Ferreira, J.; Speisky, H. Polyphenols and mitochondria: An update on their increasingly emerging ROS-scavenging independent actions. *Arch. Biochem. Biophys.* **2014**, *559*, 75–90. [CrossRef]

67. León-González, A.J.; Auger, C.; Schini-Kerth, V.B. Pro-oxidant activity of polyphenols and its implication on cancer chemoprevention and chemotherapy. *Biochem. Pharmacol.* **2015**, *98*, 371–380. [CrossRef]

68. Hadi, S.; Bhat, S.; Azmi, A.; Hanif, S.; Shamim, U.; Ullah, M. Oxidative breakage of cellular DNA by plant polyphenols: A putative mechanism for anticancer properties. *Semin. Cancer Biol.* **2007**, *17*, 370–376. [CrossRef]

69. Nakazato, T.; Ito, K.; Miyakawa, Y.; Kinjo, K.; Yamada, T.; Hozumi, N.; Ikeda, Y.; Kizaki, M. Catechin, a green tea component, rapidly induces apoptosis of myeloid leukemic cells via modulation of reactive oxygen species production in vitro and inhibits tumor growth in vivo. *Haematologica* **2005**, *90*, 317–325.

70. Jeong, J.C.; Jang, S.W.; Kim, T.H.; Kwon, C.H.; Kim, Y.K. Mulberry Fruit (*Moris fructus*) Extracts Induce Human Glioma Cell Death In Vitro Through ROS-Dependent Mitochondrial Pathway and Inhibits Glioma Tumor Growth In Vivo. *Nutr. Cancer* **2010**, *62*, 402–412. [CrossRef]

71. Gundala, S.R.; Yang, C.; Mukkavilli, R.; Paranjpe, R.; Brahmbhatt, M.; Pannu, V.; Cheng, A.; Reid, M.D.; Aneja, R. Hydroxychavicol, a betel leaf component, inhibits prostate cancer through ROS-driven DNA damage and apoptosis. *Toxicol. Appl. Pharmacol.* **2014**, *280*, 86–96. [CrossRef] [PubMed]

72. Li, G.-X.; Chen, Y.-K.; Hou, Z.; Xiao, H.; Jin, H.; Lu, G.; Lee, M.-J.; Liu, B.; Guan, F.; Yang, Z.; et al. Pro-oxidative activities and dose-response relationship of (-)-epigallocatechin-3-gallate in the inhibition of lung cancer cell growth: A comparative study in vivo and in vitro. *Carcinogenesis* **2010**, *31*, 902–910. [CrossRef] [PubMed]

73. Halliwell, B. Are polyphenols antioxidants or pro-oxidants? What do we learn from cell culture and in vivo studies? *Arch. Biochem. Biophys.* **2008**, *476*, 107–112. [CrossRef] [PubMed]

74. Leslie, T.K.; James, A.D.; Zaccagna, F.; Grist, J.T.; Deen, S.; Kennerley, A.; Riemer, F.; Kaggie, J.D.; Gallagher, F.A.; Gilbert, F.J.; et al. Sodium homeostasis in the tumour microenvironment. *Biochim. Biophys. Acta Rev. Cancer* **2019**, *1872*, 188304. [CrossRef] [PubMed]

75. Cardone, R.A.; Alfarouk, K.O.; Elliott, R.L.; Alqahtani, S.S.; Ahmed, S.B.M.; Aljarbou, A.N.; Greco, M.R.; Cannone, S.; Reshkin, S.J. The Role of Sodium Hydrogen Exchanger 1 in Dysregulation of Proton Dynamics and Reprogramming of Cancer Metabolism as a Sequela. *Int. J. Mol. Sci.* **2019**, *20*, 3694. [CrossRef]

76. Amith, S.R.; Fong, S.; Baksh, S.; Fliegel, L. Na (+)/H (+)exchange in the tumour microenvironment: Does NHE1 drive breast cancer carcinogenesis? *Int. J. Dev. Biol.* **2015**, *59*, 367–377. [CrossRef]

77. Reshkin, S.J.; Bellizzi, A.; Caldeira, S.; Albarani, V.; Malanchi, I.; Poignee, M.; Alunni-Fabbroni, M.; Casavola, V.; Tommasino, M. Na+/H+ exchanger-dependent intracellular alkalinization is an early event in malignant transformation and plays an essential role in the development of subsequent transformation-associated phenotypes. *FASEB J.* **2000**, *14*, 2185–2197. [CrossRef]

78. Amith, S.R.; Fliegel, L. Na+/H+ exchanger-mediated hydrogen ion extrusion as a carcinogenic signal in triple-negative breast cancer etiopathogenesis and prospects for its inhibition in therapeutics. *Semin. Cancer Biol.* **2017**, *43*, 35–41. [CrossRef]

79. Stock, C.; Pedersen, S.F. Roles of pH and the Na+/H+ exchanger NHE1 in cancer: From cell biology and animal models to an emerging translational perspective? *Semin. Cancer Biol.* **2017**, *43*, 5–16. [CrossRef]

80. McIntyre, A.; Hulikova, A.; Ledaki, I.; Snell, C.; Singleton, D.; Steers, G.; Seden, P.; Jones, D.; Bridges, E.; Wigfield, S.; et al. Disrupting Hypoxia-Induced Bicarbonate Transport Acidifies Tumor Cells and Suppresses Tumor Growth. *Cancer Res.* **2016**, *76*, 3744–3755. [CrossRef]

81. Flinck, M.; Kramer, S.H.; Pedersen, S.F. Roles of pH in control of cell proliferation. *Acta Physiol.* **2018**, *223*, e13068. [CrossRef] [PubMed]

82. Prevarskaya, N.; Skryma, R.; Shuba, Y. Ion Channels in Cancer: Are Cancer Hallmarks Oncochannelopathies? *Physiol. Rev.* **2018**, *98*, 559–621. [CrossRef] [PubMed]

83. White, K.A.; Grillo-Hill, B.K.; Barber, D.L. Cancer cell behaviors mediated by dysregulated pH dynamics at a glance. *J. Cell Sci.* **2017**, *130*, 663–669. [CrossRef] [PubMed]

84. Webb, B.A.; Chimenti, M.; Jacobson, M.P.; Barber, D.L. Dysregulated pH: A perfect storm for cancer progression. *Nat. Rev. Cancer* **2011**, *11*, 671–677. [CrossRef]

85. Daniel, C.; Bell, C.; Burton, C.; Harguindey, S.; Reshkin, S.J.; Rauch, C. The role of proton dynamics in the development and maintenance of multidrug resistance in cancer. *Biochim. Biophys. Acta* **2013**, *1832*, 606–617. [CrossRef]

86. Harguindey, S.; Orive, G.; Luis Pedraz, J.; Paradiso, A.; Reshkin, S.J. The role of pH dynamics and the Na+/H+ antiporter in the etiopathogenesis and treatment of cancer. Two faces of the same coin–one single nature. *Biochim. Biophys. Acta* **2005**, *1756*, 1–24. [CrossRef]

87. Lagadic-Gossmann, D.; Huc, L.; Lecureur, V. Alterations of intracellular pH homeostasis in apoptosis: Origins and roles. *Cell Death Differ.* **2004**, *11*, 953–961. [CrossRef]

88. Matsuyama, S.; Reed, J.C. Mitochondria-dependent apoptosis and cellular pH regulation. *Cell Death Differ.* **2000**, *7*, 1155–1165. [CrossRef]

89. Li, X.; Tsauo, J.; Geng, C.; Zhao, H.; Lei, X.; Li, X. Ginsenoside Rg3 Decreases NHE1 Expression via Inhibiting EGF-EGFR-ERK1/2-HIF-1 α Pathway in Hepatocellular Carcinoma: A Novel Antitumor Mechanism. *Am. J. Chin. Med.* **2018**, *46*, 1915–1931. [CrossRef]

90. Abdelazeem, K.N.M.; Singh, Y.; Lang, F.; Salker, M.S. Negative Effect of Ellagic Acid on Cytosolic pH Regulation and Glycolytic Flux in Human Endometrial Cancer Cells. *Cell. Physiol. Biochem.* **2017**, *41*, 2374–2382. [CrossRef]

91. Hwang, E.; Sim, S.; Park, S.H.; Song, K.D.; Lee, H.-K.; Heo, T.-H.; Jun, H.S.; Kim, S.-J. Anti-proliferative effect of *Zea mays* L. cob extract on rat C6 glioma cells through regulation of glycolysis, mitochondrial ROS, and apoptosis. *Biomed. Pharm.* **2018**, *98*, 726–732. [CrossRef] [PubMed]

92. Terada, H. Uncouplers of oxidative phosphorylation. *Environ. Health Perspect.* **1990**, *87*, 213–218. [CrossRef] [PubMed]

93. Romaschenko, V.P.; Zinovkin, R.A.; Galkin, I.I.; Zakharova, V.V.; Panteleeva, A.A.; Tokarchuk, A.V.; Lyamzaev, K.G.; Pletjushkina, O.Y.; Chernyak, B.V.; Popova, E.N. Low Concentrations of Uncouplers of Oxidative Phosphorylation Prevent Inflammatory Activation of Endothelial Cells by Tumor Necrosis Factor. *Biochem. Mosc.* **2015**, *80*, 610–619. [CrossRef] [PubMed]

94. Reis, F.H.Z.; Pardo-Andreu, G.L.; Nuñez-Figueredo, Y.; Cuesta-Rubio, O.; Marín-Prida, J.; Uyemura, S.A.; Curti, C.; Alberici, L.C. Clusianone, a naturally occurring nemorosone regioisomer, uncouples rat liver mitochondria and induces HepG2 cell death. *Chem. Biol. Interact.* **2014**, *212*, 20–29. [CrossRef]

95. Datta, S.; Li, J.; Mahdi, F.; Jekabsons, M.B.; Nagle, D.G.; Zhou, Y.-D. Glycolysis inhibitor screening identifies the bis-geranylacylphloroglucinol protonophore moronone from Moronobea coccinea. *J. Nat. Prod.* **2012**, *75*, 2216–2222. [CrossRef]

96. Sell, T.S.; Belkacemi, T.; Flockerzi, V.; Beck, A. Protonophore properties of hyperforin are essential for its pharmacological activity. *Sci. Rep.* **2014**, *4*, 7500. [CrossRef]

97. Barnes, J.; Anderson, L.A.; Phillipson, J.D. St John's wort (*Hypericum perforatum* L.): A review of its chemistry, pharmacology and clinical properties. *J. Pharm. Pharm.* **2001**, *53*, 583–600. [CrossRef]

98. Zirak, N.; Shafiee, M.; Soltani, G.; Mirzaei, M.; Sahebkar, A. *Hypericum perforatum* in the treatment of psychiatric and neurodegenerative disorders: Current evidence and potential mechanisms of action. *J. Cell. Physiol.* **2019**, *234*, 8496–8508. [CrossRef]

99. Singer, A.; Wonnemann, M.; Müller, W.E. Hyperforin, a major antidepressant constituent of St. John's Wort, inhibits serotonin uptake by elevating free intracellular Na+1. *J. Pharm. Exp.* **1999**, *290*, 1363–1368.

100. Ng, Q.X.; Venkatanarayanan, N.; Ho, C.Y.X. Clinical use of *Hypericum perforatum* (St John's wort) in depression: A meta-analysis. *J. Affect. Disord.* **2017**, *210*, 211–221. [CrossRef]

101. Apaydin, E.A.; Maher, A.R.; Shanman, R.; Booth, M.S.; Miles, J.N.V.; Sorbero, M.E.; Hempel, S. A systematic review of St. John's wort for major depressive disorder. *Syst. Rev.* **2016**, *5*, 148. [CrossRef]

102. Napoli, E.; Siracusa, L.; Ruberto, G.; Carrubba, A.; Lazzara, S.; Speciale, A.; Cimino, F.; Saija, A.; Cristani, M. Phytochemical profiles, phototoxic and antioxidant properties of eleven *Hypericum* species—A comparative study. *Phytochemistry* **2018**, *152*, 162–173. [CrossRef] [PubMed]

103. Bruni, R.; Sacchetti, G. Factors affecting polyphenol biosynthesis in wild and field grown St. John's Wort (*Hypericum perforatum* L. Hypericaceae/Guttiferae). *Molecules* **2009**, *14*, 682–725. [CrossRef] [PubMed]

104. Seyis, F.; Yurteri, E.; Özcan, A.; Cirak, C. Altitudinal impacts on chemical content and composition of *Hypericum perforatum*, a prominent medicinal herb. *S. Afr. J. Bot.* **2020**, *135*, 391–403. [CrossRef]

105. Krammer, B.; Verwanger, T. Molecular response to hypericin-induced photodamage. *Curr. Med. Chem.* **2012**, *19*, 793–798. [CrossRef] [PubMed]

106. Zheng, Y.; Yin, G.; Le, V.; Zhang, A.; Lu, Y.; Yang, M.; Fei, Z.; Liu, J. Hypericin-based Photodynamic Therapy Induces a Tumor-Specific Immune Response and an Effective DC-based cancer Immunotherapy. *Biochem. Pharm.* **2014**. [CrossRef] [PubMed]

107. Albert, D.; Zündorf, I.; Dingermann, T.; Müller, W.E.; Steinhilber, D.; Werz, O. Hyperforin is a dual inhibitor of cyclooxygenase-1 and 5-lipoxygenase. *Biochem. Pharm.* **2002**, *64*, 1767–1775. [CrossRef]

108. Feisst, C.; Pergola, C.; Rakonjac, M.; Rossi, A.; Koeberle, A.; Dodt, G.; Hoffmann, M.; Hoernig, C.; Fischer, L.; Steinhilber, D.; et al. Hyperforin is a novel type of 5-lipoxygenase inhibitor with high efficacy in vivo. *Cell. Mol. Life Sci.* **2009**, *66*, 2759–2771. [CrossRef] [PubMed]

109. Novelli, M.; Menegazzi, M.; Beffy, P.; Porozov, S.; Gregorelli, A.; Giacopelli, D.; De Tata, V.; Masiello, P. St. John's wort extract and hyperforin inhibit multiple phosphorylation steps of cytokine signaling and prevent inflammatory and apoptotic gene induction in pancreatic β cells. *Int. J. Biochem. Cell Biol.* **2016**, *81*, 92–104. [CrossRef]

110. Biber, A.; Fischer, H.; Römer, A.; Chatterjee, S.S. Oral bioavailability of hyperforin from hypericum extracts in rats and human volunteers. *Pharmacopsychiatry* **1998**, *31* (Suppl. 1), 36–43. [CrossRef]

111. Schulz, H.-U.; Schürer, M.; Bässler, D.; Weiser, D. Investigation of the bioavailability of hypericin, pseudohypericin, hyperforin and the flavonoids quercetin and isorhamnetin following single and multiple oral dosing of a hypericum extract containing tablet. *Arzneimittelforschung* **2005**, *55*, 15–22. [CrossRef]

112. Kraus, B.; Wolff, H.; Elstner, E.F.; Heilmann, J. Hyperforin is a modulator of inducible nitric oxide synthase and phagocytosis in microglia and macrophages. *Naunyn. Schmiedebergs Arch. Pharm.* **2010**, *381*, 541–553. [CrossRef] [PubMed]

113. Novelli, M.; Beffy, P.; Gregorelli, A.; Porozov, S.; Mascia, F.; Vantaggiato, C.; Masiello, P.; Menegazzi, M. Persistence of STAT-1 inhibition and induction of cytokine resistance in pancreatic β cells treated with St John's wort and its component hyperforin. *J. Pharm. Pharm.* **2019**, *71*, 93–103. [CrossRef] [PubMed]

114. Orth, H.C.; Rentel, C.; Schmidt, P.C. Isolation, purity analysis and stability of hyperforin as a standard material from *Hypericum perforatum* L. *J. Pharm. Pharm.* **1999**, *51*, 193–200. [CrossRef] [PubMed]

115. Gobbi, M.; Moia, M.; Funicello, M.; Riva, A.; Morazzoni, P.; Mennini, T. In vitro effects of the dicyclohexylammonium salt of hyperforin on interleukin-6 release in different experimental models. *Planta Med.* **2004**, *70*, 680–682. [CrossRef] [PubMed]

116. Donà, M.; Dell'Aica, I.; Pezzato, E.; Sartor, L.; Calabrese, F.; Della Barbera, M.; Donella-Deana, A.; Appendino, G.; Borsarini, A.; Caniato, R.; et al. Hyperforin inhibits cancer invasion and metastasis. *Cancer Res.* **2004**, *64*, 6225–6232. [CrossRef] [PubMed]

117. Gartner, M.; Müller, T.; Simon, J.C.; Giannis, A.; Sleeman, J.P. Aristoforin, a novel stable derivative of hyperforin, is a potent anticancer agent. *Chembiochem* **2005**, *6*, 171–177. [CrossRef]

118. Hammer, K.D.P.; Hillwig, M.L.; Solco, A.K.S.; Dixon, P.M.; Delate, K.; Murphy, P.A.; Wurtele, E.S.; Birt, D.F. Inhibition of prostaglandin E(2) production by anti-inflammatory hypericum perforatum extracts and constituents in RAW264.7 Mouse Macrophage Cells. *J. Agric. Food Chem.* **2007**, *55*, 7323–7331. [CrossRef]

119. Koeberle, A.; Rossi, A.; Bauer, J.; Dehm, F.; Verotta, L.; Northoff, H.; Sautebin, L.; Werz, O. Hyperforin, an Anti-Inflammatory Constituent from St. John's Wort, Inhibits Microsomal Prostaglandin E(2) Synthase-1 and Suppresses Prostaglandin E(2) Formation in vivo. *Front. Pharm.* **2011**, *2*, 7. [CrossRef]

120. Novelli, M.; Beffy, P.; Menegazzi, M.; De Tata, V.; Martino, L.; Sgarbossa, A.; Porozov, S.; Pippa, A.; Masini, M.; Marchetti, P.; et al. St. John's wort extract and hyperforin protect rat and human pancreatic islets against cytokine toxicity. *Acta Diabetol.* **2014**, *51*, 113–121. [CrossRef]

121. Novelli, M.; Masiello, P.; Beffy, P.; Menegazzi, M. Protective Role of St. John's Wort and Its Components Hyperforin and Hypericin against Diabetes through Inhibition of Inflammatory Signaling: Evidence from In Vitro and In Vivo Studies. *Int. J. Mol. Sci.* **2020**, *21*, 8108. [CrossRef]

122. Müller, M.; Essin, K.; Hill, K.; Beschmann, H.; Rubant, S.; Schempp, C.M.; Gollasch, M.; Boehncke, W.H.; Harteneck, C.; Müller, W.E.; et al. Specific TRPC6 channel activation, a novel approach to stimulate keratinocyte differentiation. *J. Biol. Chem.* **2008**, *283*, 33942–33954. [CrossRef] [PubMed]

123. Benedí, J.; Arroyo, R.; Romero, C.; Martín-Aragón, S.; Villar, A.M. Antioxidant properties and protective effects of a standardized extract of Hypericum perforatum on hydrogen peroxide-induced oxidative damage in PC12 cells. *Life Sci.* **2004**, *75*, 1263–1276. [CrossRef] [PubMed]

124. Feisst, C.; Werz, O. Suppression of receptor-mediated Ca2+ mobilization and functional leukocyte responses by hyperforin. *Biochem. Pharm.* **2004**, *67*, 1531–1539. [CrossRef] [PubMed]

125. Lee, S.-K.; Kim, J.-E.; Kim, Y.-J.; Kim, M.-J.; Kang, T.-C. Hyperforin attenuates microglia activation and inhibits p65-Ser276 NFκB phosphorylation in the rat piriform cortex following status epilepticus. *Neurosci. Res.* **2014**, *85*, 39–50. [CrossRef] [PubMed]

126. Wang, H.; Shao, B.; Yu, H.; Xu, F.; Wang, P.; Yu, K.; Han, Y.; Song, M.; Li, Y.; Cao, Z. Neuroprotective role of hyperforin on aluminum maltolate-induced oxidative damage and apoptosis in PC12 cells and SH-SY5Y cells. *Chem. Biol. Interact.* **2019**, *299*, 15–26. [CrossRef]

127. Menegazzi, M.; Di Paola, R.; Mazzon, E.; Muià, C.; Genovese, T.; Crisafulli, C.; Suzuki, H.; Cuzzocrea, S. Hypericum perforatum attenuates the development of carrageenan-induced lung injury in mice. *Free Radic. Biol. Med.* **2006**, *40*, 740–753. [CrossRef] [PubMed]

128. Di Paola, R.; Mazzon, E.; Muià, C.; Crisafulli, C.; Genovese, T.; Di Bella, P.; Esposito, E.; Menegazzi, M.; Meli, R.; Suzuki, H.; et al. Protective effect of Hypericum perforatum in zymosan-induced multiple organ dysfunction syndrome: Relationship to its inhibitory effect on nitric oxide production and its peroxynitrite scavenging activity. *Nitric Oxide* **2007**, *16*, 118–130. [CrossRef]

129. Genovese, T.; Mazzon, E.; Di Paola, R.; Muià, C.; Crisafulli, C.; Menegazzi, M.; Malleo, G.; Suzuki, H.; Cuzzocrea, S. Hypericum perforatum attenuates the development of cerulein-induced acute pancreatitis in mice. *Shock* **2006**, *25*, 161–167. [CrossRef]

130. De Paola, R.; Muià, C.; Mazzon, E.; Genovese, T.; Crisafulli, C.; Menegazzi, M.; Caputi, A.P.; Suzuki, H.; Cuzzocrea, S. Effects of Hypericum perforatum extract in a rat model of ischemia and reperfusion injury. *Shock* **2005**, *24*, 255–263. [CrossRef]

131. Genovese, T.; Mazzon, E.; Menegazzi, M.; Di Paola, R.; Muià, C.; Crisafulli, C.; Bramanti, P.; Suzuki, H.; Cuzzocrea, S. Neuroprotection and enhanced recovery with hypericum perforatum extract after experimental spinal cord injury in mice. *Shock* **2006**, *25*, 608–617. [CrossRef]

132. Raso, G.M.; Pacilio, M.; Di Carlo, G.; Esposito, E.; Pinto, L.; Meli, R. In-vivo and in-vitro anti-inflammatory effect of Echinacea purpurea and Hypericum perforatum. *J. Pharm. Pharm.* **2002**, *54*, 1379–1383. [CrossRef] [PubMed]

133. Savikin, K.; Dobrić, S.; Tadić, V.; Zdunić, G. Antiinflammatory activity of ethanol extracts of *Hypericum perforatum* L., *H. barbatum* Jacq., *H. hirsutum* L., *H. richeri* Vill. and *H. androsaemum* L. in rats. *Phytother. Res.* **2007**, *21*, 176–180. [CrossRef] [PubMed]

134. Abdel-Salam, O.M.E. Anti-inflammatory, antinociceptive, and gastric effects of Hypericum perforatum in rats. *Sci. World J.* **2005**, *5*, 586–595. [CrossRef] [PubMed]

135. Zdunić, G.; Godevac, D.; Milenković, M.; Vucićević, D.; Savikin, K.; Menković, N.; Petrović, S. Evaluation of Hypericum perforatum oil extracts for an antiinflammatory and gastroprotective activity in rats. *Phytother. Res.* **2009**, *23*, 1559–1564. [CrossRef]

136. Dost, T.; Ozkayran, H.; Gokalp, F.; Yenisey, C.; Birincioglu, M. The effect of *Hypericum perforatum* (St. John's Wort) on experimental colitis in rat. *Dig. Dis. Sci.* **2009**, *54*, 1214–1221. [CrossRef]

137. Hu, Z.-P.; Yang, X.-X.; Chan, S.Y.; Xu, A.-L.; Duan, W.; Zhu, Y.-Z.; Sheu, F.-S.; Boelsterli, U.A.; Chan, E.; Zhang, Q.; et al. St. John's wort attenuates irinotecan-induced diarrhea via down-regulation of intestinal pro-inflammatory cytokines and inhibition of intestinal epithelial apoptosis. *Toxicol. Appl. Pharm.* **2006**, *216*, 225–237. [CrossRef]

138. Hadzhiiliev, V.; Dimov, D. Separate isolation of hyperforin from *Hypericum perforatum* (St. John's Wort) pursuant to the coefficents LOG Kow, PKa and densities of the included compounds. *TJS* **2015**, *13*, 19–23. [CrossRef]

139. Roz, N.; Rehavi, M. Hyperforin inhibits vesicular uptake of monoamines by dissipating pH gradient across synaptic vesicle membrane. *Life Sci.* **2003**, *73*, 461–470. [CrossRef]

140. Froestl, B.; Steiner, B.; Müller, W.E. Enhancement of proteolytic processing of the beta-amyloid precursor protein by hyperforin. *Biochem. Pharm.* **2003**, *66*, 2177–2184. [CrossRef]

141. Friedland, K.; Harteneck, C. Hyperforin: To Be or Not to Be an Activator of TRPC(6). In *Reviews of Physiology, Biochemistry and Pharmacology Vol. 169*; Nilius, B., Gudermann, T., Jahn, R., Lill, R., Petersen, O.H., de Tombe, P.P., Eds.; Springer: Cham, Switzerland, 2015; Volume 169, pp. 1–24. ISBN 978-3-319-26563-6.

142. Tu, P.; Gibon, J.; Bouron, A. The TRPC6 channel activator hyperforin induces the release of zinc and calcium from mitochondria. *J. Neurochem.* **2010**, *112*, 204–213. [CrossRef]

143. Werz, O.; Steinhilber, D. Therapeutic options for 5-lipoxygenase inhibitors. *Pharmacol. Ther.* **2006**, *112*, 701–718. [CrossRef] [PubMed]

144. Imreova, P.; Feruszova, J.; Kyzek, S.; Bodnarova, K.; Zduriencikova, M.; Kozics, K.; Mucaji, P.; Galova, E.; Sevcovicova, A.; Miadokova, E.; et al. Hyperforin Exhibits Antigenotoxic Activity on Human and Bacterial Cells. *Molecules* **2017**, *22*, 167. [CrossRef] [PubMed]

145. Meinke, M.C.; Schanzer, S.; Haag, S.F.; Casetti, F.; Müller, M.L.; Wölfle, U.; Kleemann, A.; Lademann, J.; Schempp, C.M. In vivo photoprotective and anti-inflammatory effect of hyperforin is associated with high antioxidant activity in vitro and ex vivo. *Eur. J. Pharm. Biopharm.* **2012**, *81*, 346–350. [CrossRef] [PubMed]

146. Ševčovičová, A.; Šemeláková, M.; Plšíková, J.; Loderer, D.; Imreová, P.; Gálová, E.; Kožurková, M.; Miadoková, E.; Fedoročko, P. DNA-protective activities of hyperforin and aristoforin. *Toxicol. In Vitro* **2015**, *29*, 631–637. [CrossRef]

147. Hecht, S.M. Bleomycin: New perspectives on the mechanism of action. *J. Nat. Prod.* **2000**, *63*, 158–168. [CrossRef]

148. Kurze, A.-K.; Buhs, S.; Eggert, D.; Oliveira-Ferrer, L.; Müller, V.; Niendorf, A.; Wagener, C.; Nollau, P. Immature O-glycans recognized by the macrophage glycoreceptor CLEC10A (MGL) are induced by 4-hydroxy-tamoxifen, oxidative stress and DNA-damage in breast cancer cells. *Cell Commun. Signal.* **2019**, *17*, 107. [CrossRef]

149. Lagadic-Gossmann, D.; Hardonnière, K.; Mograbi, B.; Sergent, O.; Huc, L. Disturbances in H+ dynamics during environmental carcinogenesis. *Biochimie* **2019**, *163*, 171–183. [CrossRef]

150. Manna, S.K.; Golla, S.; Golla, J.P.; Tanaka, N.; Cai, Y.; Takahashi, S.; Krausz, K.W.; Matsubara, T.; Korboukh, I.; Gonzalez, F.J. St. John's Wort Attenuates Colorectal Carcinogenesis in Mice through Suppression of Inflammatory Signaling. *Cancer Prev. Res.* **2015**, *8*, 786–795. [CrossRef]

151. Kuai, Y.; Liu, H.; Liu, D.; Liu, Y.; Sun, Y.; Xie, J.; Sun, J.; Fang, Y.; Pan, H.; Han, W. An ultralow dose of the NADPH oxidase inhibitor diphenyleneiodonium (DPI) is an economical and effective therapeutic agent for the treatment of colitis-associated colorectal cancer. *Theranostics* **2020**, *10*, 6743–6757. [CrossRef]

152. Rahimi, R.; Abdollahi, M. An update on the ability of St. John's wort to affect the metabolism of other drugs. *Expert Opin. Drug Metab. Toxicol.* **2012**, *8*, 691–708. [CrossRef]

153. He, X.; Feng, S. Role of Metabolic Enzymes P450 (CYP) on Activating Procarcinogen and their Polymorphisms on the Risk of Cancers. *Curr. Drug Metab.* **2015**, *16*, 850–863. [CrossRef] [PubMed]

154. Reed, L.; Arlt, V.M.; Phillips, D.H. The role of cytochrome P450 enzymes in carcinogen activation and detoxication: An in vivo–in vitro paradox. *Carcinogenesis* **2018**, *39*, 851–859. [CrossRef] [PubMed]

155. Beyerle, J.; Holowatyj, A.N.; Haffa, M.; Frei, E.; Gigic, B.; Schrotz-King, P.; Boehm, J.; Habermann, N.; Stiborova, M.; Scherer, D.; et al. Expression Patterns of Xenobiotic-Metabolizing Enzymes in Tumor and Adjacent Normal Mucosa Tissues among Patients with Colorectal Cancer: The ColoCare Study. *Cancer Epidemiol. Biomark. Prev.* **2020**, *29*, 460–469. [CrossRef] [PubMed]

156. Satia, J.A.; Littman, A.; Slatore, C.G.; Galanko, J.A.; White, E. Associations of herbal and specialty supplements with lung and colorectal cancer risk in the VITamins and Lifestyle study. *Cancer Epidemiol. Biomark. Prev.* **2009**, *18*, 1419–1428. [CrossRef] [PubMed]

157. Reichling, J.; Weseler, A.; Saller, R. A Current Review of the Antimicrobial Activity of *Hypericum perforatum* L. *Pharmacopsychiatry* **2001**, *34*, 116–118. [CrossRef]

158. Boyanova, L. Comparative evaluation of the activity of plant infusions against *Helicobacter pylori* strains by three methods. *World J. Microbiol. Biotechnol.* **2014**, *30*, 1633–1637. [CrossRef]

159. Pang, R.; Tao, J.; Zhang, S.; Zhu, J.; Yue, X.; Zhao, L.; Ye, P.; Zhu, Y. In vitro anti-hepatitis B virus effect of *Hypericum perforatum* L. *J. Huazhong Univ. Sci. Technol.* **2010**, *30*, 98–102. [CrossRef]

160. Schempp, C.M.; Kirkin, V.; Simon-Haarhaus, B.; Kersten, A.; Kiss, J.; Termeer, C.C.; Gilb, B.; Kaufmann, T.; Borner, C.; Sleeman, J.P.; et al. Inhibition of tumour cell growth by hyperforin, a novel anticancer drug from St. John's wort that acts by induction of apoptosis. *Oncogene* **2002**, *21*, 1242–1250. [CrossRef]

161. Hsu, F.-T.; Chen, W.-T.; Wu, C.-T.; Chung, J.-G. Hyperforin induces apoptosis through extrinsic/intrinsic pathways and inhibits EGFR/ERK/NF-κB-mediated anti-apoptotic potential in glioblastoma. *Environ. Toxicol.* **2020**, *35*, 1058–1069. [CrossRef]

162. Wiechmann, K.; Müller, H.; Fischer, D.; Jauch, J.; Werz, O. The acylphloroglucinols hyperforin and myrtucommulone A cause mitochondrial dysfunctions in leukemic cells by direct interference with mitochondria. *Apoptosis* **2015**, *20*, 1508–1517. [CrossRef]

163. Chiang, I.-T.; Chen, W.-T.; Tseng, C.-W.; Chen, Y.-C.; Kuo, Y.-C.; Chen, B.-J.; Weng, M.-C.; Lin, H.-J.; Wang, W.-S. Hyperforin Inhibits Cell Growth by Inducing Intrinsic and Extrinsic Apoptotic Pathways in Hepatocellular Carcinoma Cells. *Anticancer Res.* **2017**, *37*, 161–167. [CrossRef] [PubMed]

164. Naven, R.T.; Swiss, R.; Klug-McLeod, J.; Will, Y.; Greene, N. The development of structure-activity relationships for mitochondrial dysfunction: Uncoupling of oxidative phosphorylation. *Toxicol. Sci.* **2013**, *131*, 271–278. [CrossRef] [PubMed]

165. Pardo-Andreu, G.L.; Nuñez-Figueredo, Y.; Tudella, V.G.; Cuesta-Rubio, O.; Rodrigues, F.P.; Pestana, C.R.; Uyemura, S.A.; Leopoldino, A.M.; Alberici, L.C.; Curti, C. The anti-cancer agent nemorosone is a new potent protonophoric mitochondrial uncoupler. *Mitochondrion* **2011**, *11*, 255–263. [CrossRef] [PubMed]

166. Han, Y.H.; Kim, S.W.; Kim, S.H.; Kim, S.Z.; Park, W.H. 2,4-dinitrophenol induces G1 phase arrest and apoptosis in human pulmonary adenocarcinoma Calu-6 cells. *Toxicol. In Vitro* **2008**, *22*, 659–670. [CrossRef] [PubMed]

167. De Graaf, A.O.; van den Heuvel, L.P.; Dijkman, H.B.P.M.; de Abreu, R.A.; Birkenkamp, K.U.; de Witte, T.; van der Reijden, B.A.; Smeitink, J.A.M.; Jansen, J.H. Bcl-2 prevents loss of mitochondria in CCCP-induced apoptosis. *Exp. Cell Res.* **2004**, *299*, 533–540. [CrossRef] [PubMed]

168. Chaudhari, A.A.; Seol, J.-W.; Kang, S.-J.; Park, S.-Y. Mitochondrial transmembrane potential and reactive oxygen species generation regulate the enhanced effect of CCCP on TRAIL-induced SNU-638 cell apoptosis. *J. Vet. Med. Sci.* **2008**, *70*, 537–542. [CrossRef] [PubMed]

169. Zakharova, V.V.; Pletjushkina, O.Y.; Galkin, I.I.; Zinovkin, R.A.; Chernyak, B.V.; Krysko, D.V.; Bachert, C.; Krysko, O.; Skulachev, V.P.; Popova, E.N. Low concentration of uncouplers of oxidative phosphorylation decreases the TNF-induced endothelial permeability and lethality in mice. *Biochim. Biophys. Acta Mol. Basis Dis.* **2017**, *1863*, 968–977. [CrossRef]

170. Alp, H.; Tutun, H.; Kaplan, H.; ŞïNgïRïK, E.; Altintaş, L. Meme Kanseri Hücre Hattında Hypericum perforatum Ekstresinin Apoptotik Etkilerinin Araştırılması. *Harran Üniversitesi Vet. Fakültesi Derg.* **2019**, *8*, 198–202. [CrossRef]

171. Orhan, I.E.; Kartal, M.; Gülpinar, A.R.; Yetkin, G.; Orlikova, B.; Diederich, M.; Tasdemir, D. Inhibitory effect of St. John's Wort oil macerates on TNFα-induced NF-κB activation and their fatty acid composition. *J. Ethnopharmacol.* **2014**, *155*, 1086–1092. [CrossRef]

172. Liu, Y.-C.; Lin, K.-H.; Hsieh, J.-H.; Chung, J.-G.; Tan, Z.-L.; Hsu, F.-T.; Chiang, C.-H. Hyperforin Induces Apoptosis Through Extrinsic/Intrinsic Pathways and Inhibits NF-kB-modulated Survival and Invasion Potential in Bladder Cancer. *In Vivo* **2019**, *33*, 1865–1877. [CrossRef]

173. You, M.-K.; Kim, H.-J.; Kook, J.H.; Kim, H.-A. St. John's Wort Regulates Proliferation and Apoptosis in MCF-7 Human Breast Cancer Cells by Inhibiting AMPK/mTOR and Activating the Mitochondrial Pathway. *Int. J. Mol. Sci.* **2018**, *19*, 966. [CrossRef] [PubMed]

174. Merhi, F.; Tang, R.; Piedfer, M.; Mathieu, J.; Bombarda, I.; Zaher, M.; Kolb, J.-P.; Billard, C.; Bauvois, B. Hyperforin inhibits Akt1 kinase activity and promotes caspase-mediated apoptosis involving Bad and Noxa activation in human myeloid tumor cells. *PLoS ONE* **2011**, *6*, e25963. [CrossRef] [PubMed]

175. Hostanska, K.; Reichling, J.; Bommer, S.; Weber, M.; Saller, R. Hyperforin a constituent of St John's wort (Hypericum perforatum L.) extract induces apoptosis by triggering activation of caspases and with hypericin synergistically exerts cytotoxicity towards human malignant cell lines. *Eur. J. Pharm. Biopharm.* **2003**, *56*, 121–132. [CrossRef]

176. Quiney, C.; Billard, C.; Faussat, A.M.; Salanoubat, C.; Ensaf, A.; Naït-Si, Y.; Fourneron, J.D.; Kolb, J.-P. Pro-apoptotic properties of hyperforin in leukemic cells from patients with B-cell chronic lymphocytic leukemia. *Leukemia* **2006**, *20*, 491–497. [CrossRef] [PubMed]

177. Zaher, M.; Tang, R.; Bombarda, I.; Merhi, F.; Bauvois, B.; Billard, C. Hyperforin induces apoptosis of chronic lymphocytic leukemia cells through upregulation of the BH3-only protein Noxa. *Int. J. Oncol.* **2012**, *40*, 269–276. [CrossRef] [PubMed]

178. Smit, L.A.; Hallaert, D.Y.H.; Spijker, R.; de Goeij, B.; Jaspers, A.; Kater, A.P.; van Oers, M.H.J.; van Noesel, C.J.M.; Eldering, E. Differential Noxa/Mcl-1 balance in peripheral versus lymph node chronic lymphocytic leukemia cells correlates with survival capacity. *Blood* **2007**, *109*, 1660–1668. [CrossRef] [PubMed]

179. Lim, H.W.; Lim, H.Y.; Wong, K.P. Uncoupling of oxidative phosphorylation by curcumin: Implication of its cellular mechanism of action. *Biochem. Biophys. Res. Commun.* **2009**, *389*, 187–192. [CrossRef]

180. Ghosh, A.K.; Kay, N.E.; Secreto, C.R.; Shanafelt, T.D. Curcumin inhibits prosurvival pathways in chronic lymphocytic leukemia B cells and may overcome their stromal protection in combination with EGCG. *Clin. Cancer Res.* **2009**, *15*, 1250–1258. [CrossRef] [PubMed]

181. Nieminen, A.I.; Eskelinen, V.M.; Haikala, H.M.; Tervonen, T.A.; Yan, Y.; Partanen, J.I.; Klefström, J. Myc-induced AMPK-phospho p53 pathway activates Bak to sensitize mitochondrial apoptosis. *Proc. Natl. Acad. Sci. USA* **2013**, *110*, E1839–E1848. [CrossRef]

182. Martínez-Poveda, B.; Verotta, L.; Bombardelli, E.; Quesada, A.R.; Medina, M.A. Tetrahydrohyperforin and octahydrohyperforin are two new potent inhibitors of angiogenesis. *PLoS ONE* **2010**, *5*, e9558. [CrossRef]

183. Martínez-Poveda, B.; Quesada, A.R.; Medina, M.A. Hyperforin, a bio-active compound of St. John's Wort, is a new inhibitor of angiogenesis targeting several key steps of the process. *Int. J. Cancer* **2005**, *117*, 775–780. [CrossRef]

184. Lorusso, G.; Vannini, N.; Sogno, I.; Generoso, L.; Garbisa, S.; Noonan, D.M.; Albini, A. Mechanisms of Hyperforin as an anti-angiogenic angioprevention agent. *Eur. J. Cancer* **2009**, *45*, 1474–1484. [CrossRef] [PubMed]

185. Rothley, M.; Schmid, A.; Thiele, W.; Schacht, V.; Plaumann, D.; Gartner, M.; Yektaoglu, A.; Bruyère, F.; Noël, A.; Giannis, A.; et al. Hyperforin and aristoforin inhibit lymphatic endothelial cell proliferation in vitro and suppress tumor-induced lymphangiogenesis in vivo. *Int. J. Cancer* **2009**, *125*, 34–42. [CrossRef] [PubMed]

186. Zhao, Y.; Adjei, A.A. Targeting Angiogenesis in Cancer Therapy: Moving Beyond Vascular Endothelial Growth Factor. *Oncology* **2015**, *20*, 660–673. [CrossRef] [PubMed]

187. Fukai, T.; Ushio-Fukai, M. Cross-Talk between NADPH Oxidase and Mitochondria: Role in ROS Signaling and Angiogenesis. *Cells* **2020**, *9*, 1849. [CrossRef] [PubMed]

188. Roebuck, K.A.; Carpenter, L.R.; Lakshminarayanan, V.; Page, S.M.; Moy, J.N.; Thomas, L.L. Stimulus-specific regulation of chemokine expression involves differential activation of the redox-responsive transcription factors AP-1 and NF-kappaB. *J. Leukoc. Biol.* **1999**, *65*, 291–298. [CrossRef]

189. Shi, Q.; Le, X.; Wang, B.; Abbruzzese, J.L.; Xiong, Q.; He, Y.; Xie, K. Regulation of vascular endothelial growth factor expression by acidosis in human cancer cells. *Oncogene* **2001**, *20*, 3751–3756. [CrossRef]

190. Wyder, L.; Suply, T.; Ricoux, B.; Billy, E.; Schnell, C.; Baumgarten, B.U.; Maira, S.M.; Koelbing, C.; Ferretti, M.; Kinzel, B.; et al. Reduced pathological angiogenesis and tumor growth in mice lacking GPR4, a proton sensing receptor. *Angiogenesis* **2011**, *14*, 533–544. [CrossRef]

191. Robey, I.F.; Baggett, B.K.; Kirkpatrick, N.D.; Roe, D.J.; Dosescu, J.; Sloane, B.F.; Hashim, A.I.; Morse, D.L.; Raghunand, N.; Gatenby, R.A.; et al. Bicarbonate increases tumor pH and inhibits spontaneous metastases. *Cancer Res.* **2009**, *69*, 2260–2268. [CrossRef]

192. Quiney, C.; Billard, C.; Mirshahi, P.; Fourneron, J.-D.; Kolb, J.-P. Hyperforin inhibits MMP-9 secretion by B-CLL cells and microtubule formation by endothelial cells. *Leukemia* **2006**, *20*, 583–589. [CrossRef]

193. Chung, T.; Lee, Y.; Kim, C. Hepatitis B viral HBx induces matrix metalloproteinase-9 gene expression through activation of ERKs and PI-3K/AKT pathways: Involvement of invasive potential. *FASEB J.* **2004**, *18*, 1123–1125. [CrossRef] [PubMed]

194. Karaarslan, S.; Cokmert, S.; Cokmez, A. Does St. John's Wort cause regression in gastrointestinal system adenocarcinomas? *World J. Gastrointest. Oncol.* **2015**, *7*, 369–374. [CrossRef] [PubMed]

195. Heerdt, B.G.; Houston, M.A.; Augenlicht, L.H. The intrinsic mitochondrial membrane potential of colonic carcinoma cells is linked to the probability of tumor progression. *Cancer Res.* **2005**, *65*, 9861–9867. [CrossRef] [PubMed]

196. Kandel, B.A.; Ekins, S.; Leuner, K.; Thasler, W.E.; Harteneck, C.; Zanger, U.M. No activation of human pregnane X receptor by hyperforin-related phloroglucinols. *J. Pharm. Exp.* **2014**, *348*, 393–400. [CrossRef] [PubMed]

 *antioxidants* 

Article

Low Dose Combined Treatment with Ultraviolet-C and Withaferin a Enhances Selective Killing of Oral Cancer Cells

Sheng-Yao Peng [1], Yen-Yun Wang [2,3,4], Ting-Hsun Lan [2,5], Li-Ching Lin [6,7,8], Shyng-Shiou F. Yuan [3,4,9], Jen-Yang Tang [10,11,*] and Hsueh-Wei Chang [1,3,4,12,*]

[1] Department of Biomedical Science and Environmental Biology, PhD Program in Life Science, College of Life Science, Kaohsiung Medical University, Kaohsiung 80708, Taiwan; u109851101@kmu.edu.tw
[2] School of Dentistry, College of Dental Medicine, Kaohsiung Medical University, Kaohsiung 80708, Taiwan; wyy@kmu.edu.tw (Y.-Y.W.); tinghsun.lan@gmail.com (T.-H.L.)
[3] Center for Cancer Research, Kaohsiung Medical University, Kaohsiung 80708, Taiwan; yuanssf@kmu.edu.tw
[4] Cancer Center, Kaohsiung Medical University Hospital, Kaohsiung 80708, Taiwan
[5] Division of Prosthodontics, Department of Dentistry, Kaohsiung Medical University Hospital, Kaohsiung Medical University, Kaohsiung 80708, Taiwan
[6] Department of Radiation Oncology, Chi-Mei Foundation Medical Center, Tainan 71004, Taiwan; 8508a6@mail.chimei.org.tw
[7] School of Medicine, Taipei Medical University, Taipei 11031, Taiwan
[8] Chung Hwa University Medical Technology, Tainan 71703, Taiwan
[9] Translational Research Center, Kaohsiung Medical University Hospital, Kaohsiung 80708, Taiwan
[10] Department of Radiation Oncology, Faculty of Medicine, College of Medicine, Kaohsiung Medical University, Kaohsiung 80708, Taiwan
[11] Department of Radiation Oncology, Kaohsiung Medical University Hospital, Kaohsiung 80708, Taiwan
[12] Institute of Medical Science and Technology, National Sun Yat-sen University, Kaohsiung 80424, Taiwan
* Correspondence: reyata@kmu.edu.tw (J.-Y.T.); changhw@kmu.edu.tw (H.-W.C.); Tel.: +886-7-312-1101 (ext. 8105) (J.-Y.T.); +886-7-312-1101 (ext. 2691) (H.-W.C.)

Received: 15 October 2020; Accepted: 11 November 2020; Published: 13 November 2020

Abstract: Withaferin A (WFA), a *Withania somnifera*-derived triterpenoid, is an anticancer natural product. The anticancer effect of nonionizing radiation such as ultraviolet-C (UVC) as well as the combined treatment of UVC and WFA is rarely investigated. Low dose UVC and/or WFA treatments ($12 J/m^2$ and/or $1 \mu M$) were chosen to evaluate antioral cancer cell line effects by examining cytotoxicity, cell cycle disruption, apoptosis induction, and DNA damage. For two cancer cell lines (Ca9-22 and HSC-3), single treatment (UVC or WFA) showed about 80% viability, while a combined treatment of UVC/WFA showed about 40% viability. In contrast, there was noncytotoxicity to normal oral cell lines (HGF-1). Compared to single treatment and control, low dose UVC/WFA shows high inductions of apoptosis in terms of flow cytometric detections for subG1, annexin V, pancaspase changes as well as Western blotting for detecting cleaved poly (ADP-ribose) polymerase (c-PARP) and caspase 3 (c-Cas 3) and luciferase assay for detecting Cas 3/7 activity. Low dose UVC/WFA also showed high inductions of oxidative stress and DNA damage in terms of flow cytometric detections of reactive oxygen species (ROS), mitochondrial superoxide (MitoSOX) generation, and membrane potential (MitoMP) destruction, γH2AX and 8-oxo-2'deoxyguanosine (8-oxodG) types of DNA damages. For comparison, low dose UVC/WFA show rare inductions of annexin V, Cas 3/7 activity, ROS, MitoSOX, and MitoMP changes to normal oral HGF-1 cells. Therefore, low dose UVC/WFA provides a novel selectively killing mechanism to oral cancer cells, suggesting that WFA is a UVC sensitizer to inhibit the proliferation of oral cancer cells.

Keywords: ultraviolet-C (UVC); withanolide; combined treatment; oral cancer; DNA damage

---

Antioxidants **2020**, *9*, 1120; doi:10.3390/antiox9111120　　　　　www.mdpi.com/journal/antioxidants

1. Introduction

Radiotherapy and chemotherapy are commonly used for curing oral cancer [1]. However, they also raise the problems of radio- and chemo-resistance [2]. Recently, a novel strategy was developed to reduce chemoresistance by using low dose treatment. For example, in the clinical drug doxorubicin, its low dose treatment improved survival by suppressing the patient-derived chemoresistant leukemia stem cells in an animal model study [3]. Accordingly, low dose drug treatments have a potential for reducing chemoresistance in oral cancer therapy.

Similarly, high dose radiation is commonly associated with radioresistance to cancer such as prostate cancer cells [4]. Low dose radiation treatments have a potential for reducing radioresistance in cancer therapy. In addition to dose adjustments of radiation, a combined treatment provides an alternative strategy to overcome radioresistance in cancer therapy. For example, DNA-dependent protein kinase (DNA-PK) inhibitors KU57788 and IC87361 [5] and multikinase inhibitors sorafenib and sunitinib [6] displayed radiosensitization to head and neck cancer cells.

In addition to X-ray radiotherapy for oral cancer, several types of nonionizing radiation were developed for dental curing, including light-emitting diodes (LED), quartz-tungsten-halogen (QTH), argon lasers, and plasma arc curing (PAC) [7]. Except for dental curing, another nonionizing radiation such as ultraviolet-C (UVC) is commonly used for its germicidal properties on dental instruments [8]. Based on Google scholar and PubMed searching (retrieval on 7 September 2020), the hits for "X-ray resistance" studies were higher than "UVC resistance". Therefore, the combined treatment of UVC and drugs may have potential to improve the anticancer therapy. Moreover, UVC is also reported to have an anticancer effect against pancreatic [9] and colon [10] cancer cells. However, high cytotoxic doses were chosen in the above studies and may have some side effects. Accordingly, low doses of UVC may reduce the side effects to normal tissues.

Withaferin A (WFA), a triterpenoid isolated from *Withania somnifera*, shows antioxidant properties [11,12]. In addition to antioxidant effect, WFA also exhibits anticancer effects by showing cytotoxicity and inducing apoptosis to oral [13,14] and other types of cancer cells [15–18]. However, the observed cytotoxic effects of WFA in anticancer studies were based on high doses. Recently, combined treatments of WFA with other drugs have been reported. For example, the combined treatment of WFA and oxaliplatin showed synergistic antitumor activity in pancreatic cancer cells [16]. WFA was also combined with caffeic acid phenethyl ester (CAPE) for selective toxicity for ovarian and cervical cancer cells [19]. However, the combined treatment of WFA with UVC irradiation is rarely investigated.

We hypothesized that low dose WFA had an UVC radiosensitizing effect to oral cancer cells. To address this hypothesis, a low dose combined treatment (UVC/WFA) and individual treatments of oral cancer cells were compared for their changes of cytotoxicity, cell cycle, apoptosis, oxidative stress, and DNA damage.

2. Materials and Methods

2.1. Cell Cultures and Chemical

Two human oral cancer cell lines (Ca9-22 and HSC-3) and normal oral cell lines (HGF-1) were commercially obtained from the Japanese Collection of Research Bioresources (JCRB) Cell Bank (National Institute of Biomedical Innovation, Osaka, Japan) and the American Type Culture Collection (ATCC; Manassas, VA, USA). Ca9-22 cells were derived from gingival squamous carcinoma. HSC-3 cells were derived from tongue squamous carcinoma with high metastatic potential. HGF-1 cells were derived from gingival fibroblasts. The relative proliferation rate for Ca9-22, HSC-3, and HGF-1 cells was 1.9:1.6:1.0, respectively [20]. Their routinely cultured methods were described as indicated in [21]. WFA was obtained from Sigma-Aldrich (St. Louis, MO, USA). Cisplatin (Selleckchem; Houston, TX, USA) and H_2O_2 (Sigma-Aldrich; St. Louis, MO, USA) were used as a positive control treatment.

2.2. Cytotoxicity, ATP Content Determination, and Cell Morphology

Cytotoxicity was evaluated by tetrazolium-based 3-(4,5-dimethylthiazol-2-yl)-5-(3-carboxymethoxyphenyl)-2-(4-sulfophenyl)-2H-tetrazolium (MTS) kit (Promega Corporation, Madison, WI, USA) as described previously [22]. The cell viability in terms of cellular ATP content was evaluated by an ATP-lite assay kit (PerkinElmer Life Sciences, Boston, MA, USA) [23]. The drug interaction for the UVC/WFA combined treatment was analyzed as previously described [24]. In brief, the formula for determining the synergy (α), i.e., additive, synergistic, or antagonistic for $\alpha = 1$, >1, and <1, respectively, was listed as follows: $\alpha =$ survival fraction (SF) for UVC alone treatment $\times$ SF for WFA alone treatment/SF for UVC/WFA combined treatment. Cells were photographed at 100$\times$ magnification for morphology analysis.

2.3. UVC and/or WFA Treatments

After removing medium, cells were treated with a germicidal UVC lamp (254 nm) in a laminar flow chamber at a rate of 2 J/m^2/sec [25] for 6 sec to archive the designed dose 12 J/m^2. Subsequently, medium containing WFA was added for post-treatment. The control contained DMSO only without UVC and WFA treatments. All experiments including DMSO only, UVC and/or WFA treatments had the same concentration of 0.01% DMSO.

2.4. Cell Cycle

DNA was stained with 7-Aminoactinmycin D (7AAD) (Biotium, Inc., Hayward, CA, USA) (1 µg/mL, 37 °C, 30 min) for a cell cycle assay [26]. Flow cytometer (Guava easyCyte; Luminex, TX, USA) and FlowJo software (Becton-Dickinson; Franklin Lakes, NJ, USA) were applied. Both subG1 and >4 N populations were individually counted. G1, S, and G2/M populations were totally adjusted to 100%.

2.5. Apoptosis

A flow cytometer (Guava easyCyte) applying FlowJo software (Becton-Dickinson) was used to measure apoptosis in terms of annexin V (Strong Biotect Corporation, Taipei, Taiwan)/7AAD [27] and a pancaspase activity assay kit (Abcam, Cambridge, UK) [23] assays. Apoptosis was also detected by Western blotting applying primary antibodies for the cleaved forms of poly (ADP-ribose) polymerase (c-PARP) and caspase 3 (c-Cas 3) (Cell Signalling Technology Inc., Danvers, MA, USA) (diluted 1:1000) and internal control mAb-β actin (Sigma-Aldrich, St. Louis, MO, USA) as described previously [14]. Cas 3/7 activity was determined by a Caspase-Glo® 3/7 Assay (Promega; Madison, WI, USA) based on luminescent detection as described previously [28].

2.6. Reactive Oxygen Species (ROS)

Reactive species-detecting probe 2′,7′-dichlorodihydrofluorescein diacetate (DCFH-DA) (10 µM, 37 °C, 30 min) was interacting with ROS to generate fluorescence, which was determined by a flow cytometer (Guava easyCyte) applying FlowJo software (Becton-Dickinson) [29].

2.7. Mitochondrial Superoxide (MitoSOX)

Reactive species-detecting probe MitoSOX Red (Molecular Probes, Invitrogen, Eugene, OR, USA) (50 nM, 37 °C, 30 min) was interacting with MitoSOX to generate fluorescence, which was determined by a flow cytometer (Guava easyCyte) applying FlowJo software (Becton-Dickinson) [30].

2.8. Mitochondrial Membrane Potential (MitoMP)

MitoMP-sensitive probe DiOC$_2$(3) (Invitrogen, Eugene, OR, USA) (5 nM, 37 °C, 30 min) was used to determine MitoMP by a flow cytometer (Guava easyCyte) applying FlowJo software (Becton-Dickinson) [31].

2.9. γH2AX

After cell harvesting and fixation, γH2AX primary antibody (Santa Cruz Biotechnology; Santa Cruz, CA, USA) (1:500 dilution), secondary antibody conjugated Alexa Fluor®488 (Cell Signaling Technology) (1:10000 dilution), and 7AAD (5 µg/mL) were chosen for flow cytometry reaction [32] and analyzed by a flow cytometer (Guava easyCyte) applying FlowJo software (Becton-Dickinson). By Western blotting, γH2AX was also probed using primary antibodies for γH2AX (Santa Cruz Biotechnology; Santa Cruz, CA, USA) (diluted 1:1000).

2.10. 8-Oxo-2′-Deoxyguanosine (8-OxodG)

After cell harvesting and fixation, fluorescein isothiocyanate (FITC) conjugated 8-oxodG antibody (1:10,000 dilution) (Santa Cruz Biotechnology) were chosen for flow cytometry reaction at 4 °C for 1 h and analyzed by a flow cytometer (Guava easyCyte) applying FlowJo software (Becton-Dickinson).

2.11. Statistics

The significance for multiple comparisons were tested by One-way ANOVA with the Tukey Honestly Significant Difference (HSD) post hoc test using JMP12 software (SAS Institute, Cary, NC, USA). Groups showing no overlapping letters indicate significant differences.

3. Results

3.1. WFA Shows UVC Sensitizing Effects on Cytotoxicity of Oral Cancer Cells

Based on the 24-h MTS assay, low cytotoxic UVC and WFA (12 J/m^2 and 1 µM) around 80% viability were used to evaluate the cytotoxic effect of a combined treatment (UVC/WFA) of oral cancer Ca9-22 and HSC-3 cells and normal oral HGF-1 cells (Figure 1). As shown in Figure 1A, UVC/WFA-treated Ca9-22 cells show a lower viability of 42.2% than UVC or WFA alone (73.7% or 83.4%) in a 24-h MTS assay. As shown in Figure 1B, UVC/WFA-treated HSC-3 cells show lower viability for 40.6% than UVC or WFA alone (82.7% or 79.4%) in a 24-h MTS assay. In contrast, UVC/WFA shows no cytotoxicity towards normal oral HGF-1 cells (Figure 1C).

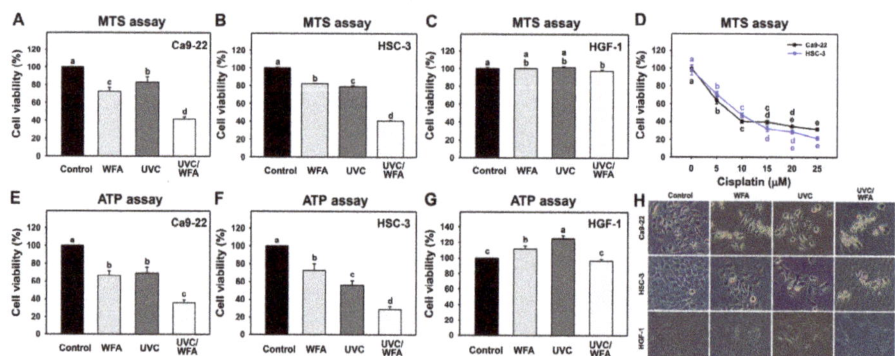

Figure 1. MTS and ATP assays for Withaferin A (WFA) and/or UV treatments. Human oral cancer Ca9-22 and HSC-3 cells and normal oral HGF-1 cells were treated with control (0.01% DMSO), WFA (1 µM), ultraviolet-C (UVC) (12 J/m^2), and a combined treatment (UVC/WFA) or cisplatin for 24 h. (**A–C**) MTS assay for WFA treatment. (**D**) MTS assay for a positive cisplatin control. (**E–G**) ATP assay for WFA treatment. Groups showing no overlapping letters (a–e) indicate significant differences ($p < 0.05$~0.0001). Data are the mean ± SD ($n = 3$ independent experiments, each experiment was performed with three replications). (**H**) Cell morphology. Cell images were photographed at 100× magnification.

The interaction effects of UVC and WFA for Figure 1A,B (MTS assay (α values were 1.46 ± 0.14 and 1.11 ± 0.03 for Ca9-22 and HSC-3 cells, respectively)) and for Figure 1E,F (ATP assay (α values were 1.28 ± 0.09 and 1.39 ± 0.15 for Ca9-22 and HSC-3 cells, respectively)) show synergistic behavior. For comparison, cisplatin shows cytotoxicity towards oral cancer cells with less drug sensitivity than UVC/WFA (Figure 1D). Similarly, UVC/WFA-treated oral cancer Ca9-22 and HSC-3 cells show lower viability than UVC or WFA alone in a 24-h ATP assay (Figure 1E,F). Similarly to the MTS assay, UVC/WFA shows no cytotoxicity to normal oral HGF-1 cells in terms of ATP assay (Figure 1G). Furthermore, the UVC and/or WFA treatments show the apoptosis-like morphology such as cell shrinkages for oral cancer cells but not for oral normal cells (Figure 1H).

3.2. WFA Shows UVC Sensitizing Effect on Cell Cycle Disturbance of Oral Cancer Cells

Figure 2A shows the cell cycle assays of oral cancer Ca9-22 and HSC-3 cells following 24-h treatments with control, WFA (1 μM), UVC (12 J/m^2), or UVC/WFA. For Ca9-22 cells (Figure 2B), 24-h UVC/WFA treatment induces higher sub-G1, G2/M, and > 4N populations (%) than UVC, WFA, and the control. For HSC-3 cells, 24-h UVC/WFA treatment induces higher sub-G1 and S populations (%) than UVC, WFA, and the control. In contrast, G1 populations (%) of UVC/WFA for oral cancer Ca9-22 and HSC-3 cells are lower than UVC, WFA, and the control. For comparison, H$_2$O$_2$ shows G2/M arrest in oral cancer cells as a positive control (Figure 2C,D).

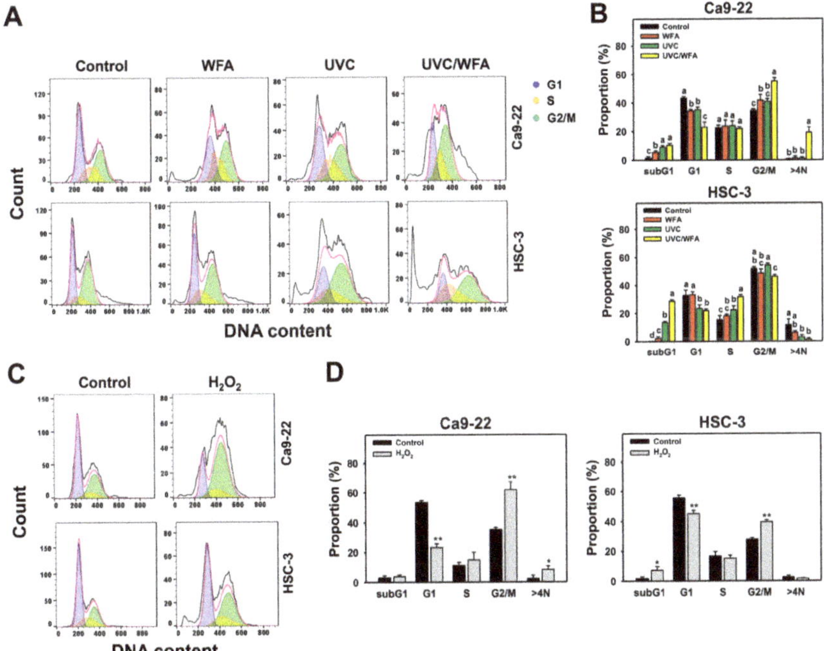

Figure 2. Cell cycle assay for WFA and/or UV treatments. Human oral cancer Ca9-22 and HSC-3 cells were treated with control (0.01% DMSO), WFA (1 μM), UVC (12 J/m^2), and a combined treatment (UVC/WFA) for 24 h. (**A,B**) Typical cell cycle patterns and statistics. Groups showing no overlapping letters (a–d) indicate significant differences ($p < 0.05{\sim}0.0001$). Data are the mean $\pm$ SD ($n = 3$ independent experiments, each experiment collected with 5000 gated cell counts). (**C,D**) Cell cycle patterns for a positive control of G2/M arrest. Cells were treated with H$_2$O$_2$ for 0 and 200 μM for 24 h. *** $p < 0.05{\sim}0.0001$. Data are the mean $\pm$ SD ($n = 3$ independent experiments, each experiment collected with 5000 gated cell counts).

3.3. WFA Shows UVC Sensitizing Effect on Annexin V Expression and Caspase Activation of Oral Cancer Cells

The apoptosis-like status for increasing subG1 (Figure 2) was further examined by other apoptosis analyses as follows. According to an annexin V/7AAD assay (Figure 3A), 24-h UVC/WFA treatment induces higher annexin V (+) (%) populations in oral cancer Ca9-22 and HSC-3 cells than UVC, WFA, and control treatments (Figure 3B). In contrast, UVC and/or WFA treatments to normal oral HGF-1 cells show little annexin V (+) (%) populations.

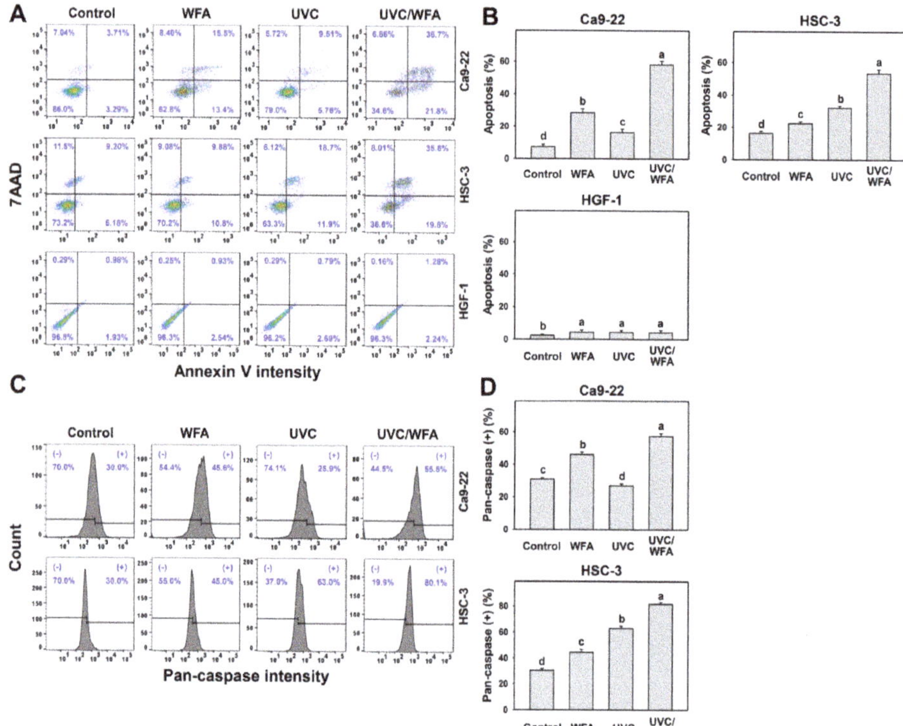

Figure 3. Annexin V and pancaspase assays of WFA and/or UV treatments. Human oral cancer Ca9-22 and HSC-3 cells and normal oral HGF-1 cells were treated with control (0.01% DMSO), WFA (1 µM), UVC (12 J/m^2), and a combined treatment (UVC/WFA) for 24 h. (**A,B**) Typical annexin V/7AAD patterns and statistics. Apoptosis (%) is the percentage of annexin V-positive population. (**C,D**) Typical pancaspase pattern and statistics. (+) is the percentage for pancaspase-positive populations. Groups showing no overlapping letters (a–d) indicate significant differences ($p < 0.05$~0.0001). Data are the mean ± SD ($n = 3$ independent experiments, each experiment collected with 5000 gated cell counts).

According to a pancaspase assay (Figure 3C), UVC/WFA induces higher pancaspase (+) (%) populations in oral cancer Ca9-22 and HSC-3 cells than UVC, WFA, and the control (Figure 3D). Based on Cas 3/7 activity, UVC/WFA also induces higher Cas 3/7 activity in oral cancer and normal oral cells than UVC, WFA, and the control (Figure 4A–C). It is noted that UVC/WFA shows higher Cas 3/7 activity in oral cancer Ca9-22 and HSC-3 cells than normal oral HGF-1 cells.

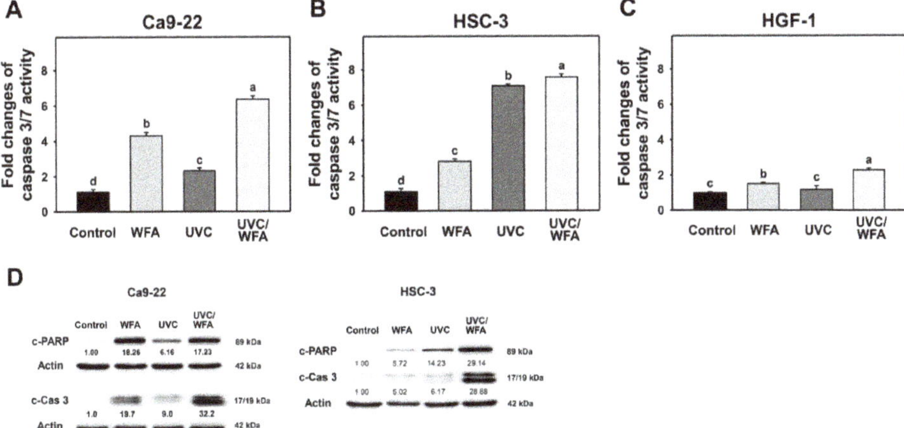

Figure 4. Caspase and PARP activations of WFA and/or UV treatments. Human oral cancer Ca9-22 and HSC-3 cells and normal oral HGF-1 cells were treated with control (0.01% DMSO), WFA (1 μM), UVC (12 J/m^2), and a combined treatment (UVC/WFA) for 24 h. (**A–C**) Caspase 3/7 activity for Ca9-22, HSC-3, and HGF-1 cells. Groups showing no overlapping letters (a–d) indicate significant differences ($p < 0.05 \sim 0.0001$). Data are the mean ± SD ($n = 3$ independent experiments, each experiment was performed with three replications). (**D**) Western blotting for apoptotic protein c-PARP and c-Cas 3 expressions of oral cancer (Ca9-22 and HSC-3) cells.

According to Western blotting (Figure 4D), UVC/WFA induces higher expressions for the apoptotic protein such as the cleaved form of c-PARP and c-Cas 3 in oral cancer Ca9-22 and HSC-3 cells than UVC, WFA, and control.

3.4. WFA Shows UVC Sensitizing Effect on ROS Generation of Oral Cancer Cells

Many factors such as oxidative stresses may induce apoptosis [33]. Because ROS generation following UVC [34], WFA [14], or UVC/natural product [35] treatments are fast, the ROS detection time of WFA/UVC was chosen as 12 h. Using DCFH-DA staining, the ROS generation was detected as oxidative stress by flow cytometry. According to an ROS assay for oral cancer Ca9-22 and HSC-3 cells (top and middle, Figure 5A), 12-h UVC/WFA treatment induces higher ROS (+) (%) populations than UVC, WFA, and control treatments (Figure 5B). In contrast, a ROS assay for normal oral HGF-1 cells (bottom, Figure 5A) shows no significant difference between UVC/WFA and single treatment, indicating that ROS is unable to be induced in WFA and/or UVC-treated normal oral HGF-1 cells.

3.5. WFA Shows UVC Sensitizing Effect on MitoSOX Generation of Oral Cancer Cells

Using MitoSOX red staining, the MitoSOX generation was detected as oxidative stress by flow cytometry. According to a MitoSOX assay for oral cancer Ca9-22 and HSC-3 cells (top and medium, Figure 6A), 24-h UVC/WFA treatment induces higher MitoSOX (+) (%) populations than UVC, WFA, and control treatments (Figure 6B). In contrast, a MitoSOX assay for normal oral HGF-1 cells (bottom, Figure 6A) show no significant difference between UVC/WFA and single treatment. Moreover, WFA/UVC induces higher MitoSOX in oral cancer Ca9-22 and HSC-3 cells than in normal oral HGF-1 cells.

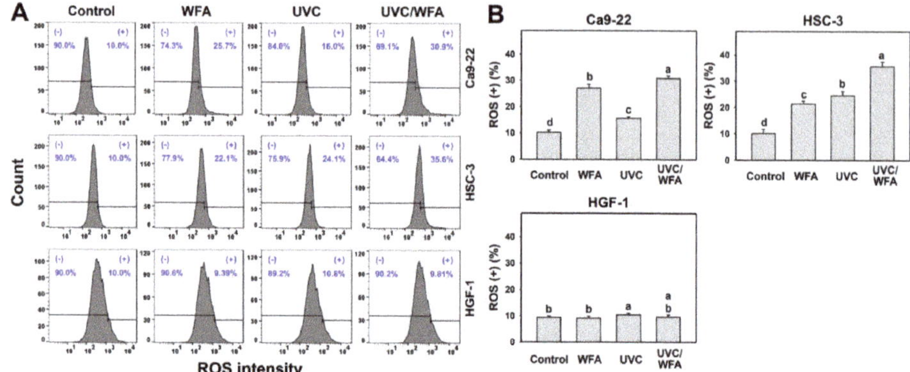

Figure 5. Reactive oxygen species (ROS) assays of WFA and/or UV treatments. Human oral cancer Ca9-22 and HSC-3 and normal oral HGF-1 cells were treated with control (0.01% DMSO), WFA (1 μM), UVC (12 J/m^2), and a combined treatment (UVC/WFA) for 12 h. (**A,B**) Typical ROS patterns and statistics. (+) is the percentage for ROS-positive populations. Groups showing no overlapping letters (a–d) indicate significant differences ($p < 0.05\sim0.0001$). Data are the mean ± SD ($n = 3$ independent experiments, each experiment collected with 5000 gated cell counts).

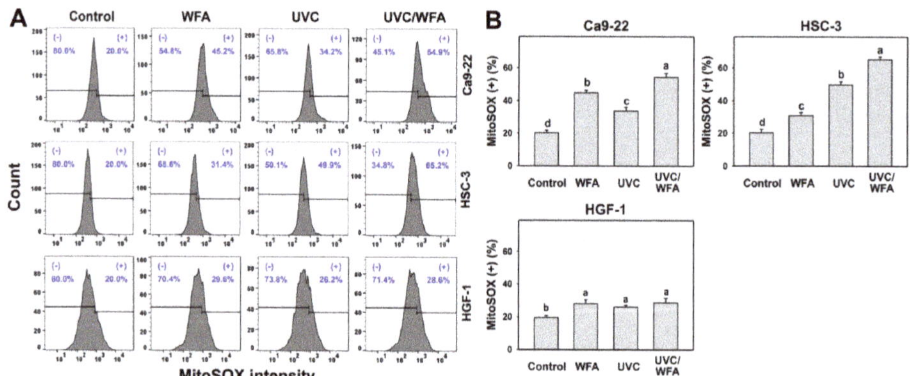

Figure 6. Mitochondrial superoxide (MitoSOX) assays of WFA and/or UV treatments. Human oral cancer Ca9-22 and HSC-3 cells and normal oral HGF-1 cells were treated with control (0.01% DMSO), WFA (1 μM), UVC (12 J/m^2), and a combined treatment (UVC/WFA) for 24 h. (**A,B**) Typical MitoSOX patterns and statistics. (+) is the percentage for MitoSOX-positive populations. Groups showing no overlapping letters (a–d) indicate significant differences ($p < 0.01\sim0.0001$). Data are the mean ± SD ($n = 3$ independent experiments, each experiment collected with 5000 gated cell counts).

3.6. WFA Shows UVC Sensitizing Effect on MitoMP Destruction of Oral Cancer Cells

Using DiOC$_2$(3) staining, the MitoMP depletion was detected as oxidative stress by flow cytometry. According to a MitoMP assay for oral cancer Ca9-22 and HSC-3 cells (top and medium, Figure 7A), 24-h UVC/WFA treatment induces higher MitoMP (-) (%) populations than UVC, WFA, and control treatments (Figure 7B). In contrast, a MitoMP assay for normal oral HGF-1 cells (bottom, Figure 7A) shows no significant difference between UVC/WFA and single treatment. Moreover, WFA/UVC induces higher MitoMP (-) (%) populations in oral cancer Ca9-22 and HSC-3 cells than in normal oral HGF-1 cells.

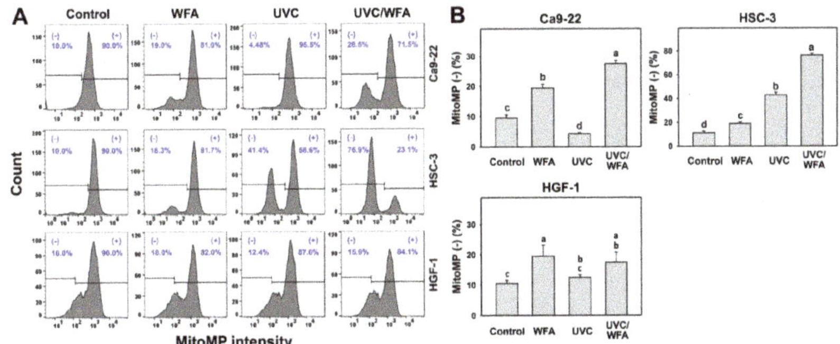

Figure 7. Membrane potential (MitoMP) assays of WFA and/or UV treatments. Human oral cancer Ca9-22 and HSC-3 cells and normal oral HGF-1 cells were treated with control (0.01% DMSO), WFA (1 μM), UVC (12 J/m^2), and a combined treatment (UVC/WFA) for 24 h. (**A,B**) Typical MitoMP patterns and statistics. (-) is the percentage for MitoMP-negative populations. Groups showing no overlapping letters (a–d) indicate significant differences ($p < 0.05$~0.0001). Data are the mean ± SD ($n = 3$ independent experiments, each experiment collected with 5000 gated cell counts).

3.7. WFA Shows UVC Sensitizing Effect on γH2AX and 8-oxodG Expressions of Oral Cancer Cells

Since oxidative stress is prone to induce DNA damage [33], the changes of DNA damage following UVC and/or WFA treatment were detected by flow cytometry and Western blotting. According to a γH2AX assay (Figure 8A), 24-h UVC/WFA treatment induces higher γH2AX (+) (%) populations in oral cancer Ca9-22 and HSC-3 cells than UVC, WFA, and control treatments (Figure 8B). According to Western blotting (Figure 8C), UVC/WFA induces higher expressions for γH2AX in oral cancer Ca9-22 and HSC-3 cells than UVC, WFA, and the control. According to an 8-oxodG assay (Figure 9A), 24-h UVC/WFA treatment induces higher 8-oxodG (+) (%) populations in oral cancer Ca9-22 and HSC-3 cells than UVC, WFA, and the control (Figure 9B).

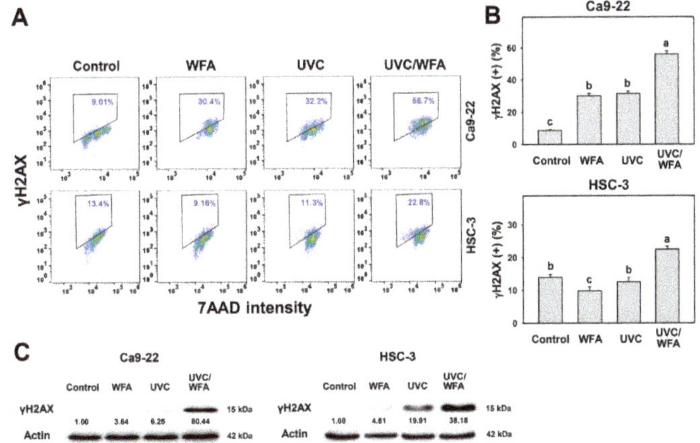

Figure 8. γH2AX assays of WFA and/or UV treatments. Human oral cancer Ca9-22 and HSC-3 cells were treated with control (0.01% DMSO), WFA (1 μM), UVC (12 J/m^2), and a combined treatment (UVC/WFA) for 24 h. (**A,B**) Typical γH2AX patterns and statistics. Box indicates the percentage for γH2AX-positive populations. Groups showing no overlapping letters (a–c) indicate significant differences ($p < 0.01$~0.0001). Data are the mean ± SD ($n = 3$ independent experiments, each experiment collected with 5000 gated cell counts). (**C**) Western blotting for γH2AX expression.

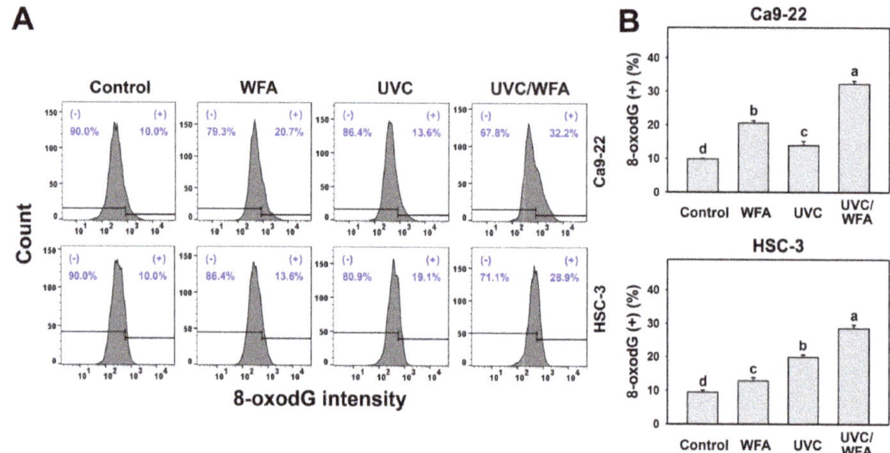

Figure 9. 8-oxo-2′deoxyguanosine (8-oxodG) assays of WFA and/or UV treatments. Human oral cancer Ca9-22 and HSC-3 cells. Cells were treated with control (0.01% DMSO), WFA (1 µM), UVC (12 J/m^2), and a combined treatment (UVC/WFA) for 24 h. (**A**,**B**) Typical 8-oxodG patterns and statistics. (+) is the percentage for 8-oxodG-positive populations. Groups showing no overlapping letters (a–d) indicate significant differences ($p < 0.0001$). Data are the mean ± SD ($n = 3$ independent experiments, each experiment collected with 5000 gated cell counts).

4. Discussion

WFA demonstrated X-ray radiosensitizing ability in many cancer cells [36–38] but its UVC sensitizing effect is rarely reported. UVC generation is more easy to handle with a user-friendly device than X-ray. Therefore, the present study focused on exploring the possible UVC sensitizing effect of WFA with cell line models. To avoid the potential resistance of high dose WFA and UVC, this study was evaluating low dose combined effects of UVC/WFA to oral cancer cells.

4.1. Combined Treatment UVC/WFA Effectively Kills Oral Cancer Cells without Side Effect to Normal Oral Cells

Low dose ionizing radiation and low dose drug treatment are novel strategies to reduce radioresistance and chemoresistance [3,4]. A combined treatment of radiation with drugs would also provide a strategy to overcome radioresistance and would improve cancer therapy [5,6].

Recently, the radiosensitizing studies using low dose clinical drugs and UVC were reported. For example, low dose cisplatin (10 µM) and UVC (10 J/m^2) jointed to induce antiproliferation of colon cancer SW480 and DLD-1 cells [39]. Combined treatment with low dose methanolic extracts of *Cryptocarya concinna* (10 µg/mL; 80.4% viability) and UVC (14 J/m^2; 83.2% viability) improve antiproliferation against oral cancer cells compared to single treatment [25].

WFA was previously shown to have no cytotoxicity towards normal oral HGF-1 cells at a low dose below 3 µM [14], while it showed cytotoxicity to oral cancer Ca9-22 and CAL 27 cells below 3 µM. Therefore, oral cancer therapy would benefit by the selective killing effect of WFA. Alternatively, in the present study this low dose treatment was combined with nonionizing radiation, i.e., UVC. Combining these strategies, we demonstrated that a low dose combined treatment of UVC/WFA dramatically decreased cell viability (~40%) of two oral cancer cell lines (Ca9-22 and HSC-3) without cytotoxic effects on normal oral cells in a 24-h MTS assay (Figure 1).

Consistently, UVC/WFA shows substantially decreased ATP content in oral cancer cells but does not decrease in oral normal cells. Moreover, the half maximal inhibitory concentration (IC$_{50}$) values of cisplatin to oral cancer Ca9-22 and HSC-3 cells in a 24-h MTS assay are 7.9 and 9.6 µM, respectively.

At a low dose, UVC (12 J/m^2) and WFA (1 µM) were used in the present study to demonstrate a higher antiproliferation ability to oral cancer Ca9-22 and HSC-3 cells (42.2% and 40.6% viability, respectively) than cisplatin alone. It is noted that the oral cancer Ca9-22 cell line is recently described for being problematic in terms of showing contamination with the cell line MSK922 (according to the Cellosaurus database [40]). It warrants detailed investigation by applying other oral cancer cell lines to support the UVC radiosensitizing effect of WFA in the future. Moreover, the HGF-1 cell line is derived from fibroblasts rather than from normal epithelial cells. The human oral keratinocytes would be a better control to show that the suppression of cell viability and induction of apoptosis after UVC/WFA treatment is cancer cell specific.

WFA shows antioxidant properties such as 1,1-diphenyl-2-picrylhydrazyl (DPPH) radical scavenging ability in vitro [12]. WFA also showed in vivo antioxidant properties with activation of superoxide dismutase (SOD), catalase (CAT), and glutathione peroxidase (GPX) in the frontal cortex of the rat brain [11]. Antioxidants are commonly reported to prevent oxidative stress-related damage [41]. In addition to WFA, the roots of *Withania somnifera* also contain several WFA analogues such as 1-oxo-5beta and 6beta-epoxy-witha-2-enolide, which prevent the UVB-induced skin carcinoma in rat [42,43]. In contrast, WFA inhibited proliferation of oral [14] and lung [17] cancer cells as well as leukemia cells [44] through oxidative stress induction. Similarly, our finding shows that WFA in combination with UVC can inhibit more cell proliferation than its single treatment. This controversy can be partly explained by the concept that antioxidants are bifunctional to regulate oxidative stress, i.e., low concentration inhibits oxidative stress but high concentration induces oxidative stress [32,45].

4.2. Oxidative Stresses Are Higher in a Combined UVC/WFA Than in Single Treatment

ROS modulating is one of the strategies for antiproliferation of cancer cells [33]. Both high dose UVC (200 J/m^2) [46] and WFA [14,18] trigger ROS in cancer cells. For low dose treatment, UVC (14 J/m^2 [25] and 12 J/m^2 (Figure 1); ~80% viability) induces ROS generation of oral cancer Ca9-22 and HSC-3 cells. For low dose treatment, WFA (0.5 µM [47] for 100% viability and 1 µM (Figure 1) for ~80% viability) also triggers ROS generation of oral cancer Ca-22 and HSC-3 cells. At a low dose combined treatment, UVC/WFA triggers more ROS generation than single treatment. Similarly, UVC/WFA triggers more MitoSOX generation and MitoMP destruction than single treatment (Figures 5–7). These results suggest that UVC/WFA cooperatively triggers oxidative stress in oral cancer cells compared to single treatment. Moreover, UVC/WFA triggers more ROS and MitoSOX generation as well as MitoMP destruction in oral cancer cells than in normal oral HGF-1 cells. Although some of these oxidative stress changes for UVC/WFA were only slightly higher than at single treatment, the combined differential oxidative stresses (ROS/MitoSOX generations and MitoMP depletion) may cooperatively contribute to the cancer cell specific effects of a UVC/WFA combined treatment.

4.3. DNA Damages and Apoptosis Are Higher in UVC/WFA than Single Treatment

Cellular oxidative stress enhances DNA damages [33]. High dose UVC causes γH2AX-detecting DNA double strand break damage [48] and 8-oxodG-detecting oxidative DNA damage [49]. High dose WFA causes γH2AX DNA damage [14]. In the present study, low dose UVC/WFA triggers more DNA damages (γH2AX and 8-oxodG) in oral cancer Ca9-22 and HSC-3 cells than single treatment (Figures 8 and 9). These results suggest that UVC/WFA cooperatively trigger DNA damage in oral cancer cells compared to single treatment. In addition to DNA damage induction, γH2AX also functions for senescence induction [50]. Our present flow cytometry approach could not discriminate between these two effects. A promising alternative would be a detailed investigation by immunofluorescence microscopy analyzing DNA damage foci formation in the future.

High dose UVC [25,51] or WFA [14,18] causes apoptosis in cancer cells. Similarly, low dose UVC/WFA triggers more apoptosis than single treatment as evidenced by flow cytometry or luminescent detection indicating an increasing subG1 population, annexin V expression, pancaspase activation, and Cas 3/7 activation as well as by Western blotting indicating c-PARP and c-Cas 3 overexpression in oral cancer cells (Figures 2–4). Moreover, UVC/WFA triggers more Cas 3/7 activation in oral cancer cells than in HGF-1 cells (Figure 4). Additionally, the loss of the mitochondrial membrane potential is often associated with early stages of apoptosis [52]. Similarly, a combination of low dose UVC/WFA triggers more mitochondrial membrane depolarization in oral cancer cells than single treatment (Figure 7), supporting also that low dose UVC/WFA triggers more apoptosis than single treatment.

Furthermore, the DNA repair system may be inhibited for drug radiosensitization. For example, withanolide D is helpful in radiosensitization of ovarian cancer SKOV3 cells by suppressing DNA repair such as non-homologous end joining (NHEJ). This warrants a detailed examination of the role of DNA repair in UVC/WFA synergistic effects to oral cancer cells in the future.

4.4. Cell Cycle Changes at UVC and/or WFA Treatment

UVC/WFA accumulated a larger subG1 population than single treatment in both oral cancer Ca9-22 and HSC-3 cells. However, UVC/WFA induced more G2/M arrest in Ca9-22 cells than upon single treatment but no G2/M arrest appeared in HSC-3 cells. Because Ca9-22 and HSC-3 cells are gingiva and tongue squamous cell carcinoma cell lines [53], it is possible that UVC/WFA-induced cell cycle G2/M arrest is tissue-dependent response.

5. Conclusions

The present study examined the combined effect of low dose UVC/WFA on regulating cell proliferation of oral cancer cells. Single treatment at low dose UVC or WFA showed low cytotoxicity to oral cancer Ca9-22 and HSC-3 cells. Combined treatment of low dose UVC/WFA highly induces oxidative stress and resulted in apoptosis and DNA damage of oral cancer cells. Moreover, low dose UVC/WFA showed highly cytotoxic, oxidative stress, and apoptosis induction in oral cancer cells without effects on normal oral cells. Therefore, low dose UVC/WFA shows a selective killing potential and effectively inhibits oral cancer cell proliferation with no cytotoxic side effects to normal oral cells. In the future, detailed investigations in vivo, such as those using mouse models, should be conducted to further validate the promising effects of a low dose UVC/WFA combined treatment in oral cancer therapy.

Author Contributions: Conceptualization, J.-Y.T. and H.-W.C.; Data curation, S.-Y.P.; Formal analysis, S.-Y.P.; Methodology, Y.-Y.W., L.-C.L., S.-S.F.Y.; Supervision, J.-Y.T. and H.-W.; Writing—original draft, S.-Y.P., T.-H.L., and H.-W.C.; Writing—review and editing, J.-Y.T. and H.-W.C. All authors have read and agreed to the published version of the manuscript.

Funding: This work was partly supported by funds of the Ministry of Science and Technology (MOST 108-2320-B-037-015-MY3 and MOST 108-2314-B-037-020), the National Sun Yat-sen University-KMU Joint Research Project (#NSYSUKMU 109-I002), the Kaohsiung Medical University Hospital (KMUH108-8R67), the Kaohsiung Medical University Research Center (KMU-TC108A04), and the Health and welfare surcharge of tobacco products, the Ministry of Health and Welfare, Taiwan (MOHW109-TDU-B-212-134016).

Acknowledgments: The authors thank our colleague Hans-Uwe Dahms for editing the manuscript.

Conflicts of Interest: The authors declare no conflict of interest.

References

1. Huang, S.H.; O'Sullivan, B. Oral cancer: Current role of radiotherapy and chemotherapy. *Med. Oral Patol. Oral Cir. Bucal* **2013**, *18*, e233–e240. [CrossRef]
2. Moeller, B.J.; Richardson, R.A.; Dewhirst, M.W. Hypoxia and radiotherapy: Opportunities for improved outcomes in cancer treatment. *Cancer Metastasis Rev.* **2007**, *26*, 241–248. [CrossRef]
3. Perry, J.M.; Tao, F.; Roy, A.; Lin, T.; He, X.C.; Chen, S.; Lu, X.; Nemechek, J.; Ruan, L.; Yu, X.; et al. Overcoming Wnt-beta-catenin dependent anticancer therapy resistance in leukaemia stem cells. *Nat. Cell Biol.* **2020**, *22*, 689–700. [CrossRef]

4. Saga, R.; Matsuya, Y.; Takahashi, R.; Hasegawa, K.; Date, H.; Hosokawa, Y. Analysis of the high-dose-range radioresistance of prostate cancer cells, including cancer stem cells, based on a stochastic model. *J. Radiat. Res.* **2019**, *60*, 298–307. [CrossRef]

5. Lee, T.W.; Wong, W.W.; Dickson, B.D.; Lipert, B.; Cheng, G.J.; Hunter, F.W.; Hay, M.P.; Wilson, W.R. Radiosensitization of head and neck squamous cell carcinoma lines by DNA-PK inhibitors is more effective than PARP-1 inhibition and is enhanced by SLFN11 and hypoxia. *Int. J. Radiat. Biol.* **2019**, *95*, 1597–1612. [CrossRef]

6. Affolter, A.; Samosny, G.; Heimes, A.S.; Schneider, J.; Weichert, W.; Stenzinger, A.; Sommer, K.; Jensen, A.; Mayer, A.; Brenner, W.; et al. Multikinase inhibitors sorafenib and sunitinib as radiosensitizers in head and neck cancer cell lines. *Head Neck* **2017**, *39*, 623–632. [CrossRef]

7. Omidi, B.R.; Gosili, A.; Jaber-Ansari, M.; Mahdkhah, A. Intensity output and effectiveness of light curing units in dental offices. *J. Clin. Exp. Dent.* **2018**, *10*, e555–e560. [CrossRef]

8. Coohill, T.P.; Sagripanti, J.L. Bacterial inactivation by solar ultraviolet radiation compared with sensitivity to 254 nm radiation. *Photochem. PhotoBiol.* **2009**, *85*, 1043–1052. [CrossRef]

9. Yamauchi, T.; Adachi, S.; Yasuda, I.; Nakashima, M.; Kawaguchi, J.; Yoshioka, T.; Hirose, Y.; Kozawa, O.; Moriwaki, H. Ultra-violet irradiation induces apoptosis via mitochondrial pathway in pancreatic cancer cells. *Int. J. Oncol.* **2011**, *39*, 1375–1380.

10. Adachi, S.; Yasuda, I.; Nakashima, M.; Yamauchi, T.; Kawaguchi, J.; Shimizu, M.; Itani, M.; Nakamura, M.; Nishii, Y.; Yoshioka, T.; et al. Ultraviolet irradiation can induce evasion of colon cancer cells from stimulation of epidermal growth factor. *J. Biol. Chem.* **2011**, *286*, 26178–26187. [CrossRef] [PubMed]

11. Bhattacharya, S.K.; Satyan, K.S.; Ghosal, S. Antioxidant activity of glycowithanolides from *Withania somnifera*. *Indian J. Exp. Biol.* **1997**, *35*, 236–239. [PubMed]

12. Devkar, S.T.; Jagtap, S.D.; Katyare, S.S.; Hegde, M.V. Estimation of antioxidant potential of individual components present in complex mixture of *Withania somnifera* (Ashwagandha) root fraction by thin-layer chromatography-2,2-diphenyl-1-picrylhdrazyl method. *J. Planar Chromatogr.* **2014**, *27*, 157–161. [CrossRef]

13. Yang, I.-H.; Kim, L.-H.; Shin, J.-A.; Cho, S.-D. Chemotherapeutic effect of withaferin A in human oral cancer cells. *J. Cancer Ther.* **2015**, *6*, 735–742. [CrossRef]

14. Chang, H.W.; Li, R.N.; Wang, H.R.; Liu, J.R.; Tang, J.Y.; Huang, H.W.; Chan, Y.H.; Yen, C.Y. Withaferin A induces oxidative stress-mediated apoptosis and DNA damage in oral cancer cells. *Front. Physiol.* **2017**, *8*, 634. [CrossRef] [PubMed]

15. Munagala, R.; Kausar, H.; Munjal, C.; Gupta, R.C. Withaferin A induces p53-dependent apoptosis by repression of HPV oncogenes and upregulation of tumor suppressor proteins in human cervical cancer cells. *Carcinogenesis* **2011**, *32*, 1697–1705. [CrossRef]

16. Li, X.; Zhu, F.; Jiang, J.; Sun, C.; Wang, X.; Shen, M.; Tian, R.; Shi, C.; Xu, M.; Peng, F.; et al. Synergistic antitumor activity of withaferin A combined with oxaliplatin triggers reactive oxygen species-mediated inactivation of the PI3K/AKT pathway in human pancreatic cancer cells. *Cancer Lett.* **2015**, *357*, 219–230. [CrossRef]

17. Hsu, J.H.; Chang, P.M.; Cheng, T.S.; Kuo, Y.L.; Wu, A.T.; Tran, T.H.; Yang, Y.H.; Chen, J.M.; Tsai, Y.C.; Chu, Y.S.; et al. Identification of withaferin A as a potential candidate for anti-cancer therapy in non-small cell lung cancer. *Cancers* **2019**, *11*, 1003. [CrossRef]

18. Xia, S.; Miao, Y.; Liu, S. Withaferin A induces apoptosis by ROS-dependent mitochondrial dysfunction in human colorectal cancer cells. *Biochem. Biophys. Res. Commun.* **2018**, *503*, 2363–2369. [CrossRef]

19. Sari, A.N.; Bhargava, P.; Dhanjal, J.K.; Putri, J.F.; Radhakrishnan, N.; Shefrin, S.; Ishida, Y.; Terao, K.; Sundar, D.; Kaul, S.C.; et al. Combination of withaferin-A and CAPE provides superior anticancer potency: Bioinformatics and experimental evidence to their molecular targets and mechanism of action. *Cancers* **2020**, *12*, 1160. [CrossRef]

20. Tsugeno, Y.; Sato, F.; Muragaki, Y.; Kato, Y. Cell culture of human gingival fibroblasts, oral cancer cells and mesothelioma cells with serum-free media, STK1 and STK2. *Biomed. Rep.* **2014**, *2*, 644–648. [CrossRef]

21. Wang, H.R.; Tang, J.Y.; Wang, Y.Y.; Farooqi, A.A.; Yen, C.Y.; Yuan, S.F.; Huang, H.W.; Chang, H.W. Manoalide preferentially provides antiproliferation of oral cancer cells by oxidative stress-mediated apoptosis and DNA damage. *Cancers* **2019**, *11*, 1303. [CrossRef]

22. Yeh, C.C.; Tseng, C.N.; Yang, J.I.; Huang, H.W.; Fang, Y.; Tang, J.Y.; Chang, F.R.; Chang, H.W. Antiproliferation and induction of apoptosis in Ca9-22 oral cancer cells by ethanolic extract of *Gracilaria tenuistipitata*. *Molecules* **2012**, *17*, 10916–10927. [CrossRef] [PubMed]

23. Chen, C.Y.; Yen, C.Y.; Wang, H.R.; Yang, H.P.; Tang, J.Y.; Huang, H.W.; Hsu, S.H.; Chang, H.W. Tenuifolide B from *Cinnamomum tenuifolium* stem selectively inhibits proliferation of oral cancer cells via apoptosis, ROS generation, mitochondrial depolarization, and DNA damage. *Toxins* **2016**, *8*, 319. [CrossRef] [PubMed]

24. Liu, P.F.; Tsai, K.L.; Hsu, C.J.; Tsai, W.L.; Cheng, J.S.; Chang, H.W.; Shiau, C.W.; Goan, Y.G.; Tseng, H.H.; Wu, C.H.; et al. Drug repurposing screening identifies tioconazole as an ATG4 inhibitor that suppresses autophagy and sensitizes cancer cells to chemotherapy. *Theranostics* **2018**, *8*, 830–845. [CrossRef] [PubMed]

25. Chang, H.W.; Tang, J.Y.; Yen, C.Y.; Chang, H.S.; Huang, H.W.; Chung, Y.A.; Chen, I.S.; Huang, M.Y. Synergistic anti-oral cancer effects of UVC and methanolic extracts of *Cryptocarya concinna* roots via apoptosis, oxidative stress and DNA damage. *Int. J. Radiat. Biol.* **2016**, *92*, 263–272. [CrossRef]

26. Vignon, C.; Debeissat, C.; Georget, M.T.; Bouscary, D.; Gyan, E.; Rosset, P.; Herault, O. Flow cytometric quantification of all phases of the cell cycle and apoptosis in a two-color fluorescence plot. *PLoS ONE* **2013**, *8*, e68425. [CrossRef]

27. Tang, J.Y.; Xu, Y.H.; Lin, L.C.; Ou-Yang, F.; Wu, K.H.; Tsao, L.Y.; Yu, T.J.; Huang, H.W.; Wang, H.R.; Liu, W.; et al. LY303511 displays antiproliferation potential against oral cancer cells in vitro and in vivo. *Environ. Toxicol.* **2019**, *34*, 958–967. [CrossRef]

28. Wang, S.C.; Wang, Y.Y.; Lin, L.C.; Chang, M.Y.; Yuan, S.F.; Tang, J.Y.; Chang, H.W. Combined treatment of sulfonyl chromen-4-ones (CHW09) and ultraviolet-C (UVC) enhances proliferation inhibition, apoptosis, oxidative stress, and DNA damage against oral cancer cells. *Int. J. Mol. Sci.* **2020**, *21*, 6443. [CrossRef]

29. Chang, Y.T.; Huang, C.Y.; Tang, J.Y.; Liaw, C.C.; Li, R.N.; Liu, J.R.; Sheu, J.H.; Chang, H.W. Reactive oxygen species mediate soft corals-derived sinuleptolide-induced antiproliferation and DNA damage in oral cancer cells. *OncoTargets Ther.* **2017**, *10*, 3289–3297. [CrossRef]

30. Huang, H.W.; Tang, J.Y.; Ou-Yang, F.; Wang, H.R.; Guan, P.Y.; Huang, C.Y.; Chen, C.Y.; Hou, M.F.; Sheu, J.H.; Chang, H.W. Sinularin selectively kills breast cancer cells showing G2/M arrest, apoptosis, and oxidative DNA damage. *Molecules* **2018**, *23*, 849. [CrossRef]

31. Tang, J.Y.; Wu, C.Y.; Shu, C.W.; Wang, S.C.; Chang, M.Y.; Chang, H.W. A novel sulfonyl chromen-4-ones (CHW09) preferentially kills oral cancer cells showing apoptosis, oxidative stress, and DNA damage. *Environ. Toxicol.* **2018**, *33*, 1195–1203. [CrossRef] [PubMed]

32. Yen, C.Y.; Hou, M.F.; Yang, Z.W.; Tang, J.Y.; Li, K.T.; Huang, H.W.; Huang, Y.H.; Lee, S.Y.; Fu, T.F.; Hsieh, C.Y.; et al. Concentration effects of grape seed extracts in anti-oral cancer cells involving differential apoptosis, oxidative stress, and DNA damage. *BMC Complement. Altern. Med.* **2015**, *15*, 94. [CrossRef] [PubMed]

33. Tang, J.Y.; Farooqi, A.A.; Ou-Yang, F.; Hou, M.F.; Huang, H.W.; Wang, H.R.; Li, K.T.; Fayyaz, S.; Shu, C.W.; Chang, H.W. Oxidative stress-modulating drugs have preferential anticancer effects—Involving the regulation of apoptosis, DNA damage, endoplasmic reticulum stress, autophagy, metabolism, and migration. *Semin. Cancer Biol.* **2018**, *58*, 109–117. [CrossRef] [PubMed]

34. Huang, C.; Li, J.; Ding, M.; Leonard, S.S.; Wang, L.; Castranova, V.; Vallyathan, V.; Shi, X. UV Induces phosphorylation of protein kinase B (Akt) at Ser-473 and Thr-308 in mouse epidermal Cl 41 cells through hydrogen peroxide. *J. Biol. Chem.* **2001**, *276*, 40234–40240. [CrossRef]

35. Peng, S.Y.; Lin, L.C.; Yang, Z.W.; Chang, F.R.; Cheng, Y.B.; Tang, J.Y.; Chang, H.W. Combined treatment with low cytotoxic ethyl acetate Nepenthes extract and ultraviolet-C improves antiproliferation to oral cancer cells via oxidative stress. *Antioxidants* **2020**, *9*, 876. [CrossRef]

36. Sharada, A.C.; Solomon, F.E.; Devi, P.U.; Udupa, N.; Srinivasan, K.K. Antitumor and radiosensitizing effects of withaferin A on mouse Ehrlich ascites carcinoma in vivo. *Acta Oncol.* **1996**, *35*, 95–100. [CrossRef]

37. Devi, P.U.; Kamath, R.; Rao, B.S. Radiosensitization of a mouse melanoma by withaferin A: In vivo studies. *Indian J. Exp. Biol.* **2000**, *38*, 432–437.

38. Yang, E.S.; Choi, M.J.; Kim, J.H.; Choi, K.S.; Kwon, T.K. Combination of withaferin A and X-ray irradiation enhances apoptosis in U937 cells. *Toxicol. Vitr.* **2011**, *25*, 1803–1810. [CrossRef]

39. Kawaguchi, J.; Adachi, S.; Yasuda, I.; Yamauchi, T.; Nakashima, M.; Ohno, T.; Shimizu, M.; Yoshioka, T.; Itani, M.; Kozawa, O.; et al. Cisplatin and ultra-violet-C synergistically down-regulate receptor tyrosine kinases in human colorectal cancer cells. *Mol. Cancer* **2012**, *11*, 45. [CrossRef]

40. Robin, T.; Capes-Davis, A.; Bairoch, A. CLASTR: The Cellosaurus STR similarity search tool—A precious help for cell line authentication. *Int. J. Cancer* **2020**, *146*, 1299–1306. [CrossRef]

41. Tan, B.L.; Norhaizan, M.E.; Liew, W.P.; Sulaiman Rahman, H. Antioxidant and oxidative stress: A mutual interplay in age-related diseases. *Front. Pharmacol.* **2018**, *9*, 1162. [CrossRef] [PubMed]

42. Mathur, S.; Kaur, P.; Sharma, M.; Katyal, A.; Singh, B.; Tiwari, M.; Chandra, R. The treatment of skin carcinoma, induced by UV B radiation, using 1-oxo-5beta, 6beta-epoxy-witha-2-enolide, isolated from the roots of *Withania somnifera*, in a rat model. *Phytomedicine* **2004**, *11*, 452–460. [CrossRef] [PubMed]

43. Braun, L.; Cohen, M. *Herbs and Natural Supplements, Volume 2: An. Evidence-Based Guide*; Elsevier Health Sciences: Melbourne, Australia, 2015; Volume 2.

44. Malik, F.; Kumar, A.; Bhushan, S.; Khan, S.; Bhatia, A.; Suri, K.A.; Qazi, G.N.; Singh, J. Reactive oxygen species generation and mitochondrial dysfunction in the apoptotic cell death of human myeloid leukemia HL-60 cells by a dietary compound withaferin A with concomitant protection by N-acetyl cysteine. *Apoptosis* **2007**, *12*, 2115–2133. [CrossRef] [PubMed]

45. Bouayed, J.; Bohn, T. Exogenous antioxidants–Double-edged swords in cellular redox state: Health beneficial effects at physiologic doses versus deleterious effects at high doses. *Oxid. Med. Cell Longev.* **2010**, *3*, 228–237. [CrossRef] [PubMed]

46. Chan, W.H.; Yu, J.S. Inhibition of UV irradiation-induced oxidative stress and apoptotic biochemical changes in human epidermal carcinoma A431 cells by genistein. *J. Cell Biochem.* **2000**, *78*, 73–84. [CrossRef]

47. Yu, T.J.; Tang, J.Y.; Ou-Yang, F.; Wang, Y.Y.; Yuan, S.F.; Tseng, K.; Lin, L.C.; Chang, H.W. Low concentration of withaferin A inhibits oxidative stress-mediated migration and invasion in oral cancer cells. *Biomolecules* **2020**, *10*, 777. [CrossRef]

48. Cadet, J.; Sage, E.; Douki, T. Ultraviolet radiation-mediated damage to cellular DNA. *Mutat. Res.* **2005**, *571*, 3–17. [CrossRef]

49. Evans, M.D.; Cooke, M.S.; Podmore, I.D.; Zheng, Q.; Herbert, K.E.; Lunec, J. Discrepancies in the measurement of UVC-induced 8-oxo-2′-deoxyguanosine: Implications for the analysis of oxidative DNA damage. *Biochem. Biophys. Res. Commun.* **1999**, *259*, 374–378. [CrossRef]

50. Turinetto, V.; Giachino, C. Multiple facets of histone variant H2AX: A DNA double-strand-break marker with several biological functions. *Nucleic Acids Res.* **2015**, *43*, 2489–2498. [CrossRef]

51. Dunkern, T.R.; Fritz, G.; Kaina, B. Ultraviolet light-induced DNA damage triggers apoptosis in nucleotide excision repair-deficient cells via Bcl-2 decline and caspase-3/-8 activation. *Oncogene* **2001**, *20*, 6026–6038. [CrossRef]

52. Ly, J.D.; Grubb, D.R.; Lawen, A. The mitochondrial membrane potential (deltapsi(m)) in apoptosis; an update. *Apoptosis* **2003**, *8*, 115–128. [CrossRef] [PubMed]

53. Bairoch, A. The Cellosaurus, a Cell-Line Knowledge Resource. *J. Biomol. Tech.* **2018**, *29*, 25–38. [CrossRef] [PubMed]

Publisher's Note: MDPI stays neutral with regard to jurisdictional claims in published maps and institutional affiliations.

 *antioxidants*

Review

Oxidative Stress-Inducing Anticancer Therapies: Taking a Closer Look at Their Immunomodulating Effects

Jinthe Van Loenhout [1,*], Marc Peeters [1,2], Annemie Bogaerts [3], Evelien Smits [1,†] and Christophe Deben [1,†]

1   Center for Oncological Research (CORE), Integrated Personalized and Precision Oncology Network (IPPON), University of Antwerp, 2610 Wilrijk, Belgium; Marc.Peeters@uza.be (M.P.); evelien.smits@uza.be (E.S.); christophe.deben@uantwerpen.be (C.D.)
2   Department of Oncology, Multidisciplinary Oncological Center Antwerp, Antwerp University Hospital, 2650 Edegem, Belgium
3   Plasma Lab for Applications in Sustainability and Medicine ANTwerp (PLASMANT), University of Antwerp, 2610 Wilrijk, Belgium; annemie.bogaerts@uantwerpen.be
*   Correspondence: jinthe.vanloenhout@uantwerpen.be
†   Shared senior author.

Received: 29 October 2020; Accepted: 25 November 2020; Published: 27 November 2020

Abstract: Cancer cells are characterized by higher levels of reactive oxygen species (ROS) compared to normal cells as a result of an imbalance between oxidants and antioxidants. However, cancer cells maintain their redox balance due to their high antioxidant capacity. Recently, a high level of oxidative stress is considered a novel target for anticancer therapy. This can be induced by increasing exogenous ROS and/or inhibiting the endogenous protective antioxidant system. Additionally, the immune system has been shown to be a significant ally in the fight against cancer. Since ROS levels are important to modulate the antitumor immune response, it is essential to consider the effects of oxidative stress-inducing treatments on this response. In this review, we provide an overview of the mechanistic cellular responses of cancer cells towards exogenous and endogenous ROS-inducing treatments, as well as the indirect and direct antitumoral immune effects, which can be both immunostimulatory and/or immunosuppressive. For future perspectives, there is a clear need for comprehensive investigations of different oxidative stress-inducing treatment strategies and their specific immunomodulating effects, since the effects cannot be generalized over different treatment modalities. It is essential to elucidate all these underlying immune effects to make oxidative stress-inducing treatments effective anticancer therapy.

Keywords: cancer therapy; oxidative stress; immune system

1. Introduction

Reactive oxygen species (ROS) is a collective term referring to unstable, reactive, partially reduced oxygen derivatives that are produced during metabolic processes within the mitochondria, peroxisomes and the endoplasmic reticulum (ER). A subset of ROS is also continuously generated by enzymatic reactions involving cyclooxygenases, nicotinamide adenine dinucleotide phosphate (NADPH) oxidases (NOX), xanthine oxidases, lipogenesis and through the iron-catalyzed Fenton reaction [1]. Examples of ROS include hydrogen peroxide (H_2O_2), superoxide anion ($O_2^{\bullet-}$), singlet oxygen ($^1O_2$) and hydroxyl radical ($^{\bullet}OH$) [2]. Tight regulation of these ROS levels is crucial for cellular life. Therefore, cells benefit from a complex scavenging system based on different antioxidants, including superoxide dismutase (SOD), glutathione (GSH) peroxidase, peroxiredoxin, thioredoxin (Trx) and catalase [1]. Additionally to the strong antioxidant activity of the beforementioned enzymes, various non-enzymatic small-molecule antioxidants such as glutathione, ascorbic acid, vitamin E, polyphenolic compounds also act as scavengers for different types of ROS [3].

Cancer cells are characterized by increased production of ROS compared to normal cells. The persistently high levels of ROS can be explained by the imbalance between oxidants and antioxidants in cancer cells and the ongoing aerobic glycolysis by pyruvate oxidation in the mitochondria, also known as the Warburg-effect [4]. This is a consequence of hypoxia in the tumor microenvironment (TME) resulting from an imbalance between oxygen supply and consumption due to uncontrollable cell proliferation, altered metabolism and abnormal tumor blood vessel growth [5]. Cancer cells evolved mechanisms to protect themselves from this intrinsic oxidative stress and developed an adaptation mechanism by upregulation of pro-survival molecules and their antioxidant defense system to maintain the redox balance [6]. For instance, nuclear factor erythroid 2-related factor 2 (Nrf2), which is a transcription factor in the first line of antioxidant defense against oxidative stress, is often upregulated in cancer cells and supports cancer cell proliferation [7].

A low to moderate increase in intracellular ROS levels may result in activation of oncogenes (such as Akt), which are involved in cell proliferation, and inactivation of tumor suppressor genes, angiogenesis and mitochondrial dysfunction, thereby serving as a signaling molecule in cancer survival [4]. Conversely, when the levels of ROS are further elevated, they can overcome the defensive antioxidant system of cancer cells, causing cell death [8].

Consequently, there are two different approaches based on the redox balance to counteract cancer cells. In the first approach, oxidative stress can be decreased via scavenging intracellular ROS. For example, increasing intake of antioxidants (e.g., vitamin C and E) can deplete oxidative stress, subsequently causing growth inhibition and increased susceptibility to cell death in cancer cells, due to a crisis in energy production [9]. However, this antioxidant supplementation remains controversial [10]. Increasing evidence has shown that antioxidant supplementation fails to provide cancer protection and can even affect cancer mortality [11–13]. These observations are further supported and rationalized by recent studies demonstrating that oxidative stress can inhibit cancer progression and metastasis and that the GSH and Trx antioxidant systems, which are under the transcriptional regulation of Nrf2, may promote tumorigenesis and resistance to therapy [14].

The second approach is by increasing ROS levels in cancer cells and thereby crossing the threshold of cancer cell death. This can be done either by direct production of ROS via exogenous approaches or indirectly by increasing intracellular ROS concentrations via targeted inhibition of previously mentioned endogenous antioxidant systems in cancer cells. Several investigations are suggestive of the fact that the underlying mechanism of action and efficacy of conventional therapies (e.g., radiotherapy and chemotherapy) inducing cancer cell death, is the generation of elevated ROS levels during treatment [15–17].

In this review, we will focus on therapies related to this second approach and how they influence the TME, more specifically the immune cell compartment, to provide an overview of the effect of ROS induction on the antitumor immune response.

1.1. Exogenous ROS Generation

One mechanism of enhancing oxidative stress levels to target cancer cells is via exogenous delivery of ROS using different physical modalities (Figure 1).

Ionizing radiation is widely used to treat many types of cancer. During radiation, cancer cells are eradicated through free radicals such as superoxide and hydroxyl radicals which are generated by radiolysis of water in extracellular environments and indirectly damage critical targets, such as DNA [15]. In addition, radiotherapy can also alter mitochondrial membrane permeability and activate NADPH oxidase, which in turn further stimulates ROS production [18]. Besides radiotherapy, other physical modalities that can induce a substantial increase in ROS levels are being investigated in cancer research, including photodynamic therapy (PDT) and cold atmospheric plasma (CAP) [19–22].

PDT is a light-based oncological intervention. Here, a photosensitizer is applied and subsequently activated by light. Upon activation, exogenously produced ROS is generated [23]. CAP is an ionized gas that can be produced at atmospheric pressure near room temperature. It is composed of reactive oxygen and nitrogen species, excited molecules, ions, electrons and other physical factors, such as electromagnetic fields and ultraviolet radiation [24].

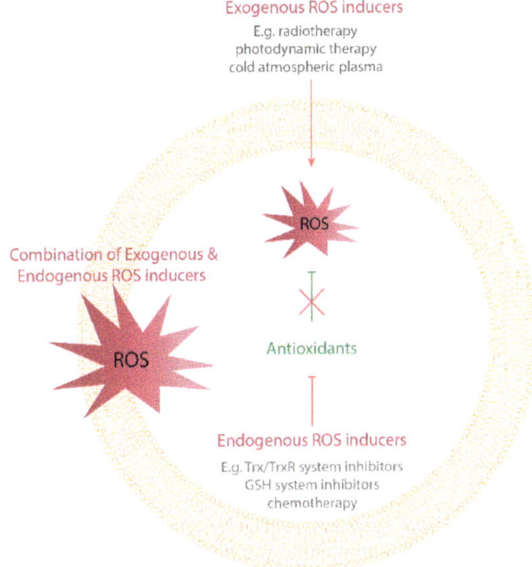

Figure 1. Oxidative stress-inducing treatment strategies. Oxidative stress can be induced by exogenous delivery of reactive oygen species (ROS) using physical treatment modalities, as well as by targeting the endogenous antioxidant system causing an intracellular accumulation of ROS. Cancer cells counteract exogenous delivery of high ROS levels by enhancing their antioxidant capacity. Therefore, a combination of both exogenous and endogenous ROS delivery by targeting the antioxidants, can be a promising anticancer strategy.

1.2. Endogenous ROS Generation

The second mechanism of enhancing oxidative stress levels is via intracellular ROS accumulation through chemotherapy or targeted inhibition of the elevated antioxidant system (Figure 1). A lot of chemotherapeutic agents enhance intracellular levels of ROS and can alter the redox homeostasis of cancer cells. This amplification of ROS levels towards cytotoxic levels is one of the proposed mechanisms by which multiple chemotherapeutics induce tumor regression. The level of ROS generation is different among several compounds. Agents that generate high levels of ROS include anthracyclines (e.g., doxorubicin), platinum coordination complexes (e.g., cisplatin), alkylating agents (e.g., cyclophosphamide), camptothecins, arsenic agents and topoisomerase inhibitors, while nucleoside, nucleotide analogs, antifolates, taxanes and vinca alkaloids only generate low levels of ROS [25].

There are two mechanisms for elevated ROS production during chemotherapy, namely through mitochondrial ROS generation and inhibition of the cellular antioxidant system and thereby interfering with ROS metabolism in cancer cells [25]. Several agents, including arsenic trioxide, doxorubicin and cisplatin, have been reported to induce a loss of mitochondrial membrane potential and to inhibit respiratory complexes, leading to the disruption of mitochondrial electron transport chain (ECT) and electron leakage, which is a major source of elevated ROS levels [26–28].

The other mechanism for intracellular ROS accumulation is the inhibition of the antioxidant system during chemotherapy. For instance, imexon, a small-molecule used to treat advanced cancer of the breast, lung or prostate, binds to thiols such as GSH, causing a depletion of cellular GSH and consequently an accumulation of oxidative stress in cancer cells [29]. For some chemotherapeutics, more than one target site for ROS generation in cancer cells has been identified. For example, in addition to mitochondrial respiration, NADPH oxidase and thioredoxin reductase (TrxR) are other targets of arsenic trioxide induced oxidative stress, inducing apoptosis [30–32].

Besides chemotherapy, selective inhibitors that block components of the cellular antioxidant system are being studied as antitumor agents that enhance endogenous ROS production. For instance, depletion of the GSH antioxidant system can also be achieved by targeting its synthesis through buthionine sulfoximine (BSO), which has been shown to exhibit anticancer activities in various types of cancer. Furthermore, inhibitors of the Xc-cystine/glutamate antiporter (e.g., sulfasalazine) may also cause GSH depletion by inhibiting the uptake of cystine, the precursor of cysteine, which is a substrate for GSH synthesis [33]. Another antioxidant is the thioredoxin/thioredoxin reductase (Trx/TrxR) system, which is shown to be upregulated in cancer cells and is correlated with cancer aggressiveness and drug resistance [30]. TrxR is required to convert oxidized Trx into its functional reductive form, which can scavenge ROS [34]. TrxR activity can effectively be blocked by the gold compound auranofin that is clinically used as an antirheumatic drug and functions as a thioredoxin inhibitor. In different cancer cells, it has been preclinically shown to induce ROS-mediated cell death, since the ROS scavenger N-acetylcysteine prevents this cytotoxic effect [35,36]. This has led to the use of auranofin in several clinical trials involving non-small cell lung and ovarian cancer (NCT01737502 and NCT03456700). The small-molecule PX-12 is another example of an antioxidant inhibitor, since it inhibits Trx and is being used as therapy for advanced cancers in clinical trials [37,38].

1.3. Molecular Pathways Involved in Oxidative Stress-Inducing Therapies

Whether ROS augment tumorigenesis or lead to apoptosis, critically depends on the intracellular ROS levels. At moderate concentration, ROS inactivate phosphatase and tensin homolog (PTEN) and unlock the PI3K-depedent recruitment of its downstream kinases, such as Akt, which will, in turn, activate NF-kB, subsequently activating the cancer cell survival signaling cascade [39]. For instance, hydrogen peroxide can reversibly oxidize cysteine thiol groups of PTEN, causing the loss of their activity and promoting the activation of the PI3K/Akt/mTOR survival pathway, consequently leading to tumor cell survival [4]. Abundant high concentrations of ROS originating from exogenous and endogenous sources, produce oxidative damage to the DNA, RNA, proteins, lipids and mitochondria, initiating apoptotic cell death (Figure 2) [39,40].

In line with this, it has been shown that the cellular response to exogenous sources of ROS strongly varies with the intensity of the treatment [41–43]. For example, low dosages of PDT and radiotherapy have been shown to transiently activate several kinases and NF-kB involved in survival signaling [43,44]. In these non-toxic dosages of PDT, kinases that are important to initiate autophagy were shown to be activated [44]. Higher dosages of radiation and PDT activate the mitochondrial apoptotic pathway and additionally can also produce a sustained activation of MAPK families including p38, MAPK, ERK1/2 and JNK apoptotic signaling proteins (Figure 2) [40,45,46].

Inhibition of the antioxidant system of cells could also induce apoptosis of cancer cells. Trx is a physiological inhibitor of ASK1 located upstream of the p38/MAPK pathway, and therefore disrupts the p38/MAPK dependent apoptosis. As such, an inhibitor of the Trx/TrxR system could induce apoptosis due to the phosphorylation of p38/MAPK, as well as the activation of JNK and ERK [34]. Additionally, several studies have shown that Trx/TrxR inhibitors downregulate the PI3K/Akt/mTOR survival pathway, causing apoptosis of different types of cancer cells [34,47–49]. The same effect was observed using an inhibitor of the GSH antioxidant pathway [50].

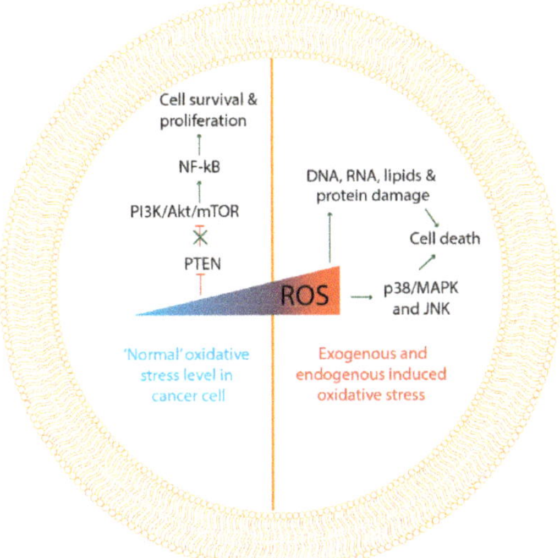

Figure 2. Molecular responses to oxidative stress. At moderate "normal" levels of oxidative stress in cancer cells (left side of the figure) ROS inactivate phosphatase and tensin homolog (PTEN) and unlock PI3K/Akt/mTOR pathway, which in turn activates NF-kB, consequently activating cancer cell survival and proliferation signaling. At high levels of oxidative stress induced by therapy, damage is produced to DNA, RNA, proteins, lipids and mitochondria, initiating apoptotic cell death. Additionally, high levels of ROS can activate p38/MAPK and JNK apoptotic signaling proteins, inducing cancer cell death.

1.4. Combinations of Different Oxidative Stress-Inducing Therapies

It has been shown that radiotherapy, PDT, as well as other ROS-inducing therapies, could induce acquired resistance to therapy. Here, NF-kB is considered to be a key component in the rise of therapy-resistant cancer [43,51]. Suppression of the NF-kB activation pathway sensitized cells to radiotherapy-induced apoptosis by increasing activation of the JNK pathway [52]. Furthermore, it has been suggested that resistance to therapies that induce intracellular ROS production, such as chemotherapy (e.g., paclitaxel and doxorubicin) and radiotherapy, is correlated with an increased antioxidant capacity of cancer cells. Here, upregulation of Nrf2 after oxidative stress contributes to the therapy resistance in cancer cells [53,54]. Due to this complexity of redox homeostasis and adaptation-mediate resistance in tumor cells, ROS-inducing treatments may not always lead to an effective antitumor effect. To overcome resistance induced by oxidative stress and to maximally exploit the ROS-mediated cell death mechanism as a therapeutic strategy, it would be beneficial to combine therapeutic strategies that exogenously induce ROS together with compounds that suppress the cellular antioxidant system.

In several preclinical studies, inhibition of GSH or Trx antioxidant systems, downstream of Nrf2 signaling, has been demonstrated to sensitize different types of tumor cells towards radiotherapy [55–57]. BSO used to inhibit the GSH production, has been shown to sensitize lung, renal and head and neck cancer to radiation. The combination of radiation and GSH depletion by BSO resulted in the activation of the JNK signaling pathway, which resulted in triggering the intrinsic apoptotic pathway [55]. A combination of other agents to disrupt endogenous redox homeostasis was also proven to improve therapeutic efficiency and overcome tumor resistance to PDT [58,59]. However, in combination with BSO, a synergistic effect with PDT was only seen when BSO alone had negligible cytotoxicity. This indicates that cancer cells with intracellular high levels of antioxidants (e.g., GSH) will be more intrinsically resistant toward antioxidant inhibitors or radiation alone, but will effectively induce cell death when these exogenous and endogenous ROS inducers are combined [59–61]. Similar effects were seen when combining BSO with a platinum-based chemotherapy-inducing ROS [62]. Additionally, inhibitors of the Trx/TrxR system (such as PX-12, auranofin and motexafin gadolinium) have shown similar effects to enhance the response against exogenously therapy-induced ROS [63,64].

Hypoxia is also one of the most important causes of exogenous oxidative stress-inducing therapy failure, because of the shortage of ROS substrate oxygen. However, it is demonstrated that more ROS is produced in hypoxic conditions compared to non-hypoxic conditions. Although the specific mechanism has not been described, it appears that the source of the increased ROS levels generated under hypoxia is the mitochondria. Hypoxia increases ROS via the transfer of electrons from ubisemiquinone to molecular oxygen at the Qo sites of complex III of the mitochondrial electron transport chain [65]. Besides mitochondria, nitric oxide synthases (NOS) and NOX have also been implicated to increase ROS production during hypoxia [66]. Moreover, NO and its derivates are a specific group of ROS synthesized by NOS. Since the inducible NOS (iNOS) is a hypoxia response gene, the generation of NO is significantly increased in tumor cells under hypoxic conditions [67]. As such, hypoxic tumor cells heavily rely on the antioxidant defense system to maintain ROS balance, making them vulnerable to inhibition of this antioxidant system [66]. For instance, BSO produces a more pronounced GSH depletion in regions of hypoxia, since GSH levels are higher in hypoxic compared to non-hypoxic regions [68]. Furthermore, auranofin was able to overcome hypoxic radiation resistance and the effect could be further amplified by combining auranofin with BSO, leading to significant tumor growth delay and increased survival rate of tumor-bearing mice [56,57]. Therefore, inhibition of the antioxidant system could be effective to counteract hypoxia-induced therapy resistance [69].

Since the upregulation of NF-kB is also a key player in acquired resistance to ROS-inducing therapy, inhibition of this transcription factor could enhance the anticancer effect. For example, auranofin has shown to decrease the expression of NF-kB, thereby overcoming acquired therapy resistance [70,71]. Similar effects were obtained when inhibiting GSH metabolism [72]. However, it should be mentioned that the activation of NF-kB is also responsible for inflammatory responses, which can induce cross-presentation of tumor antigens and stimulate antitumor immune responses [73]. This indicates that it is important to take into account the effects of exogenous and endogenous ROS-inducing therapies on the antitumoral immune system.

2. Indirect and Direct Effects of Oxidative Stress-Inducing Therapies on the Antitumoral Immune Response

In recent years, it has become clear that the immune system is a strong ally in the fight against cancer. ROS-inducing treatments have significant effects on the immune system, which can be either immunostimulatory or, in some circumstances, immunosuppressive [74,75]. Here we will discuss the direct and indirect effects of these therapies on the immune system, which are either immunostimulatory or immunosuppressive (Figure 3, Table 1).

Table 1. Overview of immunomodulating effects of different ROS-inducing therapies.

ROS-Inducing Therapy	Effect	References
Immunostimulating Effects		
Indirect Effects		
Radiotherapy, PDT, CAP, chemotherapy (e.g., oxaliplatin, doxorubicin)	Secretion of danger signals inducing ICD (e.g., ATP, IL-1β, calreticulin, HMGB1, type I IFN)	[76–83]
Radiotherapy, PDT, CAP, chemotherapy (e.g., docetaxel, doxorubicin, oxaliplatin)	Secretion of chemokines attracting T cells (e.g., CXCL9, CXCL10, CCL5)	[83–90]
Radiotherapy, PDT, CAP, chemotherapy (e.g., topotecan)	Upregulation of MHC-I molecules on tumor cells	[91–95]
Radiotherapy, PDT	Upregulation of NK cell ligands (e.g., MICA, NKG2DL)	[96,97]
Radiotherapy, GSH inhibitors (e.g., BSO)	Modulation of death receptors (e.g., Fas and CD95)	[98–100]

Table 1. *Cont.*

ROS-Inducing Therapy	Effect	References
Trx/TrxR inhibitors (e.g., butaselen)	Downregulation of PD-L1	[101]
Radiotherapy, PDT, CAP, chemotherapy (e.g., doxorubicin, oxaliplatin), Trx inhibitor	Secretion of proinflammatory cytokines (e.g., IFN-γ, TNF-α)	[78,93,102,103]
Direct Effects		
Radiotherapy, PDT, Trx/TrxR inhibitors (e.g., arsenic trioxide)	Depletion of Tregs	[103–106]
CAP, Trx inhibitor	Decrease in secretion of anti-inflammatory cytokines (e.g., IL10, TGF-β)	[78,103]
Radiotherapy, chemotherapy (e.g., cyclophosphamide), antioxidant inhibitors (e.g., noble nanoparticles)	Polarization of M2 into M1 macrophages	[107–109]
Immunosuppressive Effects		
Indirect Effects		
Radiotherapy, PDT, CAP, chemotherapy	Secretion of ATP modulating MDSCs	[110]
Radiotherapy	Secretion of chemokines attracting MDSCs (e.g., CXCL12)	[111,112]
Radiotherapy, Trx/TrxR inhibitors (e.g., auranofin, arsenic trioxide)	Upregulation of PD-L1	[101,113–115]
Radiotherapy	Secretion of anti-inflammatory cytokines (e.g., TGF-β)	[116]
Direct Effects		
Radiotherapy, PDT	Accumulation of Tregs	[116,117]
Radiotherapy	Polarization into M2 macrophages	[118]
Radiotherapy, PDT, CAP, chemotherapy (e.g., cisplatin, oxaliplatin), antioxidant inhibitors (e.g., arsenic trioxide)	Lymphocyte cytotoxicity	[87,116,119–122]
GSH inhibitor	Inhibition of DC maturation	[123]

2.1. Indirect Effects on the Antitumoral Immune Response

2.1.1. Priming of an Adaptive Immune Response

Tumor cells undergoing cell death in response to oxidative stress-induced therapy have the capacity to trigger an adaptive anticancer immune response, a concept known as immunogenic cell death (ICD). This is a unique type of cell death characterized by the release of danger signals after treatment of tumor cells, leading to the effective presentation of tumor antigens and subsequent priming of antigen-specific T cells. This process enhances the elimination of tumor cells and generates immune memory against the tumor antigens, thereby reducing the chance of recurrence [124]. Mechanistically, ICD induction requires ROS generation and further ROS-based ER stress [125]. In literature, there are already comprehensive reviews and research articles that discuss physical ROS-inducing modalities, such as radiotherapy, PDT and CAP, which have been shown to elicit effective antitumor immunity [76–80]. Additionally, chemotherapeutics which have been proven to be ICD inducers (e.g., oxaliplatin and doxorubicin) are accompanied by ROS-induced cytotoxicity [81].

Danger signals released during ICD include the release of adenosine triphosphate (ATP), which attracts dendritic cells (DCs) into the tumor and can stimulate the release of interleukin (IL)-1β, which promotes T cell priming. Moreover, calreticulin is expressed on the surface of the treated tumor cells, which promotes phagocytosis of these cells by DCs. ICD is also associated with high-mobility group box 1 (HMGB1) release, which facilitates antigen presentation and type-I interferon (IFN) secretion, mediating DC maturation [78,82,83]. Release of ATP, however, also modulates immunosuppressive properties of myeloid-derived suppressor cells (MDSCs) and can contribute to tumor growth and inhibition of antitumor immunity [110].

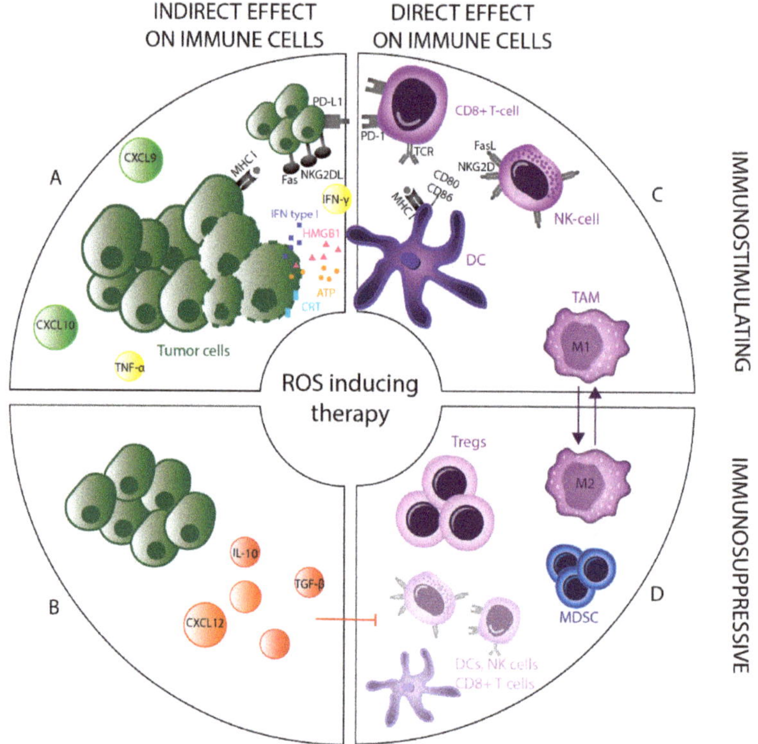

Figure 3. Direct and indirect immunomodulating effects of ROS-inducing therapy. The immunomodulating effects after ROS-inducing therapy can be divided into direct and indirect effects being immunostimulatory and/or immunosuppressive. (**A**) ROS-inducing therapy triggers recruitment and activation of DC by inducing immunogenic tumor cell death (ICD). Additionally, treated tumor cells can secrete cytokines (e.g., IFN-γ and TNF-α) and chemokines (e.g., CXCL10 and CXCL9), and can modulate their surface molecules (e.g., MHC-I, PD-L1 and NKG2DL), thereby increasing their susceptibility to T cell and NK cell-mediated cytotoxicity. (**B**) Immunosuppressive cytokines (e.g., IL10 and TGF-β) and chemokines (e.g., CXCL12) can also be secreted by tumor cells treated with ROS inducers, suppressing immunostimulatory immune cells (DCs, T cells and NK cells) and promoting immunosuppressive immune cells (Tregs and MDSCs). (**C**) Depending on the intensity, ROS-inducing treatment skews TAMs towards a more antitumoral (M1) or protumoral (M2) phenotype. (**D**) T cells and NK cells are sensitive to oxidative stress-inducing treatments, compared to Tregs which are more resistant to these toxic effects.

2.1.2. Recruitment of Leukocytes

Low infiltration of effector T cells and other leukocytes (e.g., NK cells) into the tumor represents a major obstacle for cancer immunotherapy [126]. Here, therapy can facilitate leukocyte infiltration by generating chemoattractants to induce leukocyte extravasation.

The most relevant signals regulating leukocyte infiltration are therapy-induced chemokines secreted by treated tumor cells and/or stromal components. For instance, several exogenous ROS inducers (e.g., radiotherapy, PDT and CAP) induce CXCL9 and CXCL10 secretion, which attracts T cells and thereby enhances tumor control [83–86]. By contrast, CXCL12 induced by radiotherapy can attract tumor-promoting MDSCs [111,112]. This underscores the double-edged sword of oxidative stress-induced therapy in the antitumor immune response. Additionally, a high dose of platinum-based chemotherapy is considered immunosuppressive, causing lymphopenia and neutropenia. However, complementary to other ROS-inducing treatments, it has been shown that low dose treatment enhances the T cell response with an increased number of T cells due to secreted chemokines (CXCL9, CXCL10 and CCL5) after treatment [87]. Other chemotherapy-based treatments are also able to upregulate the expression of chemokines receptors ligands in the TME, subsequently enhancing T cell recruitment [88–90].

2.1.3. Modification of the Related Surface Molecules

Susceptibility of tumor cells to T cell and NK cell-mediated cytotoxicity can be modulated by the expression pattern of surface molecules, including major histocompatibility complex (MHC)-I, MHC-II, NK cell ligands, costimulatory receptors and death receptors. The MHC class I is vital for the presentation of endogenous and potentially tumor-specific antigens to cytotoxic T cells. Radiation induced MHC-I expression on tumor cells, associated with increased susceptibility to T cell-mediated killing [91–93]. Similar to radiation, PDT and CAP use oxygen radicals and were shown to restore MHC-I expression in glioma and melanoma, respectively [94,95]. Additionally, PDT and radiotherapy induce MHC class I polypeptide–related sequence A (MICA) expression and upregulation of natural killer group 2D ligand (NKG2DL) on tumor cells. Both effects corresponded to increased NK cell-mediated killing of treated tumor cells [96,97]. By acting on the death receptors (e.g., Fas), the intrinsic immunogenic properties of the target cells can be altered after radiation, which consequently enhances their susceptibility to cytotoxic T cell-mediated killing [98]. The same effect was seen after treatment with BSO, inducing the formation of the CD95 death-inducing signaling complex [99]. Other chemotherapeutic regimens, which interfere with GSH, also increased the expression of death receptors [100].

ROS-inducing treatments also modulate programmed death-ligand 1 (PD-L1) expression. However, the interplay between ROS inducers and PD-L1 expression is complex, showing that both the up- and downregulation of PD-L1 expression can be induced. It is shown that different inhibitors of Trx/TrxR system decrease the PD-L1 protein level in tumor cells [101]. However, the opposite effect was reported with the TrxR inhibitor auranofin [113]. Like auranofin, arsenic trioxide induces PD-L1 expression in a dose-dependent manner in leukemic cells [127]. Besides antioxidant depletion, PD-L1 expression was increased through PI3K/Akt and STAT3 signaling in vivo and in vitro after conventionally fractionated radiotherapy [114,115]. In addition, PD-L1 expression may occur in response to tumor-targeting immune cells that release IFN-γ upon recognition of the antigen expressed by tumor cells [128]. Conversely, IFN-γ seems to represent the dominant effector molecule of the antitumor immune response after radiotherapy [93]. The same is true for different chemotherapeutics (such as doxorubicin and oxaliplatin) and other physical modalities inducing oxidative stress (including PDT and CAP), where IFN-γ was assessed as a reporter of T cell activity in response to treatments [102].

Many proinflammatory cytokines, including IFN-γ and tumor necrosis factor (TNF)-α, are regulated by the transcription factor NF-kB that can attract cells of the innate and adaptive immune system to mediate antitumor immune responses [129]. This highlights the paradoxical role of NF-kB, where its activation due to intermediate levels of ROS generated during lower dosages of therapeutic strategies inducing oxidative stress (e.g., radiotherapy and PDT), enhances tumor cell growth and on the other hand activates the antitumoral immunity.

None of the described indirect effects can be generalized among all different exogenous and endogenous oxidative stress-inducing therapeutic strategies. Additionally, the effects are context and dose-dependent. Further comprehensive studies are needed to fill up the gaps in the knowledge on different ROS-inducing treatments and possible combinatorial strategies concerning their specific effect on immune response priming, recruitment of leukocytes and modification of surface molecules after treatment.

2.2. Direct Effects on the Antitumoral Immune Response

2.2.1. Direct Effect on Tumor-Infiltrating Immunosuppressive Cells

Immunosuppressive cells, including tumor-associated macrophages (TAMs), regulatory T cells (Tregs) and MDSC are key components of the TME of numerous tumor types [130]. There is an interaction between tumor cells and these immune cells leading to tumor immune escape. An increased number of Tregs in tumor tissue is found in a high proportion of cancer patients and is correlated with tumor progression and poor prognosis, since Tregs help to evade host immunity. In contrast to conventional CD4+ T cells, Tregs are more resistant to oxidative stress-induced cell death [131]. This could be explained by the higher expression and secretion levels of the antioxidant molecule Trx [132]. It was shown that antioxidant Trx expression correlates with Treg representation in clinical samples of metastatic melanoma and that modulation of Trx influences the induction of Tregs and the generation of an immunotolerant cytokine profile. The addition of a Trx inhibitor decreased the number of Tregs in lung lesions. Furthermore, IFN-γ increased, whereas IL-10 and transforming growth factor (TGF)-β decreased after treatment with a Trx blocking antibody [103]. Arsenic trioxide, shown to inhibit TrxR, also induced selective depletion of Tregs and consequently increased the antitumor immune response [32,104]. Moreover, it was found that oxidative stress was the metabolic mechanism that controls tumor Treg cell functional behavior. Induction of Treg apoptosis through exogenous oxidative stress mediated the conversion of a large amount of ATP into adenosine via CD39 and CD73 and subsequently triggered an immunosuppressive cascade, tempering the therapeutic effect of immune checkpoint therapy [133]. Beyond these effects, it was shown that radiotherapy can induce TGF-β release in the TME and consequently lead to the accumulation of Tregs into the tumor tissue [116].

On the contrary, it was found that percentages of Tregs in the peripheral blood of cancer patients decreased significantly after radiotherapy [105]. It was confirmed by others that the reduction was mediated by the downregulation of CCL22 [106]. So far, there is no consensus concerning the effect of radiotherapy on Tregs, probably because the effect of radiotherapy on Tregs is context-dependent for different doses and tumor types. The same contradictory results are true for treatment with PDT [117].

Besides Tregs, certain subtypes of TAMs are also considered to have pro-tumoral functions. TAMs can differentiate from monocytes into two distinct subtypes, namely classically activated (M1) and alternatively activated (M2) macrophages with effector or suppressive function, respectively. Concerning the vulnerability of TAM to oxidative stress, M2 macrophages have lower levels of ROS compared to the M1 phenotype due to higher antioxidant activity, indicating that they will be more resistant to ROS-inducing treatments [134]. For instance, M1 macrophages were observed to be more sensitive towards radiotherapy, compared to M2 macrophages [135]. Additionally, several controversial studies have investigated the effect of chemotherapy and radiotherapy on the TAM phenotype. For example, a low dose of cyclophosphamide can promote the differentiation of M2 macrophages into M1 [107]. Similarly, low dose radiation promotes TAM skewing towards an M1 polarized phenotype and render them supportive of antitumor immunity [108]. However, higher radiation doses can polarize TAMs to an M2 phenotype promoting tumor growth, induced by factors released from irradiated cells [118]. Gold and silver nanoparticles have been shown to modulate reactive oxygen and nitrogen species production by suppressing the antioxidant system of tumor cells. When applying these nanoparticles to TAMs, there was a downregulation of TNF-α and IL-10 and an upregulation of IL-12, resulting in polarization from M2 to M1 macrophages, suggesting a radical shift from pro-tumorigenic to an anti-tumorigenic nature when TAMs undergo oxidative stress [109]. Since the polarization of TAMs is extremely dependent on the contextual signals of the TME, characterization of the oxidative stress-induced factors regulating this polarization remains to be elucidated.

2.2.2. Direct Effect on Tumor-Infiltrating Immunostimulatory Cells

Several oxidative stress-inducing treatments have the potential to increase tumor cell immunogenicity by activating ICD and secreting immunostimulatory factors that can activate innate immune responses and elicits a tumor-specific adaptive immune response. In practice, however, the toxicity of these oxidative stress-inducing treatments to T cells, NK cells and DC limits the extent of immune stimulation and can even lead to immunosuppression [136]. Consequently, oxidative stress-inducing treatments can cause severe related lymphopenia that is associated with reduced patient survival [119].

For instance, the direct effect of radiation on lymphocytes is often immunosuppressive since most subsets of lymphocytes are radiosensitive [137]. Similar direct effects have been demonstrated after chemotherapy, PTD and CAP [119–121]. Nevertheless, the various lymphocyte subtypes differ in their sensitivity to exogenous induced oxidative stress. It has been demonstrated that memory and naïve T cells, as well as NK cells, are highly sensitive, whereas effector T cells, NK-T cells and Tregs are more resistant to the toxic effects of exogenous induced oxidative stress [131,138]. Additionally, the extent of immunosuppressive properties will vary with treatment schedule and dose [136,137]. In general, activated T cells and NK cells have higher antioxidant levels (GSH and Trx), necessary to buffer the rising ROS levels upon activation and proliferation of these lymphocytes, making them less vulnerable for exogenous ROS induced cell death [139,140]. For example, IL-2 activated NK cells were more resistant to H_2O_2-induced cell death than resting NK cells due to an upregulation of the Trx system. However, H_2O_2-induced cell death was also observed in these activated NK cells in the presence of a Trx inhibitor [140]. Inhibiting the antioxidant system in T cells with arsenic trioxide also induces apoptosis in T cells by enhancing oxidative stress, decreasing intracellular GSH releasing cytochrome c, activating caspases and downregulating Bcl-2 [122]. In contrast, after Trx inhibition the expression of the activation marker CD69 was significantly increased on both CD8+ T cells and NK cells [34].

In contrast to all lymphocyte subsets, monocytes are shown to be more resistant to exogenous ROS induced cell death [75,78,121,141]. This might be explained by a stronger antioxidant defense system in phagocytes, such as monocytes and DC, which under physiological conditions protect them against self-production of ROS during oxidative burst [142]. However, depletion of the antioxidant GSH system could also inhibit DC maturation [123].

In summary, ROS-inducing treatments cause direct and indirect immune effects which can be both immunostimulatory and immunosuppressive. Current research on ROS-inducing treatments mostly focuses on one immunomodulating aspect but lacks comprehensive investigation on both stimulatory and suppressive immune effects. Additionally, it is necessary to take into account the timing and location of the effects. ROS-inducing treatments can have immediate toxic and suppressive effects on tumor-infiltrating immune cells, however, can be able to attract new systemic immune cells towards the tumor, stimulating an antitumoral immune response. Therefore, it is necessary to elucidate all these challenges when investigating oxidative stress-inducing treatment modalities as a novel anticancer strategy.

3. Conclusions

Preclinical studies have elucidated that an increase in ROS concentrations through exogenous and endogenous ROS-inducing therapies or a combination of both can be an efficient anticancer strategy. Hence, the influence of these treatments on the TME should be considered. Importantly, both the immunostimulatory as well as immunosuppressive effects have to be taken into account when investigating these anticancer modalities, because increasing ROS levels can be a double-edged sword with regards to immunomodulation and the effects cannot be generalized over different treatment modalities.

Author Contributions: Conceptualization, J.V.L., C.D., E.S.; writing—original draft preparation, J.V.L.; writing—review and editing, J.V.L., C.D., E.S. and A.B.; visualization, J.V.L.; visualization, J.V.L.; supervision, C.D., E.S., A.B., M.P., project administration, J.V.L., C.D., E.S. and M.P., funding acquisition: E.S. and M.P. All authors have read and agrees to the published version of the manuscript.

Funding: This research was funded by the Olivia Hendrickx Research Fund (21OCL06) and the University of Antwerp (FFB160231).

Conflicts of Interest: The authors declare no conflict of interest.

References

1. Perillo, B.; Di Donato, M.; Pezone, A.; Di Zazzo, E.; Giovannelli, P.; Galasso, G.; Castoria, G.; Migliaccio, A. ROS in cancer therapy: The bright side of the moon. *Exp. Mol. Med.* **2020**, *52*, 192–203. [CrossRef] [PubMed]
2. de Sa Junior, P.L.; Camara, D.A.D.; Porcacchia, A.S.; Fonseca, P.M.M.; Jorge, S.D.; Araldi, R.P.; Ferreira, A.K. The Roles of ROS in Cancer Heterogeneity and Therapy. *Oxid. Med. Cell. Longev.* **2017**, *2017*, 2467940. [CrossRef] [PubMed]
3. Kurutas, E.B. The importance of antioxidants which play the role in cellular response against oxidative/nitrosative stress: Current state. *Nutr. J.* **2016**, *15*, 71. [CrossRef] [PubMed]
4. Kumari, S.; Badana, A.K.; Malla, R. Reactive Oxygen Species: A Key Constituent in Cancer Survival. *Biomark. Insights* **2018**, *13*, 1177271918755391. [CrossRef]
5. Vaupel, P.; Harrison, L. Tumor hypoxia: Causative factors, compensatory mechanisms, and cellular response. *Oncologist* **2004**, *9* (Suppl. 5), 4–9. [CrossRef]
6. Nogueira, V.; Hay, N. Molecular pathways: Reactive oxygen species homeostasis in cancer cells and implications for cancer therapy. *Clin. Cancer Res.* **2013**, *19*, 4309–4314. [CrossRef]
7. Jaramillo, M.C.; Zhang, D.D. The emerging role of the Nrf2-Keap1 signaling pathway in cancer. *Genes Dev.* **2013**, *27*, 2179–2191. [CrossRef]
8. Galadari, S.; Rahman, A.; Pallichankandy, S.; Thayyullathil, F. Reactive oxygen species and cancer paradox: To promote or to suppress? *Free Radic. Biol. Med.* **2017**, *104*, 144–164. [CrossRef]
9. Pawlowska, E.; Szczepanska, J.; Blasiak, J. Pro- and Antioxidant Effects of Vitamin C in Cancer in correspondence to Its Dietary and Pharmacological Concentrations. *Oxid. Med. Cell. Longev.* **2019**, *2019*, 7286737. [CrossRef]
10. Singh, K.; Bhori, M.; Kasu, Y.A.; Bhat, G.; Marar, T. Antioxidants as precision weapons in war against cancer chemotherapy induced toxicity—Exploring the armoury of obscurity. *Saudi Pharm. J.* **2018**, *26*, 177–190. [CrossRef]
11. Fortmann, S.P.; Burda, B.U.; Senger, C.A.; Lin, J.S.; Whitlock, E.P. Vitamin and mineral supplements in the primary prevention of cardiovascular disease and cancer: An updated systematic evidence review for the U.S. Preventive Services Task Force. *Ann. Intern. Med.* **2013**, *159*, 824–834. [CrossRef] [PubMed]
12. Lin, J.; Cook, N.R.; Albert, C.; Zaharris, E.; Gaziano, J.M.; Van Denburgh, M.; Buring, J.E.; Manson, J.E. Vitamins C and E and beta carotene supplementation and cancer risk: A randomized controlled trial. *J. Natl. Cancer Inst.* **2009**, *101*, 14–23. [CrossRef] [PubMed]
13. Vinceti, M.; Filippini, T.; Del Giovane, C.; Dennert, G.; Zwahlen, M.; Brinkman, M.; Zeegers, M.P.; Horneber, M.; D'Amico, R.; Crespi, C.M. Selenium for preventing cancer. *Cochrane Database Syst. Rev.* **2018**, *1*, CD005195. [CrossRef]
14. Benhar, M.; Shytaj, I.L.; Stamler, J.S.; Savarino, A. Dual targeting of the thioredoxin and glutathione systems in cancer and HIV. *J. Clin. Investig.* **2016**, *126*, 1630–1639. [CrossRef] [PubMed]
15. Ozben, T. Oxidative stress and apoptosis: Impact on cancer therapy. *J. Pharm. Sci.* **2007**, *96*, 2181–2196. [CrossRef] [PubMed]
16. Kim, S.J.; Kim, H.S.; Seo, Y.R. Understanding of ROS-Inducing Strategy in Anticancer Therapy. *Oxid. Med. Cell. Longev.* **2019**, *2019*, 5381692. [CrossRef] [PubMed]
17. Zhang, B.; Wang, Y.; Su, Y. Peroxiredoxins, a novel target in cancer radiotherapy. *Cancer Lett.* **2009**, *286*, 154–160. [CrossRef]
18. Kim, W.; Lee, S.; Seo, D.; Kim, D.; Kim, K.; Kim, E.; Kang, J.; Seong, K.M.; Youn, H.; Youn, B. Cellular Stress Responses in Radiotherapy. *Cells* **2019**, *8*, 1105. [CrossRef]
19. Price, M.; Terlecky, S.R.; Kessel, D. A role for hydrogen peroxide in the pro-apoptotic effects of photodynamic therapy. *Photochem. Photobiol.* **2009**, *85*, 1491–1496. [CrossRef]
20. Moserova, I.; Truxova, I.; Garg, A.D.; Tomala, J.; Agostinis, P.; Cartron, P.F.; Vosahlikova, S.; Kovar, M.; Spisek, R.; Fucikova, J. Caspase-2 and oxidative stress underlie the immunogenic potential of high hydrostatic pressure-induced cancer cell death. *Oncoimmunology* **2017**, *6*, e1258505. [CrossRef]

21. Yamada, Y.; Takano, Y.; Satrialdi; Abe, J.; Hibino, M.; Harashima, H. Therapeutic Strategies for Regulating Mitochondrial Oxidative Stress. *Biomolecules* **2020**, *10*, 83. [CrossRef] [PubMed]

22. Turrini, E.; Laurita, R.; Stancampiano, A.; Catanzaro, E.; Calcabrini, C.; Maffei, F.; Gherardi, M.; Colombo, V.; Fimognari, C. Cold Atmospheric Plasma Induces Apoptosis and Oxidative Stress Pathway Regulation in T-Lymphoblastoid Leukemia Cells. *Oxid. Med. Cell. Longev.* **2017**, *2017*. [CrossRef] [PubMed]

23. Zhou, Z.; Song, J.; Nie, L.; Chen, X. Reactive oxygen species generating systems meeting challenges of photodynamic cancer therapy. *Chem. Soc. Rev.* **2016**, *45*, 6597–6626. [CrossRef]

24. Yan, D.; Xu, W.; Yao, X.; Lin, L.; Sherman, J.H.; Keidar, M. The Cell Activation Phenomena in the Cold Atmospheric Plasma Cancer Treatment. *Sci. Rep.* **2018**, *8*, 15418. [CrossRef] [PubMed]

25. Yang, H.; Villani, R.M.; Wang, H.; Simpson, M.J.; Roberts, M.S.; Tang, M.; Liang, X. The role of cellular reactive oxygen species in cancer chemotherapy. *J. Exp. Clin. Cancer Res.* **2018**, *37*, 266. [CrossRef] [PubMed]

26. Yen, Y.P.; Tsai, K.S.; Chen, Y.W.; Huang, C.F.; Yang, R.S.; Liu, S.H. Arsenic induces apoptosis in myoblasts through a reactive oxygen species-induced endoplasmic reticulum stress and mitochondrial dysfunction pathway. *Arch. Toxicol.* **2012**, *86*, 923–933. [CrossRef]

27. Marullo, R.; Werner, E.; Degtyareva, N.; Moore, B.; Altavilla, G.; Ramalingam, S.S.; Doetsch, P.W. Cisplatin induces a mitochondrial-ROS response that contributes to cytotoxicity depending on mitochondrial redox status and bioenergetic functions. *PLoS ONE* **2013**, *8*, e81162. [CrossRef]

28. Zhu, H.; Sarkar, S.; Scott, L.; Danelisen, I.; Trush, M.A.; Jia, Z.; Li, Y.R. Doxorubicin Redox Biology: Redox Cycling, Topoisomerase Inhibition, and Oxidative Stress. *React. Oxyg. Species* **2016**, *1*, 189–198. [CrossRef]

29. Moulder, S.; Dhillon, N.; Ng, C.; Hong, D.; Wheler, J.; Naing, A.; Tse, S.; La Paglia, A.; Dorr, R.; Hersh, E.; et al. A phase I trial of imexon, a pro-oxidant, in combination with docetaxel for the treatment of patients with advanced breast, non-small cell lung and prostate cancer. *Investig. New Drugs* **2010**, *28*, 634–640. [CrossRef]

30. Trachootham, D.; Alexandre, J.; Huang, P. Targeting cancer cells by ROS-mediated mechanisms: A radical therapeutic approach? *Nat. Rev. Drug Discov.* **2009**, *8*, 579–591. [CrossRef]

31. Chou, W.C.; Jie, C.; Kenedy, A.A.; Jones, R.J.; Trush, M.A.; Dang, C.V. Role of NADPH oxidase in arsenic-induced reactive oxygen species formation and cytotoxicity in myeloid leukemia cells. *Proc. Natl. Acad. Sci. USA* **2004**, *101*, 4578–4583. [CrossRef] [PubMed]

32. Lu, J.; Chew, E.H.; Holmgren, A. Targeting thioredoxin reductase is a basis for cancer therapy by arsenic trioxide. *Proc. Natl. Acad. Sci. USA* **2007**, *104*, 12288–12293. [CrossRef] [PubMed]

33. Desideri, E.; Ciccarone, F.; Ciriolo, M.R. Targeting Glutathione Metabolism: Partner in Crime in Anticancer Therapy. *Nutrients* **2019**, *11*, 1926. [CrossRef] [PubMed]

34. Lei, H.; Wang, G.; Zhang, J.; Han, Q. Inhibiting TrxR suppresses liver cancer by inducing apoptosis and eliciting potent antitumor immunity. *Oncol. Rep.* **2018**, *40*, 3447–3457. [CrossRef] [PubMed]

35. Sobhakumari, A.; Love-Homan, L.; Fletcher, E.V.; Martin, S.M.; Parsons, A.D.; Spitz, D.R.; Knudson, C.M.; Simons, A.L. Susceptibility of human head and neck cancer cells to combined inhibition of glutathione and thioredoxin metabolism. *PLoS ONE* **2012**, *7*, e48175. [CrossRef] [PubMed]

36. Marzano, C.; Gandin, V.; Folda, A.; Scutari, G.; Bindoli, A.; Rigobello, M.P. Inhibition of thioredoxin reductase by auranofin induces apoptosis in cisplatin-resistant human ovarian cancer cells. *Free Radic. Biol. Med.* **2007**, *42*, 872–881. [CrossRef]

37. Ramanathan, R.K.; Stephenson, J.J.; Weiss, G.J.; Pestano, L.A.; Lowe, A.; Hiscox, A.; Leos, R.A.; Martin, J.C.; Kirkpatrick, L.; Richards, D.A. A phase I trial of PX-12, a small-molecule inhibitor of thioredoxin-1, administered as a 72-hour infusion every 21 days in patients with advanced cancers refractory to standard therapy. *Investig. New Drugs* **2012**, *30*, 1591–1596. [CrossRef]

38. Baker, A.F.; Adab, K.N.; Raghunand, N.; Chow, H.; Stratton, S.P.; Squire, S.W.; Boice, M.; Pestano, L.A.; Kirkpatrick, D.L.; Dragovich, T. A phase IB trial of 24-hour intravenous PX-12, a thioredoxin-1 inhibitor, in patients with advanced gastrointestinal cancers. *Investig. New Drugs* **2013**, *31*, 631–641. [CrossRef]

39. Aggarwal, V.; Tuli, H.S.; Varol, A.; Thakral, F.; Yerer, M.B.; Sak, K.; Varol, M.; Jain, A.; Khan, M.A.; Sethi, G. Role of Reactive Oxygen Species in Cancer Progression: Molecular Mechanisms and Recent Advancements. *Biomolecules* **2019**, *9*, 735. [CrossRef]

40. Ryter, S.W.; Kim, H.P.; Hoetzel, A.; Park, J.W.; Nakahira, K.; Wang, X.; Choi, A.M. Mechanisms of cell death in oxidative stress. *Antioxid. Redox Signal.* **2007**, *9*, 49–89. [CrossRef]

41. Piette, J. Signalling pathway activation by photodynamic therapy: NF-kappa B at the crossroad between oncology and immunology. *Photochem. Photobiol. Sci.* **2015**, *14*, 1510–1517. [CrossRef] [PubMed]
42. Maier, P.; Hartmann, L.; Wenz, F.; Herskind, C. Cellular Pathways in Response to Ionizing Radiation and Their Targetability for Tumor Radiosensitization. *Int. J. Mol. Sci.* **2016**, *17*, 102. [CrossRef] [PubMed]
43. Ahmed, K.M.; Li, J.J. NF-kappa B-mediated adaptive resistance to ionizing radiation. *Free Radic. Biol. Med.* **2008**, *44*, 1–13. [CrossRef] [PubMed]
44. Reiners, J.J., Jr.; Agostinis, P.; Berg, K.; Oleinick, N.L.; Kessel, D. Assessing autophagy in the context of photodynamic therapy. *Autophagy* **2010**, *6*, 7–18. [CrossRef] [PubMed]
45. Kraus, D.; Palasuberniam, P.; Chen, B. Targeting Phosphatidylinositol 3-Kinase Signaling Pathway for Therapeutic Enhancement of Vascular-Targeted Photodynamic Therapy. *Mol. Cancer Ther.* **2017**, *16*, 2422–2431. [CrossRef] [PubMed]
46. Bundscherer, L.; Wende, K.; Ottmuller, K.; Barton, A.; Schmidt, A.; Bekeschus, S.; Hasse, S.; Weltmann, K.D.; Masur, K.; Lindequist, U. Impact of non-thermal plasma treatment on MAPK signaling pathways of human immune cell lines. *Immunobiology* **2013**, *218*, 1248–1255. [CrossRef] [PubMed]
47. Li, H.; Hu, J.; Wu, S.; Wang, L.; Cao, X.; Zhang, X.; Dai, B.; Cao, M.; Shao, R.; Zhang, R.; et al. Auranofin-mediated inhibition of PI3K/AKT/mTOR axis and anticancer activity in non-small cell lung cancer cells. *Oncotarget* **2016**, *7*, 3548–3558. [CrossRef]
48. Zheng, Z.; Fan, S.; Zheng, J.; Huang, W.; Gasparetto, C.; Chao, N.J.; Hu, J.; Kang, Y. Inhibition of thioredoxin activates mitophagy and overcomes adaptive bortezomib resistance in multiple myeloma. *J. Hematol. Oncol.* **2018**, *11*, 29. [CrossRef]
49. Duan, D.; Zhang, J.; Yao, J.; Liu, Y.; Fang, J. Targeting Thioredoxin Reductase by Parthenolide Contributes to Inducing Apoptosis of HeLa Cells. *J. Biol. Chem.* **2016**, *291*, 10021–10031. [CrossRef]
50. Hambright, H.G.; Meng, P.; Kumar, A.P.; Ghosh, R. Inhibition of PI3K/AKT/mTOR axis disrupts oxidative stress-mediated survival of melanoma cells. *Oncotarget* **2015**, *6*, 7195–7208. [CrossRef]
51. Godwin, P.; Baird, A.M.; Heavey, S.; Barr, M.P.; O'Byrne, K.J.; Gately, K. Targeting nuclear factor-kappa B to overcome resistance to chemotherapy. *Front. Oncol.* **2013**, *3*, 120. [CrossRef] [PubMed]
52. Eliseev, R.A.; Zuscik, M.J.; Schwarz, E.M.; O'Keefe, R.J.; Drissi, H.; Rosier, R.N. Increased radiation-induced apoptosis of Saos2 cells via inhibition of NFkappaB: A role for c-Jun N-terminal kinase. *J. Cell Biochem.* **2005**, *96*, 1262–1273. [CrossRef] [PubMed]
53. Telkoparan-Akillilar, P.; Suzen, S.; Saso, L. Pharmacological Applications of Nrf2 Inhibitors as Potential Antineoplastic Drugs. *Int. J. Mol. Sci.* **2019**, *20*, 2025. [CrossRef] [PubMed]
54. Wu, S.; Lu, H.; Bai, Y. Nrf2 in cancers: A double-edged sword. *Cancer Med.* **2019**, *8*, 2252–2267. [CrossRef] [PubMed]
55. Boivin, A.; Hanot, M.; Malesys, C.; Maalouf, M.; Rousson, R.; Rodriguez-Lafrasse, C.; Ardail, D. Transient alteration of cellular redox buffering before irradiation triggers apoptosis in head and neck carcinoma stem and non-stem cells. *PLoS ONE* **2011**, *6*, e14558. [CrossRef]
56. Wang, H.; Bouzakoura, S.; de Mey, S.; Jiang, H.; Law, K.; Dufait, I.; Corbet, C.; Verovski, V.; Gevaert, T.; Feron, O.; et al. Auranofin radiosensitizes tumor cells through targeting thioredoxin reductase and resulting overproduction of reactive oxygen species. *Oncotarget* **2017**, *8*, 35728–35742. [CrossRef]
57. Rodman, S.N.; Spence, J.M.; Ronnfeldt, T.J.; Zhu, Y.; Solst, S.R.; O'Neill, R.A.; Allen, B.G.; Guan, X.; Spitz, D.R.; Fath, M.A. Enhancement of Radiation Response in Breast Cancer Stem Cells by Inhibition of Thioredoxin- and Glutathione-Dependent Metabolism. *Radiat. Res.* **2016**, *186*, 385–395. [CrossRef]
58. Feng, Z.; Guo, J.; Liu, X.; Song, H.; Zhang, C.; Huang, P.; Dong, A.; Kong, D.; Wang, W. Cascade of reactive oxygen species generation by polyprodrug for combinational photodynamic therapy. *Biomaterials* **2020**, *255*, 120210. [CrossRef]
59. Lee, H.M.; Kim, D.H.; Lee, H.L.; Cha, B.; Kang, D.H.; Jeong, Y.I. Synergistic effect of buthionine sulfoximine on the chlorin e6-based photodynamic treatment of cancer cells. *Arch. Pharm. Res.* **2019**, *42*, 990–999. [CrossRef]
60. Kimani, S.G.; Phillips, J.B.; Bruce, J.I.; MacRobert, A.J.; Golding, J.P. Antioxidant inhibitors potentiate the cytotoxicity of photodynamic therapy. *Photochem. Photobiol.* **2012**, *88*, 175–187. [CrossRef]
61. Theodossiou, T.A.; Olsen, C.E.; Jonsson, M.; Kubin, A.; Hothersall, J.S.; Berg, K. The diverse roles of glutathione-associated cell resistance against hypericin photodynamic therapy. *Redox Biol.* **2017**, *12*, 191–197. [CrossRef] [PubMed]

62. Lopes-Coelho, F.; Gouveia-Fernandes, S.; Goncalves, L.G.; Nunes, C.; Faustino, I.; Silva, F.; Felix, A.; Pereira, S.A.; Serpa, J. HNF1beta drives glutathione (GSH) synthesis underlying intrinsic carboplatin resistance of ovarian clear cell carcinoma (OCCC). *Tumour Biol.* **2016**, *37*, 4813–4829. [CrossRef] [PubMed]

63. Liu, X.; Wang, W.; Yin, Y.; Li, M.; Li, H.; Xiang, H.; Xu, A.; Mei, X.; Hong, B.; Lin, W. A high-throughput drug screen identifies auranofin as a potential sensitizer of cisplatin in small cell lung cancer. *Investig. New Drugs* **2019**, *37*, 1166–1176. [CrossRef] [PubMed]

64. Smart, D.K.; Ortiz, K.L.; Mattson, D.; Bradbury, C.M.; Bisht, K.S.; Sieck, L.K.; Brechbiel, M.W.; Gius, D. Thioredoxin reductase as a potential molecular target for anticancer agents that induce oxidative stress. *Cancer Res.* **2004**, *64*, 6716–6724. [CrossRef] [PubMed]

65. Sabharwal, S.S.; Schumacker, P.T. Mitochondrial ROS in cancer: Initiators, amplifiers or an Achilles' heel? Nature reviews. *Cancer* **2014**, *14*, 709–721. [CrossRef]

66. Wang, H.; Jiang, H.; Van De Gucht, M.; De Ridder, M. Hypoxic Radioresistance: Can ROS Be the Key to Overcome It? *Cancers* **2019**, *11*, 112. [CrossRef]

67. De Ridder, M.; Verellen, D.; Verovski, V.; Storme, G. Hypoxic tumor cell radiosensitization through nitric oxide. *Nitric Oxide* **2008**, *19*, 164–169. [CrossRef]

68. Vukovic, V.; Nicklee, T.; Hedley, D.W. Differential effects of buthionine sulphoximine in hypoxic and non-hypoxic regions of human cervical carcinoma xenografts. *Radiother. Oncol.* **2001**, *60*, 69–73. [CrossRef]

69. Jiang, H.; Wang, H.; De Ridder, M. Targeting antioxidant enzymes as a radiosensitizing strategy. *Cancer Lett.* **2018**, *438*, 154–164. [CrossRef]

70. Raninga, P.V.; Di Trapani, G.; Vuckovic, S.; Tonissen, K.F. TrxR1 inhibition overcomes both hypoxia-induced and acquired bortezomib resistance in multiple myeloma through NF-small ka, Cyrillicbeta inhibition. *Cell Cycle* **2016**, *15*, 559–572. [CrossRef]

71. Nakaya, A.; Sagawa, M.; Muto, A.; Uchida, H.; Ikeda, Y.; Kizaki, M. The gold compound auranofin induces apoptosis of human multiple myeloma cells through both down-regulation of STAT3 and inhibition of NF-kappaB activity. *Leuk. Res.* **2011**, *35*, 243–249. [CrossRef]

72. Peng, L.; Linghu, R.; Chen, D.; Yang, J.; Kou, X.; Wang, X.Z.; Hu, Y.; Jiang, Y.Z.; Yang, J. Inhibition of glutathione metabolism attenuates esophageal cancer progression. *Exp. Mol. Med.* **2017**, *49*, e318. [CrossRef] [PubMed]

73. Zitvogel, L.; Apetoh, L.; Ghiringhelli, F.; Kroemer, G. Immunological aspects of cancer chemotherapy. Nature reviews. *Immunology* **2008**, *8*, 59–73. [CrossRef] [PubMed]

74. Mroz, P.; Hamblin, M.R. The immunosuppressive side of PDT. *Photochem. Photobiol. Sci.* **2011**, *10*, 751–758. [CrossRef] [PubMed]

75. Carvalho, H.A.; Villar, R.C. Radiotherapy and immune response: The systemic effects of a local treatment. *Clinics* **2018**, *73*, e557s. [CrossRef] [PubMed]

76. Adkins, I.; Fucikova, J.; Garg, A.D.; Agostinis, P.; Spisek, R. Physical modalities inducing immunogenic tumor cell death for cancer immunotherapy. *Oncoimmunology* **2014**, *3*, e968434. [CrossRef] [PubMed]

77. Zhou, J.; Wang, G.; Chen, Y.; Wang, H.; Hua, Y.; Cai, Z. Immunogenic cell death in cancer therapy: Present and emerging inducers. *J. Cell. Mol. Med.* **2019**, *23*, 4854–4865. [CrossRef]

78. Van Loenhout, J.; Flieswasser, T.; Freire Boullosa, L.; De Waele, J.; Van Audenaerde, J.; Marcq, E.; Jacobs, J.; Lin, A.; Lion, E.; Dewitte, H.; et al. Cold Atmospheric Plasma-Treated PBS Eliminates Immunosuppressive Pancreatic Stellate Cells and Induces Immunogenic Cell Death of Pancreatic Cancer Cells. *Cancers* **2019**, *11*, 1597. [CrossRef]

79. Lin, A.; Gorbanev, Y.; De Backer, J.; Van Loenhout, J.; Van Boxem, W.; Lemière, F.; Cos, P.; Dewilde, S.; Smits, E.; Bogaerts, A. Non-Thermal Plasma as a Unique Delivery System of Short-Lived Reactive Oxygen and Nitrogen Species for Immunogenic Cell Death in Melanoma Cells. *Adv. Sci.* **2019**, *6*, 1802062. [CrossRef]

80. Panzarini, E.; Inguscio, V.; Dini, L. Immunogenic cell death: Can it be exploited in PhotoDynamic Therapy for cancer? *Biomed. Res. Int.* **2013**, *2013*, 482160. [CrossRef]

81. Gebremeskel, S.; Johnston, B. Concepts and mechanisms underlying chemotherapy induced immunogenic cell death: Impact on clinical studies and considerations for combined therapies. *Oncotarget* **2015**, *6*, 41600–41619. [CrossRef]

82. Apetoh, L.; Ghiringhelli, F.; Tesniere, A.; Obeid, M.; Ortiz, C.; Criollo, A.; Mignot, G.; Maiuri, M.C.; Ullrich, E.; Saulnier, P.; et al. Toll-like receptor 4-dependent contribution of the immune system to anticancer chemotherapy and radiotherapy. *Nat. Med.* **2007**, *13*, 1050–1059. [CrossRef] [PubMed]

83. Lamberti, M.J.; Mentucci, F.M.; Roselli, E.; Araya, P.; Rivarola, V.A.; Rumie Vittar, N.B.; Maccioni, M. Photodynamic Modulation of Type 1 Interferon Pathway on Melanoma Cells Promotes Dendritic Cell Activation. *Front. Immunol.* **2019**, *10*, 2614. [CrossRef]

84. Freund, E.; Liedtke, K.R.; van der Linde, J.; Metelmann, H.R.; Heidecke, C.D.; Partecke, L.I.; Bekeschus, S. Physical plasma-treated saline promotes an immunogenic phenotype in CT26 colon cancer cells in vitro and in vivo. *Sci. Rep.* **2019**, *9*, 634. [CrossRef] [PubMed]

85. Sagwal, S.K.; Pasqual-Melo, G.; Bodnar, Y.; Gandhirajan, R.K.; Bekeschus, S. Combination of chemotherapy and physical plasma elicits melanoma cell death via upregulation of SLC22A16. *Cell Death Dis.* **2018**, *9*, 1179. [CrossRef] [PubMed]

86. Lim, J.Y.; Gerber, S.A.; Murphy, S.P.; Lord, E.M. Type I interferons induced by radiation therapy mediate recruitment and effector function of CD8(+) T cells. *Cancer Immunol. Immunother.* **2014**, *63*, 259–271. [CrossRef] [PubMed]

87. Fu, D.; Wu, J.; Lai, J.; Liu, Y.; Zhou, L.; Chen, L.; Zhang, Q. T cell recruitment triggered by optimal dose platinum compounds contributes to the therapeutic efficacy of sequential PD-1 blockade in a mouse model of colon cancer. *Am. J. Cancer Res.* **2020**, *10*, 473–490.

88. Gao, Q.; Wang, S.; Chen, X.; Cheng, S.; Zhang, Z.; Li, F.; Huang, L.; Yang, Y.; Zhou, B.; Yue, D.; et al. Cancer-cell-secreted CXCL11 promoted CD8(+) T cells infiltration through docetaxel-induced-release of HMGB1 in NSCLC. *J. Immunother. Cancer* **2019**, *7*, 42. [CrossRef]

89. Sauter, K.A.; Wood, L.J.; Wong, J.; Iordanov, M.; Magun, B.E. Doxorubicin and daunorubicin induce processing and release of interleukin-1beta through activation of the NLRP3 inflammasome. *Cancer Biol. Ther.* **2011**, *11*, 1008–1016. [CrossRef]

90. Hu, J.; Sun, C.; Bernatchez, C.; Xia, X.; Hwu, P.; Dotti, G.; Li, S. T-cell Homing Therapy for Reducing Regulatory T Cells and Preserving Effector T-cell Function in Large Solid Tumors. *Clin. Cancer Res.* **2018**, *24*, 2920–2934. [CrossRef]

91. Reits, E.A.; Hodge, J.W.; Herberts, C.A.; Groothuis, T.A.; Chakraborty, M.; Wansley, E.K.; Camphausen, K.; Luiten, R.M.; de Ru, A.H.; Neijssen, J.; et al. Radiation modulates the peptide repertoire, enhances MHC class I expression, and induces successful antitumor immunotherapy. *J. Exp. Med.* **2006**, *203*, 1259–1271. [CrossRef] [PubMed]

92. Wan, S.; Pestka, S.; Jubin, R.G.; Lyu, Y.L.; Tsai, Y.C.; Liu, L.F. Chemotherapeutics and radiation stimulate MHC class I expression through elevated interferon-beta signaling in breast cancer cells. *PLoS ONE* **2012**, *7*, e32542. [CrossRef] [PubMed]

93. Lugade, A.A.; Sorensen, E.W.; Gerber, S.A.; Moran, J.P.; Frelinger, J.G.; Lord, E.M. Radiation-induced IFN-gamma production within the tumor microenvironment influences antitumor immunity. *J. Immunol.* **2008**, *180*, 3132–3139. [CrossRef] [PubMed]

94. Zhang, S.Y.; Li, J.L.; Xu, X.K.; Zheng, M.G.; Wen, C.C.; Li, F.C. HMME-based PDT restores expression and function of transporter associated with antigen processing 1 (TAP1) and surface presentation of MHC class I antigen in human glioma. *J. Neuro-Oncol.* **2011**, *105*, 199–210. [CrossRef]

95. Bekeschus, S.; Rodder, K.; Fregin, B.; Otto, O.; Lippert, M.; Weltmann, K.D.; Wende, K.; Schmidt, A.; Gandhirajan, R.K. Toxicity and Immunogenicity in Murine Melanoma following Exposure to Physical Plasma-Derived Oxidants. *Oxid. Med. Cell. Longev.* **2017**, *2017*, 4396467. [CrossRef]

96. Belicha-Villanueva, A.; Riddell, J.; Bangia, N.; Gollnick, S.O. The effect of photodynamic therapy on tumor cell expression of major histocompatibility complex (MHC) class I and MHC class I-related molecules. *Lasers Surg. Med.* **2012**, *44*, 60–68. [CrossRef]

97. Gasser, S.; Orsulic, S.; Brown, E.J.; Raulet, D.H. The DNA damage pathway regulates innate immune system ligands of the NKG2D receptor. *Nature* **2005**, *436*, 1186–1190. [CrossRef]

98. Garnett, C.T.; Palena, C.; Chakraborty, M.; Tsang, K.Y.; Schlom, J.; Hodge, J.W. Sublethal irradiation of human tumor cells modulates phenotype resulting in enhanced killing by cytotoxic T lymphocytes. *Cancer Res.* **2004**, *64*, 7985–7994. [CrossRef]

99. Friesen, C.; Kiess, Y.; Debatin, K.M. A critical role of glutathione in determining apoptosis sensitivity and resistance in leukemia cells. *Cell Death Differ.* **2004**, *11* (Suppl. 1), S73–S85. [CrossRef]

100. Zitvogel, L.; Apetoh, L.; Ghiringhelli, F.; Andre, F.; Tesniere, A.; Kroemer, G. The anticancer immune response: Indispensable for therapeutic success? *J. Clin. Investig.* **2008**, *118*, 1991–2001. [CrossRef]

101. Bailly, C. Regulation of PD-L1 expression on cancer cells with ROS-modulating drugs. *Life Sci.* **2020**, *246*, 117403. [CrossRef] [PubMed]

102. Showalter, A.; Limaye, A.; Oyer, J.L.; Igarashi, R.; Kittipatarin, C.; Copik, A.J.; Khaled, A.R. Cytokines in immunogenic cell death: Applications for cancer immunotherapy. *Cytokine* **2017**, *97*, 123–132. [CrossRef] [PubMed]

103. Wang, X.; Dong, H.; Li, Q.; Li, Y.; Hong, A. Thioredoxin induces Tregs to generate an immunotolerant tumor microenvironment in metastatic melanoma. *Oncoimmunology* **2015**, *4*, e1027471. [CrossRef] [PubMed]

104. Thomas-Schoemann, A.; Batteux, F.; Mongaret, C.; Nicco, C.; Chereau, C.; Annereau, M.; Dauphin, A.; Goldwasser, F.; Weill, B.; Lemare, F.; et al. Arsenic trioxide exerts antitumor activity through regulatory T cell depletion mediated by oxidative stress in a murine model of colon cancer. *J. Immunol.* **2012**, *189*, 5171–5177. [CrossRef] [PubMed]

105. Napolitano, M.; D'Alterio, C.; Cardone, E.; Trotta, A.M.; Pecori, B.; Rega, D.; Pace, U.; Scala, D.; Scognamiglio, G.; Tatangelo, F.; et al. Peripheral myeloid-derived suppressor and T regulatory PD-1 positive cells predict response to neoadjuvant short-course radiotherapy in rectal cancer patients. *Oncotarget* **2015**, *6*, 8261–8270. [CrossRef]

106. Liao, C.; Xiao, W.; Zhu, N.; Liu, Z.; Yang, J.; Wang, Y.; Hong, M. Radiotherapy suppressed tumor-specific recruitment of regulator T cells via up-regulating microR-545 in Lewis lung carcinoma cells. *Int. J. Clin. Exp. Pathol.* **2015**, *8*, 2535–2544.

107. Bryniarski, K.; Szczepanik, M.; Ptak, M.; Zemelka, M.; Ptak, W. Influence of cyclophosphamide and its metabolic products on the activity of peritoneal macrophages in mice. *Pharmacol. Rep.* **2009**, *61*, 550–557. [CrossRef]

108. Klug, F.; Prakash, H.; Huber, P.E.; Seibel, T.; Bender, N.; Halama, N.; Pfirschke, C.; Voss, R.H.; Timke, C.; Umansky, L.; et al. Low-dose irradiation programs macrophage differentiation to an iNOS(+)/M1 phenotype that orchestrates effective T cell immunotherapy. *Cancer Cell* **2013**, *24*, 589–602. [CrossRef]

109. Pal, R.; Chakraborty, B.; Nath, A.; Singh, L.M.; Ali, M.; Rahman, D.S.; Ghosh, S.K.; Basu, A.; Bhattacharya, S.; Baral, R.; et al. Noble metal nanoparticle-induced oxidative stress modulates tumor associated macrophages (TAMs) from an M2 to M1 phenotype: An in vitro approach. *Int. Immunopharmacol.* **2016**, *38*, 332–341. [CrossRef]

110. Bianchi, G.; Vuerich, M.; Pellegatti, P.; Marimpietri, D.; Emionite, L.; Marigo, I.; Bronte, V.; Di Virgilio, F.; Pistoia, V.; Raffaghello, L. ATP/P2X7 axis modulates myeloid-derived suppressor cell functions in neuroblastoma microenvironment. *Cell Death Dis.* **2014**, *5*, e1135. [CrossRef]

111. Giordano, F.A.; Link, B.; Glas, M.; Herrlinger, U.; Wenz, F.; Umansky, V.; Brown, J.M.; Herskind, C. Targeting the Post-Irradiation Tumor Microenvironment in Glioblastoma via Inhibition of CXCL12. *Cancers* **2019**, *11*, 272. [CrossRef] [PubMed]

112. Eckert, F.; Schilbach, K.; Klumpp, L.; Bardoscia, L.; Sezgin, E.C.; Schwab, M.; Zips, D.; Huber, S.M. Potential Role of CXCR4 Targeting in the Context of Radiotherapy and Immunotherapy of Cancer. *Front. Immunol.* **2018**, *9*, 3018. [CrossRef] [PubMed]

113. Raninga, P.V.; Lee, A.C.; Sinha, D.; Shih, Y.Y.; Mittal, D.; Makhale, A.; Bain, A.L.; Nanayakarra, D.; Tonissen, K.F.; Kalimutho, M.; et al. Therapeutic cooperation between auranofin, a thioredoxin reductase inhibitor and anti-PD-L1 antibody for treatment of triple-negative breast cancer. *Int. J. Cancer* **2020**, *146*, 123–136. [CrossRef] [PubMed]

114. Gong, X.; Li, X.; Jiang, T.; Xie, H.; Zhu, Z.; Zhou, F.; Zhou, C. Combined Radiotherapy and Anti-PD-L1 Antibody Synergistically Enhances Antitumor Effect in Non-Small Cell Lung Cancer. *J. Thorac. Oncol.* **2017**, *12*, 1085–1097. [CrossRef] [PubMed]

115. Azad, A.; Yin Lim, S.; D'Costa, Z.; Jones, K.; Diana, A.; Sansom, O.J.; Kruger, P.; Liu, S.; McKenna, W.G.; Dushek, O.; et al. PD-L1 blockade enhances response of pancreatic ductal adenocarcinoma to radiotherapy. *EMBO Mol. Med.* **2017**, *9*, 167–180. [CrossRef] [PubMed]

116. Kachikwu, E.L.; Iwamoto, K.S.; Liao, Y.P.; DeMarco, J.J.; Agazaryan, N.; Economou, J.S.; McBride, W.H.; Schaue, D. Radiation enhances regulatory T cell representation. *Int. J. Radiat. Oncol. Biol. Phys.* **2011**, *81*, 1128–1135. [CrossRef] [PubMed]

117. Maeding, N.; Verwanger, T.; Krammer, B. Boosting Tumor-Specific Immunity Using PDT. *Cancers* **2016**, *8*, 91. [CrossRef]

118. Tsai, C.S.; Chen, F.H.; Wang, C.C.; Huang, H.L.; Jung, S.M.; Wu, C.J.; Lee, C.C.; McBride, W.H.; Chiang, C.S.; Hong, J.H. Macrophages from irradiated tumors express higher levels of iNOS, arginase-I and COX-2, and promote tumor growth. *Int. J. Radiat. Oncol. Biol. Phys.* **2007**, *68*, 499–507. [CrossRef]

119. Grossman, S.A.; Ellsworth, S.; Campian, J.; Wild, A.T.; Herman, J.M.; Laheru, D.; Brock, M.; Balmanoukian, A.; Ye, X. Survival in Patients With Severe Lymphopenia Following Treatment With Radiation and Chemotherapy for Newly Diagnosed Solid Tumors. *J. Natl. Compr. Cancer Netw.* **2015**, *13*, 1225–1231. [CrossRef]

120. Hunt, D.W.; Jiang, H.; Granville, D.J.; Chan, A.H.; Leong, S.; Levy, J.G. Consequences of the photodynamic treatment of resting and activated peripheral T lymphocytes. *Immunopharmacology* **1999**, *41*, 31–44. [CrossRef]

121. Bekeschus, S.; Kolata, J.; Winterbourn, C.; Kramer, A.; Turner, R.; Weltmann, K.D.; Broker, B.; Masur, K. Hydrogen peroxide: A central player in physical plasma-induced oxidative stress in human blood cells. *Free Radic. Res.* **2014**, *48*, 542–549. [CrossRef] [PubMed]

122. Gupta, S.; Yel, L.; Kim, D.; Kim, C.; Chiplunkar, S.; Gollapudi, S. Arsenic trioxide induces apoptosis in peripheral blood T lymphocyte subsets by inducing oxidative stress: A role of Bcl-2. *Mol. Cancer Ther.* **2003**, *2*, 711–719. [PubMed]

123. Kim, H.J.; Barajas, B.; Chan, R.C.; Nel, A.E. Glutathione depletion inhibits dendritic cell maturation and delayed-type hypersensitivity: Implications for systemic disease and immunosenescence. *J. Allergy Clin. Immunol.* **2007**, *119*, 1225–1233. [CrossRef] [PubMed]

124. Galluzzi, L.; Vitale, I.; Warren, S.; Adjemian, S.; Agostinis, P.; Martinez, A.B.; Chan, T.A.; Coukos, G.; Demaria, S.; Deutsch, E.; et al. Consensus guidelines for the definition, detection and interpretation of immunogenic cell death. *J. Immunother. Cancer* **2020**, *8*, e000337. [CrossRef] [PubMed]

125. Garg, A.D.; Dudek, A.M.; Ferreira, G.B.; Verfaillie, T.; Vandenabeele, P.; Krysko, D.V.; Mathieu, C.; Agostinis, P. ROS-induced autophagy in cancer cells assists in evasion from determinants of immunogenic cell death. *Autophagy* **2013**, *9*, 1292–1307. [CrossRef] [PubMed]

126. Melero, I.; Rouzaut, A.; Motz, G.T.; Coukos, G. T-cell and NK-cell infiltration into solid tumors: A key limiting factor for efficacious cancer immunotherapy. *Cancer Discov.* **2014**, *4*, 522–526. [CrossRef]

127. Wang, X.; Li, J.; Dong, K.; Lin, F.; Long, M.; Ouyang, Y.; Wei, J.; Chen, X.; Weng, Y.; He, T.; et al. Tumor suppressor miR-34a targets PD-L1 and functions as a potential immunotherapeutic target in acute myeloid leukemia. *Cell. Signal.* **2015**, *27*, 443–452. [CrossRef]

128. Garcia-Diaz, A.; Shin, D.S.; Moreno, B.H.; Saco, J.; Escuin-Ordinas, H.; Rodriguez, G.A.; Zaretsky, J.M.; Sun, L.; Hugo, W.; Wang, X.; et al. Interferon Receptor Signaling Pathways Regulating PD-L1 and PD-L2 Expression. *Cell Rep.* **2017**, *19*, 1189–1201. [CrossRef]

129. Broekgaarden, M.; Weijer, R.; van Gulik, T.M.; Hamblin, M.R.; Heger, M. Tumor cell survival pathways activated by photodynamic therapy: A molecular basis for pharmacological inhibition strategies. *Cancer Metastasis Rev.* **2015**, *34*, 643–690. [CrossRef]

130. Vasievich, E.A.; Huang, L. The suppressive tumor microenvironment: A challenge in cancer immunotherapy. *Mol. Pharm.* **2011**, *8*, 635–641. [CrossRef]

131. Mougiakakos, D.; Johansson, C.C.; Kiessling, R. Naturally occurring regulatory T cells show reduced sensitivity toward oxidative stress-induced cell death. *Blood* **2009**, *113*, 3542–3545. [CrossRef] [PubMed]

132. Mougiakakos, D.; Johansson, C.C.; Jitschin, R.; Bottcher, M.; Kiessling, R. Increased thioredoxin-1 production in human naturally occurring regulatory T cells confers enhanced tolerance to oxidative stress. *Blood* **2011**, *117*, 857–861. [CrossRef] [PubMed]

133. Maj, T.; Wang, W.; Crespo, J.; Zhang, H.; Wang, W.; Wei, S.; Zhao, L.; Vatan, L.; Shao, I.; Szeliga, W.; et al. Oxidative stress controls regulatory T cell apoptosis and suppressor activity and PD-L1-blockade resistance in tumor. *Nat. Immunol.* **2017**, *18*, 1332–1341. [CrossRef] [PubMed]

134. Griess, B.; Mir, S.; Datta, K.; Teoh-Fitzgerald, M. Scavenging reactive oxygen species selectively inhibits M2 macrophage polarization and their pro-tumorigenic function in part, via Stat3 suppression. *Free Radic. Biol. Med.* **2020**, *147*, 48–60. [CrossRef]

135. Leblond, M.M.; Peres, E.A.; Helaine, C.; Gerault, A.N.; Moulin, D.; Anfray, C.; Divoux, D.; Petit, E.; Bernaudin, M.; Valable, S. M2 macrophages are more resistant than M1 macrophages following radiation therapy in the context of glioblastoma. *Oncotarget* **2017**, *8*, 72597–72612. [CrossRef]

136. Wu, J.; Waxman, D.J. Immunogenic chemotherapy: Dose and schedule dependence and combination with immunotherapy. *Cancer Lett.* **2018**, *419*, 210–221. [CrossRef]

137. Schaue, D.; McBride, W.H. T lymphocytes and normal tissue responses to radiation. *Front. Oncol.* **2012**, *2*, 119. [CrossRef]

138. Takahashi, A.; Hanson, M.G.V.; Norell, H.R.; Havelka, A.M.; Kono, K.; Malmberg, K.J.; Kiessling, R.V.R. Preferential cell death of CD8(+) effector memory (CCR7(-)CD45RA(-)) T cells by hydrogen peroxide-induced oxidative stress. *J. Immunol.* **2005**, *174*, 6080–6087. [CrossRef]

139. Mak, T.W.; Grusdat, M.; Duncan, G.S.; Dostert, C.; Nonnenmacher, Y.; Cox, M.; Binsfeld, C.; Hao, Z.; Brustle, A.; Itsumi, M.; et al. Glutathione Primes T Cell Metabolism for Inflammation. *Immunity* **2017**, *46*, 1089–1090. [CrossRef]

140. Mimura, K.; Kua, L.F.; Shimasaki, N.; Shiraishi, K.; Nakajima, S.; Siang, L.K.; Shabbir, A.; So, J.; Yong, W.P.; Kono, K. Upregulation of thioredoxin-1 in activated human NK cells confers increased tolerance to oxidative stress. *Cancer Immunol. Immunother.* **2017**, *66*, 605–613. [CrossRef]

141. Falcke, S.E.; Ruhle, P.F.; Deloch, L.; Fietkau, R.; Frey, B.; Gaipl, U.S. Clinically Relevant Radiation Exposure Differentially Impacts Forms of Cell Death in Human Cells of the Innate and Adaptive Immune System. *Int. J. Mol. Sci.* **2018**, *19*, 3574. [CrossRef] [PubMed]

142. Seres, T.; Knickelbein, R.G.; Warshaw, J.B.; Johnston, R.B., Jr. The phagocytosis-associated respiratory burst in human monocytes is associated with increased uptake of glutathione. *J. Immunol.* **2000**, *165*, 3333–3340. [CrossRef] [PubMed]

Publisher's Note: MDPI stays neutral with regard to jurisdictional claims in published maps and institutional affiliations.

 *antioxidants*

Review

Radioprotective Agents and Enhancers Factors. Preventive and Therapeutic Strategies for Oxidative Induced Radiotherapy Damages in Hematological Malignancies

Andrea Gaetano Allegra [1,†], Federica Mannino [2,†], Vanessa Innao [3], Caterina Musolino [3] and Alessandro Allegra [3,*]

[1] Radiation Oncology Unit, Department of Biomedical, Experimental, and Clinical Sciences "Mario Serio", Azienda Ospedaliero-Universitaria Careggi, University of Florence, 50100 Florence, Italy; andrea.allegra@hotmail.it

[2] Department of Clinical and Experimental Medicine, University of Messina, c/o AOU Policlinico G. Martino, Via C. Valeria Gazzi, 98125 Messina, Italy; fmannino@unime.it

[3] Department of Human Pathology in Adulthood and Childhood "Gaetano Barresi", Division of Haematology, University of Messina, 98125 Messina, Italy; vinnao@unime.it (V.I.); cmusolino@unime.it (C.M.)

* Correspondence: aallegra@unime.it; Tel.: +39-090-221-2364

† These authors contributed equally.

Received: 15 October 2020; Accepted: 10 November 2020; Published: 12 November 2020

Abstract: Radiation therapy plays a critical role in the management of a wide range of hematologic malignancies. It is well known that the post-irradiation damages both in the bone marrow and in other organs are the main causes of post-irradiation morbidity and mortality. Tumor control without producing extensive damage to the surrounding normal cells, through the use of radioprotectors, is of special clinical relevance in radiotherapy. An increasing amount of data is helping to clarify the role of oxidative stress in toxicity and therapy response. Radioprotective agents are substances that moderate the oxidative effects of radiation on healthy normal tissues while preserving the sensitivity to radiation damage in tumor cells. As well as the substances capable of carrying out a protective action against the oxidative damage caused by radiotherapy, other substances have been identified as possible enhancers of the radiotherapy and cytotoxic activity via an oxidative effect. The purpose of this review was to examine the data in the literature on the possible use of old and new substances to increase the efficacy of radiation treatment in hematological diseases and to reduce the harmful effects of the treatment.

Keywords: radiotherapy; hematological malignancies; oxidative stress; lymphoma; leukemia; multiple myeloma; apoptosis; mitochondria

---

1. Introduction

Radiotherapy in Hematologic Malignancies

Radiation therapy (RT) has an essential role in the treatment of a wide range of hematologic malignancies. The ideal radiation dosage and target volume, and efficacious system of combining radiation with systemic drugs, differ in relation to the histologic varieties, stage, and patient performance status. Reducing dosages to adjacent organs without losing disease control is of leading importance as it can increase the therapeutic ratio of subjects getting RT for hematologic malignancies [1].

Lymphomas are very radiosensitive tumors and RT is the principal treatment approach, and it is the most efficacious single therapy for local control of lymphomas. Nevertheless, curative objective is only imaginable if all lymphoma tissue can be included in the zone to be irradiated with the recommended total irradiation dosage. For this reason, RT is a single mode only in early stages of Hodgkin's lymphoma and low-grade non-Hodgkin's lymphoma. However, in most subjects, RT can be employed as consolidation treatment after chemotherapy. In the last few years, irradiation treatment of lymphomas has been improved in order to save essential tissues and decrease toxicity. However, although efficacious, RT is a disregarded type of therapy because of the concern of side effects after irradiation [2].

RT is also an efficacious treatment for multiple myeloma (MM) [3]. External beam irradiation is habitually employed for palliation of bone pain and treatment of solitary plasmacytomas [4]. In MM subjects who have diffused bone disease, systemic RT with compounds such as bone-seeking radionuclides [5,6] or treatments such as intensity modulated RT and helical tomography [7,8] have been employed. For instance, bone-seeking radionuclide treatment with radioactive samarium conjugated to a tetraphosphonate chelator has been utilized in myeloablative protocols for MM [9–12]. Moreover, the combinate use of RT with chemotherapeutic drugs such as bortezomib or thalidomide [13–15] has demonstrated good clinical effects with decreased radiotoxicity to normal tissues.

Finally, as far leukemic patients, the most frequent situation in which RT therapy is employed comprises therapy of central nervous system (CNS) disease and other types of extramedullary relapses. Either whole-brain radiation treatment or cranial spinal irradiation may be evaluated in subjects with leukemia and history of CNS disease as part of therapy before allogeneic stem cell transplant (SCT), consolidation after salvage chemo treatment, or salvage therapy. Myeloid sarcoma, also recognized as granulocytic sarcoma, and leukemia cutis are extramedullary malignancies containing myeloid blasts, in most patients arising concomitantly with acute myeloid leukemia, and they are extremely sensitive to RT [16].

2. Radiotherapy and Oxidative Stress

An expanding quantity of results is aiding to explain the effects of oxidative stress in malignancies both in the onset of the disease and in response of the treatment. In several tumoral situations, reactive oxygen species (ROS) generation and removal are out of control, leading to disproportionate quantities of ROS [17]. Antioxidants are essential in the removal of ROS, so preserving the normal physiological condition is vital [18].

An oxidatively balanced redox status may influence carcinogenesis by modifying DNA and different cellular components. On the other hand, oxidative stress increases the activities of several cytostatic drugs by provoking sublethal DNA injury and so stimulates programmed cell death. Moreover, oxidative stress markers are prognostically relevant in solid cancers [19–21], but also in multiple myeloma, lymphoma, and acute and chronic leukemia [21–27].

On the other hand, it is well known that, at molecular level, the cytotoxic actions of ionizing radiation (IR) on tumor and normal cells is essentially due to ROS and reactive nitrogen species (RNS) generation such as the hydrogen peroxide, superoxide anion, hydroxyl radical, and nitrogen dioxide via the radiolysis of cellular water [28]. If not scavenged, these ROS/RNS have a main action in the cell injury by provoking lipid peroxidation, DNA strand breaks, and protein modifications. ROS/RNS also induct the onset of different biochemical and molecular signaling effects that may either fix the injury or provoke cell death, such as apoptosis [29,30].

Among the several compounds that participate in regulating the process, the proto-oncogene bcl-2 family plays a main role. B-cell lymphoma 2 (Bcl-2) is an intracellular protein that locates to endoplasmic reticulum, mitochondria, and the nuclear envelope. It blocks programmed cell death and augments cell survival under different situations in various cell types [31,32]. The expression of Bcl-2 is intensely connected with the expression of several antioxidant enzymes and the strong relationship controlling the intracellular redox concentration has an essential action in cell destiny [33].

However, after exposure of cells to RT, the stimulation of caspases is moderately faint, and then the signal is directed via the mitochondria for efficacious death execution (type 2 or intrinsic signaling).

In fact, damage of cellular DNA after RT is a dose-dependent event and happens in both the nuclear (nDNA) and extra-nuclear DNA. Thus, besides nDNA, mitochondrial DNA (mtDNA) is similarly altered [34,35]. Several experimentations reported that mtDNA can be a target for endogenous ROS and free radicals generated by IR [36,37]. The modalities of cellular response to RT with regard to mtDNA modifications are essentially two different mechanisms. Firstly, mtDNA has limited repair systems and mitochondrial function is maintained principally due to its high copy number. One potential radio-protective system is that increased replication of mtDNA decreases the mutation occurrence of total mtDNA and reduces the onset of lethal radiation injury to the mitochondria [38,39]. This hypothesis has been supported by Zhang et al. by demonstrating augmented mtDNA copy number in bone marrow (BM) of total body irradiated animals [40].

As for the second mechanism, IR generally stimulates cell apoptosis by provoking a gathering of large scale mtDNA deletions, particularly the specific 4977 bp deletion, denoted to as the "common deletion (CD)" [41]. The site of CD is bordered by two 13 bp direct repeats (ACCTCCCTCACCA) at mtDNA nucleotide site 8470 and 13,447 [42]. Findings have demonstrated that CD can be a precise indicator of oxidative damage to mtDNA [43–45]. Wen et al. studied mtDNA modifications in irradiated human peripheral lymphocytes from acute lymphoblastic leukemia (ALL) subjects [46]. Significant inverse correlation was demonstrated between CD level and mtDNA content. CD content and mtDNA may be believed as prognostic factors to radiation toxicity [46].

However, it is well known that the post-irradiation injuries in the BM and diverse organs represent a main problem and are the principal reasons of post-irradiation mortality and morbidity [47].

The purpose of this review was to examine the data in the literature on the possibility of using old and new substances to increase the efficacy of radiation treatment in hematological diseases and to reduce the harmful effects of the treatment.

3. Radioprotective Agents

Cancer management without causing wide injury to the adjacent normal cells, via the employment of radioprotectors, is of exceptional clinical importance in RT. Radioprotective factors are compounds that temper the consequences of radiation on normal tissues, maintaining the sensitivity to radiation injury in tumor cells [29]. However, no ideal radioprotector has been obtainable so far that would be non-toxic at its best, at an efficacious dosage, and that could defend normal cells against radiation damage while maintaining the radiosensitivity of tumor cells.

A fragile equilibrium exists between oxidant and antioxidant mechanisms under physiological situations, but it is compromised in pathologic conditions. Oxidative stress activation during several diseases may reduce the antioxidant ability of cells, and vulnerability of target elements to oxidative stress increases as a result. Enzymatic and nonenzymatic antioxidants avoid surplus ROS generation and counteract the ROS. Substances such as vitamin A, vitamin C, vitamin E, catechin, epicatechin, carotenoids, and in vivo low-molecular weight antioxidants (melatonin, lipoic acid, uric acid, haptaglobuline, bilirubin, and melatonin) represent the total antioxidant capacity of plasma [48].

Among the great number of natural and synthetic radioprotective substances, the phytochemical compounds, comprising phenolics (simple phenols, benzoic acid derivatives, flavonoids, stilbenes, phenylpropanoids, tannins, lignins, and lignans) have provided encouraging effects because of their capacity to scavenge free radicals and reduce toxicity [49] (Table 1). However, the possible use of substances, such as genistein and quercetin, has been impeded due to the inadequate water-solubility, and therefore their reduced bioavailability [50,51].

Table 1. Radioprotective agents able to reduce the harmful effects of the treatment.

Agent	Mechanisms of Action	Limits	Advantages	Ref.
Phenolic (flavonoids, stilbenes, tannins, lignans, lignins, quercetin, genistein)	Inhibition of apoptosis	Limited water solubility and poor availability		[49–51]
Polyphenolic—polysaccharide conjugates	Inhibition of apoptosis		Non-toxic, water-soluble polymeric compounds. They did not protect the leukemic cells against radiation-induced apoptotic death.	[52,53]
Cimetidine	Hydroxyl radical scavenging mechanism. Decrease of inflammation, and Bax/Bcl2 ratio.		Reduction of loss of bone marrow cell count, intestinal lining destruction, and fibrosis. Differential effect on both cancer cells and adjacent healthy cells.	[54,55]
Amifostine (WR2721)	Organ repair via bone marrow recruitment or dedifferentiation.	Intolerance and significant accumulative toxicity.		[56]
STW5 (Iberogast)	Anti-apoptotic effects.	Antioxidant activity and anti-inflammatory properties	Action on radiation enteritis. Preservation of the mucosal integrity of the small intestine.	[57–60]
CBLB502	Increased expression of the strong natural antioxidant superoxide dismutase and induction of radioprotective cytokines (G-CSF, IL-6, and TNF-α). Suppression of p53-dependent apoptosis. Reduction of DNA damage and chromosomal aberrations. Action on the TLR5 signaling pathway.		It protects mammals from gastrointestinal and hematopoietic acute radiation syndrome. Reduction of IR-induced oxidative stress, reduction of decline of sperm quantity and quality.	[61–65] [66,67]
Erdosteine	Protective role on the release of free oxygen radicals. Action on TNF alpha, interleukin 1, and IL-6.		Protection against radiation induced inflammatory kidney damage.	[68]
Human umbilical cord-derived mesenchymal stromal cells	Prevention of oxidative stress and increased antioxidant status. Reduction of pro-fibrotic TGF-β1, IL-6, and IL-8 levels.		Protective effects on irradiation myocardial fibrosis with increased cell viability, reduction of collagen deposition.	[69]
Calf spleen extractive injection	Regenerating action on damaged cells.		Reduction of thrombocytopenia and leucopenia.	[70–72]
Polyxydroxylated fullerenes	Anti-oxidative effects.		Prevention of radiation-induced reduction in the white cell count.	[73,74]
Platelet-rich plasma	Administration of growth factors. Reduction of oxidative stress and inhibition of the induced apoptosis.		Neuroprotection.	[75,76]

The polyphenolic-polysaccharide conjugates (PPCs) extracted from the plants of *Rosaceae* family (*Rubus plicatus*, *Sanguisorba officinalis* L., and *Fragaria vesca* L.) as well as of *Asteraceae* family are compounds for distinguishing the secondary cell wall of higher plants (Figure 1). The PPCs are water-soluble, non-toxic polymeric substances [52]. It was reported that the PPCs were capable of augmenting the post-radiation survival of peripheral blood mononuclear cells (PBMC)s by blocking programmed cell death, while they did not defend the leukemic cells against radiation-induced programmed cell death. The PPCs guarantee an efficacious defense of PBMCs via scavenging of intracellular ROS and reducing DNA injury, while they offered no decrease of the oxidative stress and DNA injury in K562 cells (a human immortalized myelogenous leukemia cell line). These results intensely propose that the PPCs, principally those extracted from *Elodea canadensis* and *S. officinalis*, can selectively defend normal cells, thus they have the benchmarks of radioprotectors for possible employment in RT [53].

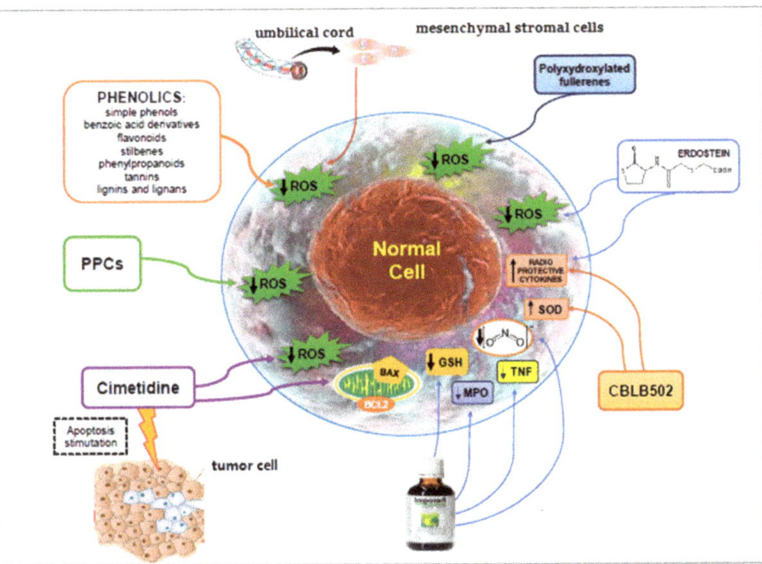

Figure 1. Natural and synthetic radioprotective substances.

A diverse possible radioprotective drug could be Cimetidine, a histamine type II receptors blocker, which has been demonstrated to have an antisuppressor action. Cimetidine has also been employed efficaciously to reestablish immune activities in subjects with malignancies, hypogammaglobulinemia, and acquired immune deficiency syndrome-related complexes [77]. Some in vitro and in vivo experiments evaluated the radioprotective activity of cimetidine on lympho-hematopoietic cells. The mechanism by which cimetidine decreases the mutagenic action of radiation is not well-known, even if a hydroxyl radical scavenging mechanism has been proposed [54]. Estaphan et al. studied the protective action of cimetidine in animals exposed to γ-irradiation and evaluated the Bcl2/Bcl2 associated X (bax) pathway as a possible fundamental mechanism [55]. Cimetidine pretreatment considerably reduced the fibrosis, the loss of BM cell count, and the intestinal lining destruction reported in the untreated irradiated animals, and drastically reduced the oxidative stress, inflammation, and Bax/Bcl2 ratio. An important relationship between Bax/Bcl2 ratio, tissue oxidative stress level, and tissue injury was demonstrated. Cimetidine could be a very encouraging radioprotective drug with a possible differential beneficial action on both cancer cells (stimulating programmed cell death) and contiguous normal cells (provoking radioprotection through blocking programmed cell death) via its antioxidant action and consequent control of type 2 apoptotic pathway [55].

However, Waller Reed Army Research Institute produced and examined over 4000 compounds that could have a radioprotective activity. WR2721 (Amifostine) was the most efficacious substance, although its use is restrained due to its accumulative toxicity [56]. So, there is still a persistent necessity to find new, efficacious, and non-toxic radio-protective substances.

Some compounds seem to be capable to exercise a protective effect on specific consequences of the RT, being able to decrease the damaging actions effects on particular organs or systems such as the intestinal mucosa, the male reproductive system, kidney, heart, or bone marrow.

Organ or system healing can be accomplished ex vivo, by administration of embryonic or adult stem cells, or in situ, by dispensing elements that will stimulate or improve the healing mechanisms of the organ itself, and several efforts have been made about the elaboration of a system for efficacious growth factors release in vivo to promote cellular repair and tissue renewal [78].

As far intestinal mucosa, radiation enteritis is a grave complication caused by the cytotoxic action of RT on the intestinal mucosa. The genesis of this situation implicates the delivery of pro-inflammatory cytokines and ROS and augments programmed cell death in the intestinal epithelium, provoking an alteration of intestinal structure and function [79,80]. Research evaluated the possible radio-protective and anti-apoptotic actions of STW5 (Iberogast®,) an herbal preparation, inclosing extracts of *Angelica archangelica, Silybum marianum, Melissa officinalis, Iberis amara, Matricaria recutita, Carum carvi, Mentha piperita, Chelidonium majus,* and *Glycyrrhiza glabra* [57].

Exposure to radiation caused apoptotic modifications associated with an augmentation in cytosolic calcium, reduction of complex I, mitochondrial cytochrome c, and B-cell lymphoma-2. Oxidative stress parameters were disturbed, while inflammation markers such as tumor necrosis factor and indicators of intestinal damage such as serum diamine oxidase were augmented. STW 5 defended against histological modifications and stabilized the deranged parameters [58].

The powerful antioxidant action of STW5 was reported in numerous other experimental situations [59] and was demonstrated by the reduction of glutathione(GSH) levels and lessening of the increased thiobarbituric acid reactive substances and nitrite concentrations in the intestinal tissue [60]. STW5 avoided the increase of myeloperoxidase and TNF levels in intestinal tissue caused by RT as well as the increase of serum level of diamine oxidase. The antioxidant and anti-inflammatory actions of STW5 have been ascribed, at least in part, to the flavonoid and phenolic carboxylic acid content of its components [57].

A different interesting compound is CBLB502, a polypeptide originated from *Salmonella* flagellin. This compound connects to the Toll-like receptor 5 (TLR5) and stimulates nuclear factor-kappa B (NFκB), a main controller of programmed cell death, inflammation, and immune response. A single administration of CBLB502 before lethal IR doses was demonstrated to defend animals from gastrointestinal and hematopoietic acute radiation syndrome, but not tumor expansion [61]. CBLB502 administration induces the expression of the powerful natural antioxidant superoxide dismutase (SOD)2 and stimulates the expression of numerous radioprotective cytokines, such as IL-6, G-CSF, and TNF-α [61]. CBLB502 may theoretically increase the therapeutic index of RT and act as a new substance stopping irradiation injury [62]. Moreover, CBLB502 displays immunotherapeutic ability by stimulating TLR5-expressing accessory immune cells [63–65].

IR can also cause damage to the male reproductive system in subjects experiencing RT [81,82]. The testis, one of the most radiosensitive organs, can be altered by low IR dosages as of the existence of replicating spermatogonial cells [83]. Abdominal irradiation during RT may cause the accrual of harmful radiation concentrations in the testes. Previous studies demonstrated that IR can alter physiologic spermatogenic metabolism, growth, and differentiation, provoking low sperm counts, sterility, and sexual alterations [84–86].

Numerous sorts of antioxidants, such as hydrogen-rich saline, have been studied for the treatment of IR-caused testicular damage, with the intention of increasing the fertility of patients [87–89]. CBLB502 was also stated to reestablish testicular antioxidant condition and decrease testicular oxidative injury caused by IR [66]. To evaluate if CBLB502 can reduce testicular damage, animals were intraperitoneally injected with CBLB502 prior to applying IR [67]. It was reported that CBLB502 administration reduced IR-provoked oxidative stress, decreased the alterations of architecture of seminiferous tubules, increased the sperm quality and quantity, and ameliorated male mouse fertility. Moreover, CBLB502 effectively decreased DNA injuries. CBLB502 was reported to stimulate the NFκB pathway and decrease the programmed cell death rate with an augmentation in anti-apoptotic B-cell lymphoma 2 concentrations. Furthermore, an IR-caused decrease in serum testosterone and superoxide dismutase (SOD) concentrations and an augmentation in malondialdehyde (MDA) concentrations were significantly inverted in CBLB502-pretreated animals. No relevant actions were reported in *TLR5* knockout mice, proposing that defense of the testis against IR by CBLB502 is essentially due to the TLR5 signaling pathway [67].

A different side effect of RT is the radiation-caused kidney damage, particularly radiation nephropathy. The frequency of RT-caused kidney injury is probably underrated because of the protracted latency and being often ascribed to diverse causes [90,91]. In RT nephropathy, renal endothelial alterations and modified hemodynamics are well-known elements [92,93].

The radiation-caused damage is principally due to the production of ROS, which causes an imbalance between pro-oxidant and antioxidant elements within the cell and, in turn, oxidation of DNA, lipids, proteins, and cell death [94]. Moreover, there is some proof for programmed cell death as the mechanism of renal tubular cell loss in RT nephropathy [95–97].

A homocysteine derivative, Erdosteine, is recognized to have protecting actions on the delivery of free oxygen radicals alongside its mucolytic effect. A study evaluated the potential protective action of Erdosteine against radiation-caused renal alterations in rats [68]. The findings demonstrated that the use of Erdosteine in rats before irradiation ameliorated the altered kidney function. Moreover, IL-1, IL-6, and TNF-alpha circulating levels were also significantly improved. Kidney glutathione peroxidase and catalase levels and reduced glutathione levels displayed almost normal concentrations with respect to the irradiated group.

Erdosteine operates in the kidney as a powerful scavenger of free radicals and might offer relevant protection against RT-caused inflammatory kidney injury [68].

Irradiation myocardial fibrosis (IMF) is a diverse complication correlated with RT for hematopoietic stem cell transplantation, and lymphoma [98]. A research evaluated the paracrine actions of human umbilical cord-derived mesenchymal stromal cells (UC-MSCs) in an experimental model of IMF [69]. Primary human cardiac fibroblasts (HCF) cells were exposed to radiation and cultured with the conditioned medium of UC-MSCs (MSCCM). MSCCM increased cell viability, decreased collagen deposition, inhibited oxidative stress, augmented antioxidant status, and diminished pro-fibrotic IL-6, IL-8, and TGF-β1 concentrations [69].

Furthermore, as regards to bone marrow toxicity, calf spleen extractive injection (CSEI), a small peptide, has been employed to ameliorate leucopenia and thrombocytopenia in subjects who experienced RT [70–72].

Polyxydroxylated fullerenes have also been recently established as exogenous redox balance controllers, able to produce anti-oxidative actions [73]. Examinations of biological consequences of fullerenol have offered confirmation for its ROS/RNS scavenger abilities in vitro and radioprotective efficacy in vivo [74].

Finally, FDA authorized Palifermin to decrease the occurrence and duration of grave oral mucositis in subjects with hematological diseases. Palifermin is a recombinant N-terminal truncated form of endogenous keratinocyte growth factor (KGF) with activity analogous to that of the KGF, but with augmented stability. Palifermin stimulates cell proliferation and increases cytoprotective systems. Thus, palifermin may inhibit epithelial cell apoptosis and block injury to the epithelial DNA, decrease the delivery of pro-inflammatory cytokines, and augment protective enzymes against free radicals [99]. Several clinical studies confirmed that the drug remarkably decreases the employ of parenteral nutrition and the duration of the mucositis, allowing an improvement in the patient's condition after chemo-radiotherapies [100,101].

4. Enhancers of Radiotherapy Activity in Hematologic Neoplasms

As well as the substances capable of carrying out a protective action against the damage caused by RT reported above, other substances have been recognized as possible enhancers of the RT and cytotoxic action (Table 2). The following section summarizes new therapeutic agents that are proposed to be inducers of oxidative stress in hematologic malignancies and may have clinical efficacy in the treatment of B-cell lymphoma and MM.

Table 2. Enhancers of radiotherapy activity in hematologic neoplasms.

Agent	Mechanisms of Action	Effects	Ref.
Natural phytochemicals (curcumin, demethoxycurcumin, quercetin, genistein)	Alteration of levels of radioprotective metabolites. Electron transfer to a radiation sensitizer.	Reduction of radio-resistance.	[100–102]
Long-chain n-3 polyunsaturated fatty acids	Their peroxidation may sensitize cells to ROS, inducing an oxidative stress. Modulation of ROS-sensitive mitogen-activated protein kinases and phosphatases, and transcription factors.	Cytotoxicity. Increased radiation-induced apoptosis.	[103,104]
Ascorbyl stearate	Augmented levels of ROS, drop in mitochondrial membrane potential and increased caspase-3 activity.	Reduction of cell proliferation, induction of apoptosis by arresting the cells at S/G2-M phase of cell cycle.	[105]
Spleen tyrosine kinase (SYK) P-site inhibitor	Increased H2O2-induced apoptosis.	Action on radio-chemotherapy resistance.	[106]
Dexamethasone	Increased superoxide and hydrogen peroxide production and augmented radiation-induced oxidative stress.	Clonogenic cell killing and apoptosis of myeloma cells.	[107]
Rituximab	Elevation in ROS generation	Increase of cell growth inhibition. Augmented apoptosis.	[108–113]
All-*trans*-retinoic acid, Metformin, IM3829	Inhibition of nuclear factor erythroid 2-related factor 2	Inhibition of cancer cell survival.	[114–116]
Gold nanoparticle-based compounds	Increased ROS levels, mitochondrial depolarization, and cell cycle redistribution.	Inhibition of protective autophagy.	[117,118]

The appearance of radio-resistance in tumor cells involves a mechanism devised by its intracellular antioxidant system to block the oxidative stress and maintain a low, steady level. It is well recognized that radio-resistance of lymphoma cells accounts for lower basal ROS levels. It is well recognized that GSH concentrations and antioxidant enzymes were greater in lymphoma cells with respect to normal lymphocytes [102].

RT has drawback of toxicity and resistance in cancer cells. A number of natural phytochemicals, such as curcumin, demethoxycurcumin, quercetin, and genistein, are demonstrated to hold radio-sensitizing potential in tumor cells [119–121].

Substances that present affinity for electrons can strengthen the biologic action of ionizing radiation [122]. Strengthening may indirectly happen by modifying concentrations of radioprotective metabolites such as glutathione, or the sensitizer may directly cooperate with cellular elements to provoke or increase cellular injury. Models of radiation sensitizers supposed to operate by the former mechanism comprise diamide (diazenedicarboxylic acid *bis*[N, N_-dimethylamide]), *tert*-butyl hydroperoxide, L-buthionine-(S,R)-sulfoximine (BSO), and other thiol depleters [123–125]. Furthermore, nitroimidazoles, tirapazamine, and oxygen directly act with DNA or other targets such as lipid or proteins [126,127]. Generally, oxygen mimetic sensitizers of this latter group are most efficacious under hypoxic situations and the activity is not covered by the competitive action of oxygen. Solvated electrons generated by the radiolysis of water may help to decrease the electron-affinic sensitizer. Nevertheless, even in the absence of ionizing radiation, electron transfer to a radiation sensitizer may happen in the presence of cellular elements that have a more negative standard reduction potential, such as NADPH, flavins, ascorbate, or reduced glutathione [128,129]. This mechanism decreases the reducing substances, which must then be restocked. Moreover, in the presence of oxygen, additional electron transfer can generate reduced oxygen species and restore the sensitizer. Generally, such redox cycling can provoke the onset of a situation of oxidative stress.

Long-chain n-3 polyunsaturated fatty acids (n-3 PUFAs), such as docosahexaenoic acid (DHA) or eicosapentaenoic acid (EPA), are easily oxidized, and n-3 PUFA-derived peroxidation products are believed essential to justify the cytotoxicity toward tumor cells [130–132]. Moreover, their peroxidation may trigger cells to ROS, causing oxidative stress [133]. Additionally, DHA- and EPA-caused oxidative stress may control ROS-sensitive molecular pathways implicated in survival signaling (Figure 2). ROS-sensitive mitogen-activated protein kinases (MAPKs), MAPK-phosphatases [134], and transcription factors [135,136] are essential elements of these pathways. Currently, however, discussion occurs if n-3 PUFAs might also exercise antioxidant actions and decrease ROS production. The paradoxical antioxidant action generated by n-3 PUFAs in several experimental patterns has been justified essentially on the basis of the stimulation of cellular antioxidant elements [103,104,137].

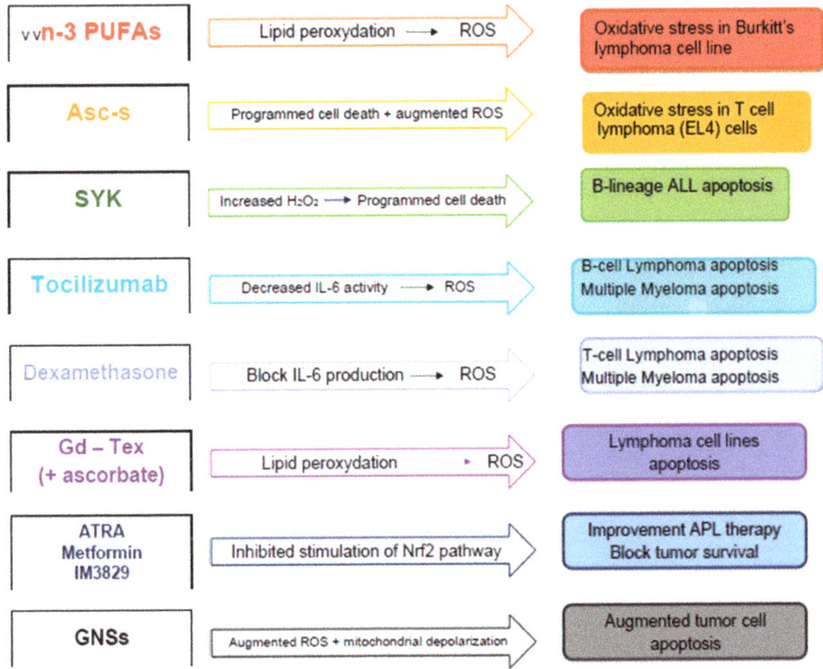

Figure 2. Reactive oxygen species (ROS)-sensitive molecular pathways implicated in survival signaling.

An increased cytotoxic action of irradiation was reported in tumor cells cultured in vitro and treated with DHA and EPA [138]. DHA was also described to block γ-radiation-provoked activation of NF-κB in Ramos cells, an extremely radiation-resistant Burkitt's lymphoma cell line, and to sensitize the cells to radiation-caused programmed cell death [139].

Encouraging results have been also reported in in vivo research in non-hematologic malignancies [105,106,140,141]. This effect was correlated to the DHA-provoked oxidative stress in tumor cells previously subjected to augmented radiation-caused production of ROS.

Ascorbyl stearate (Asc-s) is a byproduct of ascorbic acid with superior anti-tumor effectiveness with respect to ascorbic acid. In a report, authors have evaluated radio-sensitizing action of Asc-s in T cell lymphoma (EL4) cells. Asc-s and radiation administration decreased cell growth, causing programmed cell death in a dose-dependent manner by blocking the cells at S/G2-M phase of the cell cycle [142]. It also reduced the occurrence of tumor stem cells per se, with considerably greater reduction in combination with radiation therapy. Moreover, Asc-s and radiation treatment augmented ROS concentration, a drop in mitochondrial membrane potential (MMP), and augmented caspase-3 activity, thus provoking programmed cell death of EL4 cells. Furthermore, it also drastically reduced GSH/GSSG ratio due to binding of Asc-s with thiols. The augmented oxidative stress caused by Asc-s and radiation treatment was abolished by thiol antioxidants in EL4 cells. Remarkably, this redox modulation stimulated relevant augmentation in protein glutathionylation. Asc-s administration caused glutathionylation of p50-NF-kB, IκB kinase (IKK), and mutated p53, thus blocking tumor expansion during oxidative stress.

A new spleen tyrosine kinase (SYK) P-site inhibitor, 1,4-Bis (9-O dihydroquinidinyl) phthalazine/hydroquinidine 1,4-phathalazinediyl diether (C-61), significantly increased H2O2-caused programmed cell death of primary leukemia cells from each of five relapsed B-lineage acute lymphoblastic leukemia (ALL) subjects [143]. An extremely radiation-resistant subclone of the murine B-lineage leukemia cell line BCL-1 was employed to evaluate the in vivo radio-sensitizing actions of C-61. C-61 increased the antileukemia effectiveness of total-body irradiation (TBI) in a study on syngeneic bone marrow transplantation (BMT) at 20%. It is probable that the use of C-61 into the pretransplant TBI protocols of subjects with high-risk B-ALL will aid the overwhelmed radio-chemotherapy resistance of leukemia cells and thus increase survival outcome [143].

IL-6 might also increase resistance to radiotherapy. In fact, IL-6 plays an important role in promoting myeloma cell growth and chemoresistance; increased IL-6 concentrations are associated with aggressive disease and poor outcome [144,145]. IL-6 has been demonstrated to induce superoxide generation in monocytes, neutrophils, and neuronal cells [146,147]. However, surprisingly, IL-6 can provoke adaptive responses to oxidative stress in normal cells by protecting cells from hydrogen peroxide-caused cell death [148]; transgenic animals overexpressing IL-6 demonstrate resistance to hyperoxia [149]. Although MM subjects display augmented lipid peroxidation and lower concentrations of antioxidant enzymes in plasma and erythrocytes [150–156], research reported the effect of IL-6 in restoring intracellular redox homoeostasis in the setting of myeloma treatment. IL-6 employment augmented myeloma cell resistance to substances that provoke oxidative stress, comprising IR and Dex (dexamethasone) [157]. As compared to IR alone, myeloma cells treated with IL-6 plus IR showed decreased caspase-3 stimulation, annexin/propidium iodide staining, PARP [poly(ADPribose) polymerase] cleavage, and mitochondrial membrane depolarization with augmented clonogenic survival. IL-6 combined with IR or Dex augmented early intracellular pro-oxidant concentrations that were correlated to stimulation of NF-κB. In myeloma cells, upon combination with hydrogen peroxide administration, relative to TNF-α, IL-6 caused an alteration of decreased glutathione concentration and augmented NF-κB-dependent manganese superoxide dismutase (MnSOD) expression. Furthermore, knockdown of MnSOD reduced the IL-6-caused myeloma cell resistance to radiation. MitoSOX Red staining demonstrated that IL-6 administration reduced late mitochondrial oxidant generation in irradiated myeloma cells [157]. Moreover, in B-cell lymphoma, radio-resistance has been also associated with the presence of IL-6-producing tumor cells [158].

The use of substances with anti-IL-6 activity (tocilizumab) should be explored in an attempt to increase sensitivity to radiotherapy.

From this point of view, the action of steroids could be interesting. Dexamethasone has been reported to provoke oxidative cell death in T-cell lymphoma [107]; nevertheless, Dex effect in causing oxidative stress in MM has not been proved [108,159].

Bera et al. suggested a new combination of radiation plus Dex that presents greater clonogenic cell killing and programmed cell death of myeloma cells and eradicates myeloma cells when cultured with bone marrow stromal cells (BMSCs) [109]. Dex was reported to block the production of IL-6 from irradiated BMSCs. In 5TGM1 model, combined use of Dex with skeletal radiotherapy (153-Sm-EDTMP) increased median survival time and reduced radiation-caused myelosuppression. Dex augmented superoxide and hydrogen peroxide generation and increased radiation-caused oxidative stress and cell death of myeloma cells. However, Dex reduced radiation-caused augmentation in pro-oxidant concentrations and increased the clonogenic survival in progenitor cells and normal hematopoietic stem cells. Administration with either N-acetylcysteine or the combinate use of polyethylene glycol (PEG)–conjugated copper, PEG-catalase, and zinc-superoxide dismutase considerably safeguarded myeloma cells from Dex-caused clonogenic death [109]. These findings state that Dex in combination with RT increases the destroying of myeloma cells while defending normal bone marrow hematopoiesis via a system that implicates an augmentation in oxidative stress.

Magda et al. evaluated the mechanism of radiation improvement by motexafin gadolinium (Gd-Tex) in vitro [110]. Clonogenic assays were employed to evaluate radiation response in lymphoma cell lines. Gd-Tex catalyzed the oxidation of NADPH and ascorbate under aerobic conditions, generating hydrogen peroxide. Cultivation with Gd-Tex in the presence of ascorbate augmented the aerobic radiation response of E89. Gd-Tex sensitizes cells to IR by augmenting oxidative stress [110].

Finally, RT, chemotherapy, and immunotherapy are the main treatments in Non-Hodgkin's lymphoma therapy. There is the probability that the combination of chemotherapy or immunotherapy and RT may have a synergistic effect. In one report, rituximab was reported to radio-sensitize lymphoma cells controlling programmed cell death and cell cycle-related proteins [111–113,160].

The stimulation of apoptosis caused by the combined use of RT and rituximab was demonstrated to be caused by mitochondrial dissipation. ROS were described to have a role in this phenomenon [161]. Fengling et al. evaluated the action of rituximab on cell death caused by radiation in Raji lymphoma cells [114]. G2/M cell cycle arrest was reported after irradiation alone and the combination therapy. The combination therapy caused an increase in ROS production in a radiation dose-dependent manner. Moreover, rituximab increased the cell inhibition and programmed cell death caused by H2O2 [114].

An interesting field of study is the Kelch-like ECH-associated protein 1(Keap1)-nuclear factor erythroid 2-related factor 2 (Nrf2). It is the most investigated signaling pathway of cellular defense against oxidative stress [115]. When ROS concentrations are augmented, Nrf2 detaches from its inhibitor, Keap1, and translocates into the nucleus where it creates a heterodimer with small V-Maf Avian Musculoaponeurotic Fibrosarcoma Oncogene MAF proteins (sMAF) and controls the expression of genes that comprise the antioxidant response elements. Nrf2 regulates about 250 genes, essentially implicated in the endogenous antioxidant defense and detoxification of ROS. Mutation of Nrf2, or of its negative regulator Keap1, can alter their relations, which may explain the overexpression of Nrf2 signaling [115]. While normal cells are defended from DNA injury caused by ROS, tumor cells are protected against chemo- or radiotherapy [115].

Nevertheless, Nrf2 stimulation is not general. For instance, it was observed that after RT, the majority of Nrf2-targeted genes remained unchanged. Moreover, there were nine genes implicated in lipid peroxidation, which showed under expression in the case of new RT [116]. Recently, new data suggest that Nrf2 has conflicting actions in tumors. Anomalous stimulation of Nrf2 is correlated with poor prognosis as lasting stimulation of Nrf2 considerably increases the tumor cell resistance to ROS by increasing antioxidant enzymes. The increased Nrf2 in RT is recognized to be correlated with greater expression of Nrf2 downstream targets that stimulate GSH synthesis. Affecting Nrf2 activity in hematologic malignancies and tumors may be an efficacious approach to reduce radioresistance [116] and several Nrf2 inhibitors detected. For instance, All-trans-retinoic acid, employed for the therapy of acute promyelocytic leukemia, has been suggested as a specific Nrf2 inhibitor, which permits Nrf2 to form a complex with retinoid X receptor alpha (RARα), inhibiting stimulation of the Nrf2 pathway [117]. Moreover, metformin, a drug commonly used in the therapy of type 2 diabetes, decreases mRNA and protein concentrations of Nrf2 via the block of Raf/ERK/Nrf2 signaling [118], while IM3829

(4-[2-Cyclohexylethoxy] aniline) reduces Nrf2 mRNA and protein concentrations, and combination with radiation is capable to considerably block tumor survival [162].

Moreover, other radiosensitizers, which can considerably increase the radiotherapeutic efficacy, have been produced. Gold-based nanomaterials, as a novel sort of nanoparticle-based radiosensitizer, have been employed in experimentations implicating tumor RT [163]. However, therapeutic action employing the gold nanoparticle-based RT is generally not significant because of the low cellular uptake effectiveness and the autophagy-inducing capacity of these gold nanomaterials. Using gold nanospikes (GNSs), investigators built a series of thiol-poly(ethylene glycol)-modified GNSs terminated with folic acid (FA) (FA-GNSs), amine (NH_2-GNSs), methoxyl (GNSs), and the cell-penetrating peptide TAT (TAT-GNSs), and studied their actions on X-ray RT [164]. The sensitization enhancement ratio (SER), which is generally employed to assess how efficiently radiosensitizers reduce cell growth, reaches 2.30 for TAT-GNSs. The exceptionally elevated SER value for TAT-GNSs suggests the greater radiosensitization action of this nanomaterial. The radiation improvement system of these GNSs implicated the augmented ROS and mitochondrial depolarization. Surface-modified GNSs could provoke the increase of autophagy-related protein (LC3-II) and apoptosis-related protein (active caspase-3) in tumor cells. GNSs provoked the diminishing of autolysosome degradation ability and autophagosome amassing. These findings established that autophagy has a protecting action against RT, and the block of defensive autophagy with inhibitors would provoke an augmentation of cell apoptosis. As well as the above in vitro experiments, the in vivo studies also reported that X-ray + TAT-GNSs therapy had the best tumor proliferation inhibitory action, which proved the greatest radiation sensitizing action of TAT-GNSs [164].

Finally, some drugs that have long been used in the treatment of haematological neoplasms also seem to have radiosensitizing capabilities. For instance, fludarabine is also a well-known DNA damage repair inhibitor that has clinical activity against hematological cancers [165].

In an in vitro experimental study, fludarabine-P, in clinically possible dosages, is a potent radiosensitizer in a human squamous carcinoma cell line of the oropharynx (ZMK-1 cells) and has a minor action in the fetal lung fibroblasts (MRC-5) [166]. The radiosensitization of fludarabine-P appears to be over additive in the malignant cells and additive in normal fetal fibroblasts. This would suggest that fludarabine-P might increase the therapeutic ratio of radiation.

Another well-known radiosensitizer is hydroxyurea, which was used in treatment of hematological malignancies [167], and several studies demonstrated that the addition of hydroxyurea to ionizing radiation produced a synergistic effect in vitro [75,168].

5. Future Perspectives

RT is a keystone of both curative and palliative tumor treatment. However, RT is harshly limited by radiation-caused side effects. If these side effects could be decreased, a bigger dosage of radiation could be given to obtain a superior response. Pre-clinical experimentations have demonstrated that irradiation at dosage rates far exceeding those generally employed decreases radiation-induced side-effects though conserving a comparable anti-tumor effect. This is recognized as the FLASH effect [76]. To date, numerous subjects have been treated with Ultra-High Dose Rate (FLASH) RT for the therapy of lymphomas causing complete responses and reduced toxicities. The mechanism accountable for decreased tissue toxicity is yet to be clarified, but the most noticeable theory so far suggested is that acute oxygen diminution happens within the irradiated tissue [76].

In the near future, other areas of intervention will open up in the field of radiation protection, such as that of non-coding genetic material. Previous reports have recognized essential actions of specific miRNAs in radiation response in numerous malignancies [169].

miR-139-5p is a powerful controller of RT response in tumors through its modulation of genes implicated in several DNA repair and ROS protection pathways. Therapy of solid cancer cells with a miR-139-5p mimic intensely synergized with radiation both in vitro and in vivo, causing a relevant augmentation of oxidative stress and stimulation of programmed cell death. Numerous miR-139-5p target genes were also predictive of prognosis. These prognostically important miR-139-5p target genes were employed as markers to recognize radioresistant cancer xenografts extremely responsive to sensitization by treatment with a miR-139-5p mimetic [170]; one of its confirmed targets, MAT2A, has a role in ROS defense [171].

The use of specific antisense oligonucleotides, able to bind and antagonize miRNAs, could be efficacious as a therapeutic element able of provoking a diverse response to oxidative stress and RT [172].

A different field of study could be the neuroprotection. Primary Central Nervous System Lymphoma (PCNSL) is a non-Hodgkin lymphoma that occurs within the brain, spinal cord, or eyes in the absence of systemic disease. Therapy often comprises whole-brain radiotherapy. However, after therapy, some subjects may show neurologic alterations [173]. This delayed treatment-related side-effect appears to be similar to a diffuse leukoencephalopathy. Some data indicate that the mechanism is related to the disruption of frontal subcortical circuits provoked by radiation injury, probably produced by oxidative stress [174].

Platelet-rich plasma (PRP) is an autologous preparation of platelets which comprehends great amounts of growth factors such as platelet-derived growth factor, epidermal growth factor, vascular endothelial growth factor, hepatocyte growth factor, insulin-like growth factor-1, and transforming growth factor-beta 1 [175]. PRP pre-treatment considerably decreased the radiation-induced alterations. Moreover, PRP remarkably improved the state of oxidative stress and seemed to block apoptotic process [176]. A similar approach could be attempted to prevent RT-induced neurological toxicity.

In the future, the possibility of modifying the radiosensitivity of the different cell types could be decisive in guaranteeing the effectiveness of the radiotherapy treatment. The L5178Y lymphoma cells (LY-S cell line) were the first ionizing radiation-sensitive mammalian cell line to be described. Investigators isolated the LY-S line from the parental L5178Y line (later called LY-R; R for resistant) and studied its high radiosensitivity. The following studies were targeted at comprehending the cause for the radiation sensitivity difference between the LY lines [177]. The LY-R to LY-S phenotype modification was due to the oxidative shock after cell relocation from the ascitic tumor into culture medium [177].

Full knowledge of the intergenomic interactions seems crucial for understanding the cellular response to ionizing radiation. Finding therapeutic solutions for hematologic malignancies and other diseases may also depend upon such knowledge.

6. Conclusions

Ionizing radiation has a main function in modern tumor treatment due to its distinctive advantages comprising non-invasiveness and a lack of severe systemic toxicity. Even though RT has showed several degrees of success, relapse and treatment failure may happen in tumor subjects due to radioresistance. Therefore, approaches are urgently necessitated for increasing radiosensitivity of tumor cells and augmenting the radioprotection of normal cells. The elaboration of novel and clinically efficacious modulators will offer prospects for new treatment paradigms.

In this compound, planning complementary therapeutic methods founded on differences in oxidative metabolism between tumoral and normal cells could be an effective approach to create more successful treatments.

Author Contributions: Conceptualization, A.A., A.G.A., and C.M.; methodology, V.I., F.M.; formal analysis, V.I., F.M.; data curation, V.I., F.M.; writing—original draft preparation, A.A., A.G.A.; writing—review and editing, A.A., A.G.A. All authors have read and agreed to the published version of the manuscript.

Funding: This research received no external funding.

Conflicts of Interest: The authors declare no conflict of interest.

References

1. Tseng, Y.D.; Ng, A.K. Hematologic Malignancies. *Hematol. Oncol. Clin. N. Am.* **2020**, *34*, 127–142. [CrossRef] [PubMed]
2. Galunić-Bilić, L.; Šantek, F. Infradiaphragmal Radiotherapy in Patients with Lymphoma: Volume Definition and Side Effects. *Acta Clin. Croat.* **2018**, *57*, 554–560. [CrossRef]
3. Bosch, A.; Frias, Z. Radiotherapy in the treatment of multiple myeloma. *Int. J. Radiat. Oncol. Biol. Phys.* **1988**, *15*, 1363–1369. [CrossRef]
4. Kilciksiz, S.; Celik, O.K.; Pak, Y.; Demiral, A.N.; Pehlivan, M.; Orhan, O.; Tokatli, F.; Agaoglu, F.; Zincircioglu, B.; Atasoy, B.M.; et al. Clinical and prognostic features of plasmacytomas: A multicenter study of Turkish Oncology Group–Sarcoma Working Party. *Am. J. Hematol.* **2008**, *83*, 702–707. [CrossRef] [PubMed]
5. Goel, A.; Dispenzieri, A.; Witzig, T.E.; Russell, S.J. Enhancing the therapeutic index of radiation in multiple myeloma. *Drug Discov. Today Dis. Mech.* **2006**, *3*, 515–522. [CrossRef]
6. Chatterjee, M.; Chakraborty, T.; Tassone, P. Multiple myeloma: Monoclonal antibodies–based immunotherapeutic strategies and targeted radiotherapy. *Eur. J. Cancer* **2006**, *42*, 1640–1652. [CrossRef]
7. Wong, J.Y.; Liu, A.; Schultheiss, T.; Popplewell, L.; Stein, A.; Rosenthal, J.; Essensten, M.; Forman, S.; Somlo, G. Targeted total marrow irradiation using three dimensional image-guided tomographic intensity–modulated radiation therapy: An alternative to standard total body irradiation. *Biol. Blood Marrow Transplant.* **2006**, *12*, 306–315. [CrossRef]
8. Wong, J.Y.; Rosenthal, J.; Liu, A.; Schultheiss, T.; Forman, S.; Somlo, G. Image-guided total-marrow irradiation using helical tomotherapy in patients with multiple myeloma and acute leukemia undergoing hematopoietic cell transplantation. *Int. J. Radiat. Oncol. Biol. Phys.* **2009**, *73*, 273–279. [CrossRef]
9. Anderson, P.M.; Wiseman, G.A.; Dispenzieri, A.; Arndt, C.A.; Hartmann, L.C.; Smithson, W.A.; Mullan, B.P.; Bruland, O.S. High-dose samarium-153 ethylene diamine tetramethylene phosphonate: Low toxicity of skeletal irradiation in patients with osteosarcoma and bone metastases. *J. Clin. Oncol.* **2002**, *20*, 189–196. [CrossRef]
10. Dispenzieri, A.; Wiseman, G.A.; Lacy, M.Q.; Geyer, S.; Litzow, M.R.; Tefferi, A.; Inwards, D.J.; Micallef, I.N.; Ansell, S.; Gastineau, D.A.; et al. A phase II study of high dose 153-samarium EDTMP (153-sm EDMTP) and melphalan for peripheral stem cell transplantation (PBSCT) in multiple myeloma (MM). *Blood* **2003**, *102*, 982.
11. Wilky, B.A.; Loeb, D.M. Beyond Palliation: Therapeutic Applications of [153]Samarium-EDTMP. *Clin. Exp. Pharmacol.* **2013**, *3*, 1000131. [CrossRef] [PubMed]
12. Abruzzese, E.; Iuliano, F.; Trawinska, M.M.; DiMaio, M. 153Sm: Its use in multiple myeloma and report of a clinical experience. *Expert Opin. Investig. Drugs* **2008**, *17*, 1379–1387. [CrossRef] [PubMed]
13. Marchand, V.; Decaudin, D.; Servois, V.; Kirova, Y.M. Concurrent radiation therapy and lenalidomide in myeloma patient. *Radiother. Oncol.* **2008**, *87*, 152–153. [CrossRef] [PubMed]
14. Berenson, J.R.; Yellin, O.; Patel, R.; Duvivier, H.; Nassir, Y.; Mapes, R.; Abaya, C.D.; Swift, R.A. A phase I study of samarium lexidronam/bortezomib combination therapy for the treatment of relapsed or refractory multiple myeloma. *Clin. Cancer Res.* **2009**, *15*, 1069–1075. [CrossRef] [PubMed]
15. Berges, O.; Decaudin, D.; Servois, V.; Kirova, Y.M. Concurrent radiation therapy and bortezomib in myeloma patient. *Radiother. Oncol.* **2008**, *86*, 290–292. [CrossRef]
16. Bakst, R.L.; Dabaja, B.S.; Specht, L.K.; Yahalom, J. Use of radiation in extramedullary leukemia/chloroma: Guidelines from the International Lymphoma Radiation Oncology Group. *Int. J. Radiat. Oncol. Biol. Phys.* **2018**, *102*, 314–319. [CrossRef] [PubMed]
17. Karihtala, P.; Winqvist, R.; Syvaoja, J.E.; Kinnula, V.L.; Soini, Y. Increasing oxidative damage and loss of mismatch repair enzymes during breast carcinogenesis. *Eur. J. Cancer* **2006**, *42*, 2653–2659. [CrossRef] [PubMed]
18. Mates, J.M.; Segura, J.A.; Alonso, F.J.; Marquez, J. Intracellular redox status and oxidative stress: Implications for cell proliferation, apoptosis, and carcinogenesis. *Arch. Toxicol.* **2008**, *82*, 273–299. [CrossRef]
19. Karihtala, P.; Soini, Y. Reactive oxygen species and antioxidant mechanisms in human tissues and their relation to malignancies. *Acta Pathol. Microbiol. Immunol. Scand.* **2007**, *115*, 81–103. [CrossRef]
20. Karihtala, P.; Kauppila, S.; Soini, Y.; Jukkola-Vuorinen, A. Oxidative stress and counteracting mechanisms in hormone receptor positive, triple-negative and basal-like breast carcinomas. *BMC Cancer* **2011**, *11*, 262. [CrossRef]
21. Salzman, R.; Pacal, L.; Kankova, K.; Tomandl, J.; Horakova, Z.; Tothova, E.; Kostrica, R. High perioperative level of oxidative stress as a prognostic tool for identifying patients with a high risk of recurrence of head and neck squamous cell carcinoma. *Int. J. Clin. Oncol.* **2010**, *15*, 565–570. [CrossRef] [PubMed]

22. Imbesi, S.; Musolino, C.; Allegra, A.; Saija, A.; Morabito, F.; Calapai, G.; Gangemi, S. Oxidative stress in oncohematologic diseases: An update. *Expert Rev. Hematol.* **2013**, *6*, 317–325. [CrossRef]

23. Gangemi, S.; Allegra, A.; Aguennouz, M.; Alonci, A.; Speciale, A.; Cannavò, A.; Cristani, M.; Russo, S.; Spatari, G.; Alibrandi, A.; et al. Relationship between advanced oxidation protein products, advanced glycation end products, and S-nitrosylated proteins with biological risk and MDR-1 polymorphisms in patients affected by B-chronic lymphocytic leukemia. *Cancer Investig.* **2012**, *30*, 20–26. [CrossRef] [PubMed]

24. Musolino, C.; Allegra, A.; Alonci, A.; Saija, A.; Russo, S.; Cannavò, A.; Cristani, M.; Centorrino, R.; Saitta, S.; Alibrandi, A.; et al. Carbonyl group serum levels are associated with CD38 expression in patients with B chronic lymphocytic leukemia. *Clin. Biochem.* **2011**, *44*, 1487–1490. [CrossRef] [PubMed]

25. Peroja, P.; Pasanen, A.K.; Haapasaari, K.M.; Jantunen, E.; Soini, Y.; Turpeenniemi-Hujanen, T.; Bloigu, R.; Lilja, L.; Kuittinen, O.; Karihtala, P. Oxidative stress and redox state-regulating enzymes have prognostic relevance in diffuse large B-cell lymphoma. *Exp. Hematol. Oncol.* **2012**, *1*, 2. [CrossRef]

26. Zhou, F.L.; Zhang, W.G.; Wei, Y.C.; Meng, S.; Bai, G.G.; Wang, B.Y.; Yang, H.Y.; Tian, W.; Meng, X.; Zhang, H.; et al. Involvement of oxidative stress in the relapse of acute myeloid leukemia. *J. Biol. Chem.* **2010**, *285*, 15010–15015. [CrossRef]

27. Bur, H.; Haapasaari, K.M.; Turpeenniemi-Hujanen, T.; Kuittinen, O.; Auvinen, P.; Marin, K.; Koivunen, P.; Sormunen, R.; Soini, Y.; Karihtala, P. Oxidative stress markers and mitochondrial antioxidant enzyme expression are increased in aggressive Hodgkin lymphomas. *Histopathology* **2014**, *65*, 319–327. [CrossRef]

28. Redza-Dutordoir, M.; Averill-Bates, D.A. Activation of apoptosis signalling pathways by reactive oxygen species. *Biochim. Biophys. Acta* **2016**, *1863*, 2977–2992. [CrossRef]

29. Rahgoshai, S.; Mohammadi, M.; Refahi, S.; Oladghaffari, M.; Aghamiri, S. Protective effects of IMOD and cimetidine against radiation induced cellular damage. *J. Biomed. Phys. Eng.* **2017**, *8*, 133–140.

30. Azzam, E.I.; Jay-Gerin, J.P.; Pain, D. Ionizing radiation-induced metabolic oxidative stress and prolonged cell injury. *Cancer Lett.* **2012**, *327*, 48–60. [CrossRef]

31. Popgeorgiev, N.; Jabbour, L.; Gillet, G. Subcellular localization and dynamics of the Bcl-2 family of proteins. *Front. Cell Dev. Biol.* **2018**, *6*, 13. [CrossRef] [PubMed]

32. Warren, C.F.A.; Wong-Brown, M.W.; Bowden, N.A. BCL-2 family isoforms in apoptosis and cancer. *Cell Death Dis.* **2019**, *10*, 177. [CrossRef] [PubMed]

33. Pohl, S.Ö.-G.; Agostino, M.; Dharmarajan, A.; Pervaiz, S. Cross talk between cellular redox state and the antiapoptotic protein Bcl-2. *Antioxid. Redox Signal.* **2018**, *29*, 1215–1236. [CrossRef] [PubMed]

34. Iliakis, G.; Wang, H.; Perrault, A.R.; Boecker, W.; Rosidi, B.; Windhofer, F.; Wu, W.; Guan, J.; Terzoudi, G.; Antelias, G. Mechanisms of DNA double strand break repair and chromosome aberration formation. *Cytogenet. Genome Res.* **2004**, *104*, 14–20. [CrossRef] [PubMed]

35. Purkayastha, S.; Milligan, J.R.; Bernhard, W.A. On the chemical yield of base lesions, strand breaks, and clustered damage generated in plasmid DNA by the direct effect of × rays. *Radiat. Res.* **2007**, *168*, 357–366. [CrossRef] [PubMed]

36. Yakes, F.M.; van Houten, B. Mitochondrial DNA damage is more extensive and persists longer than nuclear DNA damage in human cells following oxidative stress. *Proc. Natl. Acad. Sci. USA* **1997**, *94*, 514–519. [CrossRef]

37. Rodemann, H.P.; Blaese, M.A. Responses of normal cells to ionizing radiation. *Semin. Radiat. Oncol.* **2007**, *17*, 81–88. [CrossRef]

38. Larsen, N.B.; Rasmussen, M.; Rasmussen, L.J. Nuclear and mitochondrial DNA repair: Similar pathways? *Mitochondrion* **2005**, *5*, 89–108. [CrossRef]

39. Yang, J.L.; Weissman, L.; Bohr, V.A.; Mattson, M.P. Mitochondrial DNA damage and repair in neurodegenerative disorders. *DNA Repair* **2008**, *7*, 1110–1120. [CrossRef]

40. Zhang, H.; Maguire, D.; Swarts, S.; Sun, W.; Yang, S.; Wang, W.; Liu, C.; Zhang, M.; Zhang, D.; Zhang, L.; et al. Replication of murine mitochondrial DNA following irradiation. *Adv. Exp. Med. Biol.* **2009**, *645*, 43–48.

41. Cortopassi, G.A.; Arnheim, N. Detection of a specific mitochondrial DNA deletion in tissues of older humans. *Nucleic Acids Res.* **1990**, *18*, 6927–6933. [CrossRef] [PubMed]

42. Krishnan, K.J.; Reeve, A.K.; Samuels, D.C.; Chinnery, P.F.; Blackwood, J.K.; Taylor, R.W.; Wanrooij, S.; Spelbrink, J.N.; Lightowlers, R.N.; Turnbull, D.M. What causes mitochondrial DNA deletions in human cells? *Nat. Genet.* **2008**, *40*, 275–279. [CrossRef]

43. Gerhard, G.S.; Benko, F.A.; Allen, R.G.; Tresini, M.; Kalbach, A.; Cristofalo, V.J.; Gocke, C.D. Mitochondrial DNA mutation analysis in human skin fibroblasts from fetal, young, and old donors. *Mech Ageing Dev.* **2002**, *123*, 155–166. [CrossRef]

44. Chabi, B.; Mousson de Camaret, B.; Chevrollier, A.; Boisgard, S.; Stepien, G. Random mtDNA deletions and functional consequence in aged human skeletal muscle. *Biochem. Biophys. Res. Commun.* **2005**, *332*, 542–549. [CrossRef] [PubMed]

45. Pavicic, W.H.; Richard, S.M. Correlation analysis between mtDNA 4977-bp deletion and ageing. *Mutat. Res.* **2009**, *670*, 99–102. [CrossRef]

46. Wen, Q.; Hu, Y.; Ji, F.; Qian, G. Mitochondrial DNA alterations of peripheral lymphocytes in acute lymphoblastic leukemia patients undergoing total body irradiation therapy. *Radiat. Oncol.* **2011**, *6*, 133. [CrossRef]

47. Soyal, D.; Jindal, A.; Singh, I.; Goyal, P. Protective capacity of Rosemary extract against radiation induced hepatic injury in mice. *Int. J. Radiat. Res.* **2007**, *4*, 161–168.

48. Janssen, L.; Bosman, C.B.; van Duijn, W.; Oostendorp-van de Ruit, M.M.; Kubben, F.J.; Griffioen, G.; Lamers, C.B.; van Krieken, J.H.; van de Velde, C.J.; Verspaget, H.W. Superoxide dismutases in gastric and esophageal cancer and the prognostic impact in gastric cancer. *Clin. Cancer Res.* **2000**, *6*, 3183–3192.

49. Hall, S.; Rudrawar, S.; Zunk, M.; Bernaitis, N.; Arora, D.; McDermott, C.M.; Anoopkumar-Dukie, S. Protection against Radiotherapy-Induced Toxicity. *Antioxidants* **2016**, *5*, 22. [CrossRef]

50. Xu, G.; Shi, H.; Ren, L.; Gou, H.; Gong, D.; Gao, X.; Huang, N. Enhancing the anti-colon cancer activity of quercetin by self-assembled micelles. *Int. J. Nanomed.* **2015**, *10*, 2051–2063. [CrossRef]

51. Rusin, A.; Krawczyk, Z.; Grynkiewicz, G.; Gogler, A.; Zawisza-Puchałka, J.; Szeja, W. Synthetic derivatives of genistein, their properties and possible applications. *Acta Biochim. Pol.* **2010**, *57*, 23–34. [CrossRef] [PubMed]

52. Szejk, M.; Poplawski, T.; Czubatka-Bienkowska, A.; Olejnik, A.K.; Pawlaczyk-Graja, I.P.; Gancarz, R.; Zbikowska, H.M. A comparative study on the radioprotective potential of the polyphenolic glycoconjugates from medicinal plants of Rosaceae and Asteraceae families versus their aglycones. *J. Photochem. Photobiol. B* **2017**, *171*, 50–57. [CrossRef] [PubMed]

53. Szejk-Arendt, M.; Czubak-Prowizor, K.; Macieja, A.; Poplawski, T.; Olejnik, A.K.; Pawlaczyk-Graja, I.; Gancarz, R.; Zbikowska, H.M. Polyphenolic-polysaccharide conjugates from medicinal plants of Rosaceae/Asteraceae family protect human lymphocytes but not myeloid leukemia K562 cells against radiation-induced death. *Int. J. Biol. Macromol.* **2020**, *156*, 1445–1454. [CrossRef] [PubMed]

54. Mozdarani, H.; Salimi, M.; Froughizadeh, M. Effect of cimetidine and famotidine on survival of lethally gamma irradiated mice. *Iran. J. Radiat. Res.* **2008**, *5*, 187–194.

55. Estaphan, S.; Abdel-Malek, R.; Rashed, L.; Mohamed, E.A. Cimetidine a promising radio-protective agent through modulating Bax/Bcl2 ratio: An in vivo study in male rats. *J. Cell Physiol.* **2020**. [CrossRef] [PubMed]

56. Yamini, K.; Gopal, V. Natural radioprotective agents against ionizing radiation—An overview. *Int. J. Pharm. Tech. Res.* **2010**, *2*, 1421–1426.

57. Wegener, T.; Wagner, H. The active components and the pharmacological multi-target principle of STW 5 (Iberogast). *Phytomedicine* **2006**, *13*, 20–35. [CrossRef]

58. Khayyal, M.T.; Abdel-Naby, D.H.; Abdel-Aziz, H.; El-Ghazaly, M.A. A multi-component herbal preparation, STW 5, shows anti-apoptotic effects in radiation induced intestinal mucositis in rats. *Phytomedicine* **2014**, *21*, 1390–1399. [CrossRef]

59. Wadie, W.; Abdel-Aziz, H.; Zaki, H.F.; Kelber, O.; Weiser, D.; Khayyal, M.T. STW 5 is effective in dextran sulfate sodium-induced colitis in rats. *Int. J. Colorectal Dis.* **2012**, *27*, 1445–1453. [CrossRef]

60. Michael, S.; Kelber, O.; Hauschildt, S.; Spanel-Borowski, K.; Nieber, K. 2009. Inhibition of inflammation-induced alterations in rat small intestine by the herbal preparations STW 5 and STW 6. *Phytomedicine* **2009**, *16*, 161–171. [CrossRef]

61. Burdelya, L.G.; Krivokrysenko, V.I.; Tallant, T.C.; Strom, E.; Gleiberman, A.S.; Gupta, D.; Kurnasov, O.V.; Fort, F.L.; Osterman, A.L.; Didonato, J.A.; et al. An agonist of toll-like receptor 5 has radioprotective activity in mouse and primate models. *Science* **2008**, *320*, 226–230. [CrossRef] [PubMed]

62. Toshkov, I.A.; Gleiberman, A.S.; Mett, V.L.; Hutson, A.D.; Singh, A.K.; Gudkov, A.V.; Burdelya, L.G. Mitigation of Radiation-Induced Epithelial Damage by the TLR5 Agonist Entolimod in a Mouse Model of Fractionated Head and Neck Irradiation. *Radiat. Res.* **2017**, *187*, 570–580. [CrossRef] [PubMed]

63. Leigh, N.D.; Bian, G.; Ding, X.; Liu, H.; Aygun-Sunar, S.; Burdelya, L.G.; Gudkov, A.V.; Cao, X. A flagellin-derived toll-like receptor 5 agonist stimulates cytotoxic lymphocyte-mediated tumor immunity. *PLoS ONE* **2014**, *9*, e85587. [CrossRef] [PubMed]

64. Yang, H.; Brackett, C.M.; Morales-Tirado, V.M.; Li, Z.; Zhang, Q.; Wilson, M.W.; Benjamin, C.; Harris, W.; Waller, E.K.; Gudkov, A.V.; et al. The Toll-like receptor 5 agonist entolimod suppresses hepatic metastases in a murine model of ocular melanoma via an NK cell-dependent mechanism. *Oncotarget* **2016**, *7*, 2936–2950. [CrossRef]

65. Brackett, C.M.; Kojouharov, B.; Veith, J.; Greene, K.F.; Burdelya, L.G.; Gollnick, S.O.; Abrams, S.I.; Gudkov, A.V. Toll-like receptor-5 agonist, entolimod, suppresses metastasis and induces immunity by stimulating an NK-dendritic-CD$_8$$^+$ T-cell axis. *Proc. Natl. Acad. Sci. USA* **2016**, *113*, E874–E883. [CrossRef]

66. Li, W.; Ge, C.; Yang, L.; Wang, R.; Lu, Y.; Gao, Y.; Li, Z.; Wu, Y.; Zheng, X.; Wang, Z.; et al. CBLB502, an agonist of Toll-like receptor 5, has antioxidant and scavenging free radicals activities in vitro. *Int. J. Biol. Macromol.* **2016**, *82*, 97–103. [CrossRef]

67. Bai, H.; Sun, F.; Yang, G.; Wang, L.; Zhang, Q.; Zhang, Q.; Zhan, Y.; Chen, J.; Yu, M.; Li, C.; et al. CBLB502, a Toll-like receptor 5 agonist, offers protection against radiation-induced male reproductive system damage in mice. *Biol. Reprod.* **2019**, *100*, 281–291. [CrossRef]

68. Elkady, A.A.; Ibrahim, I.M. Protective effects of erdosteine against nephrotoxicity caused by gamma radiation in male albino rats. *Hum. Exp. Toxicol.* **2016**, *35*, 21–28. [CrossRef]

69. Chen, Z.Y.; Hu, Y.Y.; Hu, X.F.; Cheng, L.X. The conditioned medium of human mesenchymal stromal cells reduces irradiation-induced damage in cardiac fibroblast cells. *J. Radiat. Res.* **2018**, *59*, 555–564. [CrossRef]

70. Zhao, F.; Wang, Z.; Zheng, R. Efficacy of calf spleen extract injection combined with chemotherapy in the treatment of advanced gastric cancer. *J. Bengbu Med. Coll.* **2010**, *779*, 781.

71. Liu, Z.; Zhang, X.; Ma, J.; Wang, L.; Yang, Y. Calf spleen extract injection combined with docetaxel Saijiakapei gemcitabine treatment of patients with advanced breast cancer program in clinical research. *Cancer Res. Prev.* **2014**, *132*, 122–130.

72. Svirnovskiĭ, A.I.; Shimanskaia, T.V.; Bakun, A.V. Protivoluchevoe deĭstvie ėkstrakta regeneriruiushcheĭ selezenki. The radioprotective action of an extract of regenerating spleen. *Radiobiologiia* **1993**, *33*, 141–147. [PubMed]

73. Djordjevic, A.; Bogdanovic, G.; Dobric, S. Fullerenes in biomedicine. *J. Buon* **2006**, *11*, 391–404. [PubMed]

74. Trajković, S.; Dobrić, S.; Jaćević, V.; Dragojević-Simić, V.; Milovanović, Z.; Dordević, A. Tissue-protective effects of fullerenol C$_{60}$(OH)$_{24}$ and amifostine in irradiated rats. *Colloids Surf. B Biointerfaces* **2007**, *58*, 39–43. [CrossRef] [PubMed]

75. Chapman, T.R.; Kinsella, T.J. Ribonucleotide reductase inhibitors: A new look at an old target for radiosensitization. *Front. Oncol.* **2012**, *1*, 56. [CrossRef]

76. Wilson, J.D.; Hammond, E.M.; Higgins, G.S.; Petersson, K. Ultra-High Dose Rate (FLASH) Radiotherapy: Silver Bullet or Fool's Gold? *Front. Oncol.* **2020**, *9*, 1563. [CrossRef]

77. Jafarzadeh, A.; Nemati, M.; Khorramdelazad, H.; Hassan, Z.M. Immunomodulatory properties of cimetidine: Its therapeutic potentials for treatment of immune-related diseases. *Int. Immunopharmacol.* **2019**, *70*, 156–166. [CrossRef]

78. Little, M.H. Regrow or repair: Potential regenerative therapies for the kidney. *J. Am. Soc. Nephrol.* **2006**, *17*, 2390–2401. [CrossRef]

79. Linard, C.; Ropenga, A.; Vozenin-Brotons, M.C.; Chapel, A.; Mathe, D. Abdominal irradiation increases inflammatory cytokine expression and activates NF-kappa B in ratileal muscularis layer. *Am. J. Physiol. Gastrointest. Liver Physiol.* **2003**, *285*, 556–565. [CrossRef]

80. Wong, P.C.; Dodd, M.J.; Miaskowski, C.; Paul, S.M.; Bank, K.A.; Shiba, G.H.; Facione, N. Mucositis pain induced by radiation therapy: Prevalence, severity and use of self-care behaviors. *J. Pain Symptom Manag.* **2006**, *32*, 27–37. [CrossRef]

81. Duca, Y.; di Cataldo, A.; Russo, G.; Cannata, E.; Burgio, G.; Compagnone, M.; Alamo, A.; Condorelli, R.A.; La Vignera, S.; Calogero, A.E. Testicular Function of Childhood Cancer Survivors: Who Is Worse? *J. Clin. Med.* **2019**, *8*, 2204. [CrossRef]

82. Meistrich, M.L. Effects of chemotherapy and radiotherapy on spermatogenesis. *Eur. Urol.* **1993**, *23*, 136–141. [CrossRef]

83. Hermann, R.M.; Henkel, K.; Christiansen, H.; Vorwerk, H.; Hille, A.; Hess, C.F.; Schmidberger, H. Testicular dose and hormonal changes after radiotherapy of rectal cancer. *Radiother. Oncol.* **2005**, *75*, 83–88. [CrossRef]

84. Shen, H.; Ong, C. Detection of oxidative DNA damage in human sperm and its association with sperm function and male infertility. *Free Radic. Biol. Med.* **2000**, *28*, 529–536. [CrossRef]

85. Ash, P. The influence of radiation on fertility in man. *Br. J. Radiol.* **1980**, *53*, 271–278. [CrossRef]

86. Albuquerque, A.V.; Almeida, F.R.; Weng, C.C.; Shetty, G.; Meistrich, M.L.; Chiarini-Garcia, H. Spermatogonial behavior in rats during radiation-induced arrest and recovery after hormone suppression. *Reproduction* **2013**, *146*, 363–376. [CrossRef]

87. Chuai, Y.; Gao, F.; Li, B.; Zhao, L.; Qian, L.; Cao, F.; Wang, L.; Sun, X.; Cui, J.; Cai, J. Hydrogen-rich saline attenuates radiation-induced male germ cell loss in mice through reducing hydroxyl radicals. *Biochem. J.* **2012**, *442*, 49–56. [CrossRef]

88. Ding, J.; Wang, H.; Wu, Z.B.; Zhao, J.; Zhang, S.; Li, W. Protection of murine spermatogenesis against ionizing radiation-induced testicular injury by a green tea polyphenol. *Biol. Reprod.* **2015**, *92*, 6. [CrossRef]

89. Kim, J.; Lee, S.; Jeon, B.; Jang, W.; Moon, C.; Kim, S. Protection of spermatogenesis against gamma ray-induced damage by granulocytecolony-stimulating factor in mice. *Andrologia* **2011**, *43*, 87–93. [CrossRef] [PubMed]

90. Nguyen, T.D.; Maquart, F.X.; Monboisse, J.C. Ionizing radiations and collagen metabolism: From oxygen free radicals to radio-induced late fibrosis. *Radiat. Phys. Chem.* **2005**, *72*, 381–386. [CrossRef]

91. Cohen, E.P.; Robbins, M.E. Radiation nephropathy. *Semin. Nephrol.* **2003**, *23*, 486–499. [CrossRef]

92. Teixeira, V.P.; Boim, M.A.; Segreto, H.R.; Schor, N. Acute, subacute, and chronic x-ray effects on glomerular hemodynamics in rats. *Ren. Fail.* **1994**, *16*, 457–470. [CrossRef]

93. Withers, H.R.; Mason, K.A.; Thames, H.D., Jr. Late radiation response of kidney assayed by tubule-cell survival. *Br. J. Radiol.* **1986**, *59*, 587–595. [CrossRef] [PubMed]

94. Flora, S.J. Role of free radicals and antioxidants in health and disease. *Cell. Mol. Biol.* **2007**, *53*, 1–2.

95. Gobé, G.C.; Axelsen, R.A.; Harmon, B.V.; Allan, D.J. Cell death by apoptosis following X-irradiation of the foetal and neonatal rat kidney. *Int. J. Radiat. Biol.* **1988**, *54*, 567–576. [CrossRef]

96. Moulder, J.E.; Fish, B.L.; Regner, K.R.; Cohen, E.P.; Raife, T.J. Retinoic acid exacerbates experimental radiation nephropathy. *Radiat. Res.* **2002**, *157*, 199–203. [CrossRef]

97. Robbins, M.E.; O'Malley, Y.; Zhao, W.; Davis, C.S.; Bonsib, S.M. The role of tubulointerstitium in radiation-induced renal fibrosis. *Radiat. Res.* **2001**, *155*, 481–489. [CrossRef]

98. Hu, S.; Chen, Y.; Li, L.; Chen, J.; Wu, B.; Zhou, X.; Zhi, G.; Li, Q.; Wang, R.; Duan, H.; et al. Effects of adenovirus-mediated delivery of the human hepatocyte growth factor gene in experimental radiation-induced heart disease. *Int. J. Radiat. Oncol. Biol. Phys.* **2009**, *75*, 1537–1544. [CrossRef]

99. Blijlevens, N.; Sonis, S. Palifermin (recombinant keratinocyte growth factor-1): A pleiotropic growth factor with multiple biological activities in preventing chemotherapy- and radiotherapy-induced mucositis. *Ann. Oncol.* **2007**, *18*, 817–826. [CrossRef]

100. Lauritano, D.; Petruzzi, M.; Di Stasio, D.; Lucchese, A. Clinical effectiveness of palifermin in prevention and treatment of oral mucositis in children with acute lymphoblastic leukaemia: A case-control study. *Int. J. Oral Sci.* **2014**, *6*, 27–30. [CrossRef]

101. Spielberger, R.; Stiff, P.; Bensinger, W.; Gentile, T.; Weisdorf, D.; Kewalramani, T.; Shea, T.; Yanovich, S.; Hansen, K.; Noga, S.; et al. Palifermin for oral mucositis after intensive therapy for hematologic cancers. *N. Engl. J. Med.* **2004**, *351*, 2590–2598. [CrossRef] [PubMed]

102. Shankar, B.; Kumar, S.S.; Sainis, K.B. Generation of reactive oxygen species and radiation response in lymphocytes and tumour cells. *Radiat. Res.* **2003**, *160*, 478–487. [CrossRef]

103. Komatsu, W.; Ishihara, K.; Murata, M.; Saito, H.; Shinohara, K. Docosahexaenoic acid suppresses nitric oxide production and inducible nitric oxide synthase expression in interferon-gamma plus lipopolysaccharide stimulated murine macrophages by inhibiting the oxidative stress. *Free Radic. Biol. Med.* **2003**, *34*, 1006–1016. [CrossRef]

104. Sneddon, A.A.; Wu, H.C.; Farquharson, A.; Grant, I.; Arthur, J.R.; Rotondo, D.; Choe, S.-N.; Wahle, K.W.J. Regulation of selenoprotein GPx4 expression and activity in human endothelial cells by fatty acids, cytokines and antioxidants. *Atherosclerosis* **2003**, *171*, 57–65. [CrossRef]

105. Wen, B.; Deutsch, E.; Opolon, P.; Auperin, A.; Frascogna, V.; Connault, E.; Bourhis, J. n-3 polyunsaturated fatty acids decrease mucosal/epidermal reactions and enhance antitumour effect of ionising radiation with inhibition of tumour angiogenesis. *Br. J. Cancer* **2003**, *89*, 1102–1107. [CrossRef]

106. Minami, Y.; Miyata, H.; Doki, Y.; Yano, M.; Yamasaki, M.; Takiguchi, S.; Fujiwara, Y.; Yasuda, T.; Monden, M. omega-3 Fatty acid-containing diet (Racol) reduces toxicity of chemoradiation therapy for patients with esophageal cancer. *Gan Kagaku Ryoho* **2008**, *35*, 437–440.

107. Briehl, M.M.; Cotgreave, I.A.; Powis, G. Downregulation of the antioxidant defence during glucocorticoid-mediated apoptosis. *Cell Death Differ.* **1995**, *2*, 41–46.

108. Burington, B.; Barlogie, B.; Zhan, F.; Crowley, J.; Shaughnessy, J.D., Jr. Tumor cell gene expression changes following short-term in vivo exposure to single agent chemotherapeutics are related to survival in multiple myeloma. *Clin. Cancer Res.* **2008**, *14*, 4821–4829. [CrossRef] [PubMed]

109. Bera, S.; Greiner, S.; Choudhury, A.; Dispenzieri, A.; Spitz, D.R.; Russell, S.J.; Apollina Goel, A. Dexamethasone-induced oxidative stress enhances myeloma cell radiosensitization while sparing normal bone marrow hematopoiesis. *Neoplasia* **2010**, *12*, 980–992. [CrossRef]

110. Magda, D.; Lepp, C.; Gerasimchuk, N.; Lee, I.; Sessler, J.L.; Lin, A.; Biaglow, J.E.; Miller, R.A. Redox cycling by motexafin gadolinium enhances cellular response to ionizing radiation by forming reactive oxygen species. *Int. J. Radiat. Oncol. Biol. Phys.* **2001**, *51*, 1025–1036. [CrossRef]

111. Skvortsova, I.; Popper, B.A.; Skvortsov, S.; Saurer, M.; Auer, T.; Moser, R.; Kamleitner, H.; Zwierzina, H.; Lukas, P. Pretreatment with rituximab enhances radiosensitivity of non-Hodgkin's lymphoma cells. *J. Radiat Res.* **2005**, *46*, 241–248. [CrossRef] [PubMed]

112. Skvortsova, I.; Skvortsov, S.; Popper, B.A.; Haidenberger, A.; Saurer, M.; Gunkel, A.R.; Zwierzina, H.; Lukas, P. Rituximab enhances radiation-triggered apoptosis in non-Hodgkin's lymphoma cells via caspase-dependent and -independent mechanisms. *J. Radiat. Res.* **2006**, *47*, 183–196. [CrossRef] [PubMed]

113. Kapadia, N.S.; Engles, J.M.; Wahl, R.L. In vitro evaluation of radioprotective and radiosensitizing effects of rituximab. *J. Nucl. Med.* **2008**, *49*, 674–678. [CrossRef] [PubMed]

114. Fengling, M.; Fenju, L.; Wanxin, W.; Lijia, Z.; Jiandong, T.; Zu, W.; Xin, Y.; Qingxiang, G. Rituximab sensitizes a Burkitt lymphoma cell line to cell killing by X-irradiation. *Radiat. Environ. Biophys.* **2009**, *48*, 371–378. [CrossRef] [PubMed]

115. Zimta, A.A.; Cenariu, D.; Irimie, A.; Magdo, L.; Nabavi, S.M.; Atanasov, A.G.; Berindan-Neagoe, I. The Role of Nrf2 Activity in Cancer Development and Progression. *Cancers* **2019**, *11*, 1755. [CrossRef]

116. Wu, S.; Lu, H.; Bai, Y. Nrf2 in cancers: A double-edged sword. *Cancer Med.* **2019**, *8*, 2252–2267. [CrossRef]

117. Wang, X.J.; Hayes, J.D.; Henderson, C.J.; Wolf, C.R. Identification of retinoic acid as an inhibitor of transcription factor Nrf2 through activation of retinoic acid receptor alpha. *Proc. Natl. Acad. Sci. USA* **2007**, *104*, 19589–19594. [CrossRef]

118. Do, M.T.; Kim, H.G.; Khanal, T.; Choi, J.H.; Kim, D.H.; Jeong, T.C.; Jeong, H.G. Metformin inhibits heme oxygenase-1 expression in cancer cells through inactivation of Raf-ERK-Nrf2 signaling and AMPK-independent pathways. *Toxicol. Appl. Pharmacol.* **2013**, *271*, 229–238. [CrossRef]

119. Allegra, A.; Innao, V.; Russo, S.; Gerace, D.; Alonci, A.; Musolino, C. Anticancer Activity of Curcumin and Its Analogues: Preclinical and Clinical Studies. *Cancer Investig.* **2017**, *35*, 1–22. [CrossRef]

120. Krishnan, S.; Sandur, S.K.; Shentu, S.; Aggarwal, B.B. Curcumin enhances colorectal cancer cell radiosensitivity by suppressing the radiation-induced nuclear factorkappaB (NF-kB) pathway. *Int. J. Radiat. Oncol. Biol. Phys.* **2006**, *66*, S547. [CrossRef]

121. Jayakumar, S.; Patwardhan, R.S.; Pal, D.; Sharma, D.; Sandur, S.K. Dimethoxycurcumin, a metabolically stable analogue of curcumin enhances the radiosensitivity of cancer cells: Possible involvement of ROS and thioredoxin reductase. *Biochem. Biophys. Res. Commun.* **2016**, *478*, 446–454. [CrossRef] [PubMed]

122. Adams, G.E.; Dewey, D.L. Hydrated electrons and radiobiological sensitization. *Biochem. Biophys. Res. Commun.* **1963**, *12*, 473–477. [CrossRef]

123. Biaglow, J.E.; Varnes, M.E.; Clark, E.P.; Epp, E.R.R. The role of thiols in cellular response to radiation and drugs. *Radiat. Res.* **1983**, *95*, 437–455. [CrossRef] [PubMed]

124. Biaglow, J.E.; Varnes, M.E.; Epp, E.R.; Clark, E.P.; Tuttle, S.; Held, K.D. Role of glutathione in the aerobic radiation response. *Int. J. Radiat. Oncol. Biol. Phys.* **1989**, *16*, 1311–1314. [CrossRef]

125. Biaglow, J.E.; Mitchell, J.B.; Held, K. The importance of peroxide and superoxide in the x-ray response. *Int. J. Radiat. Oncol. Biol. Phys.* **1992**, *22*, 665–669. [CrossRef]

126. Siim, B.G.; van Zijl, P.L.; Brown, J.M. Tirapazamine-induced DNA damage measured using the comet assay correlates with cytotoxicity towards tumour cells in vitro. *Br. J. Cancer* **1996**, *73*, 952–960. [CrossRef]

127. Coleman, C.N.; Beard, C.J.; Hlatky, L. Biochemical modifiers: Hypoxic cell sensitizers. In *Radiation Oncology: Technology and Biology*; Mauch, P.M., Loeffler, J.S., Eds.; W.B. Saunders: Philadelphia, PA, USA, 1994; pp. 56–89.

128. Biaglow, J.E.; Varnes, M.E.; Roizen-Towle, L.; Clark, E.P.; Epp, E.R.; Astor, M.B.; Hall, E.J. Biochemistry of reduction of nitro heterocycles. *Biochem. Pharmacol.* **1986**, *35*, 77–90. [CrossRef]

129. Jacobson, B.; Biaglow, J.E.; Fielden, E.M.; Adams, G.E. Respiratory effects and ascorbate reactions with misonidazole and other recently developed drugs. *Cancer Clin. Trials* **1980**, *3*, 47–53.

130. Mengeaud, V.; Nano, J.L.; Fournel, S.; Rampal, P. Effects of eicosapentaenoic acid, gamma-linolenic acid and prostaglandin E1 on three human colon carcinoma cell lines. *Prostaglandins Leukot Essent Fatty Acids* **1992**, *47*, 313–319. [CrossRef]

131. Ng, Y.; Barhoumi, R.; Tjalkens, R.B.; Fan, Y.Y.; Kolar, S.; Wang, N.; Lupton, J.R.; Chapkin, R.S. The role of docosahexaenoic acid in mediating mitochondrial membrane lipid oxidation and apoptosis in colonocytes. *Carcinogenesis* **2005**, *26*, 1914–1921. [CrossRef]

132. Dommels, Y.E.; Haring, M.M.; Keestra, N.G.; Alink, G.M.; van Bladeren, P.J.; van Ommen, B. The role of cyclooxygenase in n-6 and n-3 polyunsaturated fatty acid mediated effects on cell proliferation, PGE synthesis and cytotoxicity in human colorectal carcinoma cell lines. *Carcinogenesis* **2003**, *24*, 385–392. [CrossRef] [PubMed]

133. Maziere, C.; Conte, M.A.; Degonville, J.; Ali, D.; Maziere, J.C. Cellular enrichment with polyunsaturated fatty acids induces an oxidative stress and activates the transcription factors AP1 and NFkappaB. *Biochem. Biophys. Res. Commun.* **1999**, *265*, 116–122. [CrossRef] [PubMed]

134. Bubici, C.; Papa, S.; Dean, K.; Franzoso, G. Mutual cross-talk between reactive oxygen species and nuclear factor-kappa B: Molecular basis and biological significance. *Oncogene* **2006**, *25*, 6731–6748. [CrossRef] [PubMed]

135. Simon, H.U.; Haj-Yehia, A.; Levi-Schaffer, F. Role of reactive oxygen species (ROS) in apoptosis induction. *Apoptosis* **2000**, *5*, 415–418. [CrossRef] [PubMed]

136. Hofmanova, J.; Vaculova, A.; Lojek, A.; Kozubík, A. Interaction of polyunsaturated fatty acids and sodium butyrate during apoptosis inHT-29 human colon adenocarcinoma cells. *Eur. J. Nutr.* **2005**, *44*, 40–51. [CrossRef]

137. Sarsilmaz, M.; Songur, A.; Ozyurt, H.; Kus, I.; Ozen, O.A.; Ozyurt, B.; Söğüt, S.; Akyol, O. Potential role of dietary omega-3 essential fatty acids on some oxidant/antioxidant parameters in rats' corpus striatum. *Prostaglandins Leukot Essent Fatty Acids* **2003**, *69*, 253–259. [CrossRef]

138. Vartak, S.; Robbins, M.E.; Spector, A.A. Polyunsaturated fatty acids increase the sensitivity of 36B10 rat astrocytoma cells to radiation-induced cell kill. *Lipids* **1997**, *32*, 283–292. [CrossRef]

139. Zand, H.; Rahimipour, A.; Salimi, S.; Shafiee, S.M. Docosahexaenoic acid sensitizes Ramos cells to gamma-irradiation-induced apoptosis through involvement of PPAR-gamma activation and NF-kappaB suppression. *Mol. Cell. Biochem.* **2008**, *317*, 113–120. [CrossRef]

140. Colas, S.; Paon, L.; Denis, F.; Prat, M.; Louisot, P.; Hoinard, C.; le Floch, O.; Ogilvie, G.; Bougnoux, P. Enhanced radiosensitivity of rat autochthonous mammary tumors by dietary docosahexaenoic acid. *Int. J. Cancer* **2004**, *109*, 449–454. [CrossRef]

141. Calviello, G.; Serini, S.; Piccioni, E.; Pessina, G. Antineoplastic effects of n-3 polyunsaturated fatty acids in combination with drugs and radiotherapy: Preventive and therapeutic strategies. *Nutr. Cancer* **2009**, *61*, 287–301. [CrossRef]

142. Mane, S.D.; Kamatham, A.N. Ascorbyl stearate and ionizing radiation potentiate apoptosis through intracellular thiols and oxidative stress in murine T lymphoma cells. *Chem. Biol. Interact.* **2018**, *281*, 37–50. [CrossRef] [PubMed]

143. Uckun, F.M.; Dibirdik, I.; Qazi, S. Augmentation of the Antileukemia Potency of Total-Body Irradiation (TBI) by a Novel P-site Inhibitor of Spleen Tyrosine Kinase (SYK). *Radiat. Res.* **2010**, *174*, 526–531. [CrossRef] [PubMed]

144. Allegra, A.; Innao, V.; Allegra, A.G.; Pugliese, M.; di Salvo, E.; Ventura-Spagnolo, V.; Musolino, C.; Gangemi, S. Lymphocyte Subsets and Inflammatory Cytokines of Monoclonal Gammopathy of Undetermined Significance and Multiple Myeloma. *Int. J. Mol. Sci.* **2019**, *20*, 2822. [CrossRef] [PubMed]

145. Musolino, C.; Allegra, A.; Innao, V.; Allegra, A.G.; Pioggia, G.; Gangemi, S. Inflammatory and Anti-Inflammatory Equilibrium, Proliferative and Antiproliferative Balance: The Role of Cytokines in Multiple Myeloma. *Mediators Inflamm.* **2017**, *2017*, 1852517. [CrossRef] [PubMed]

146. Kharazmi, A.; Nielsen, H.; Rechnitzer, C.; Bendtzen, K. Interleukin 6 primes human neutrophil and monocyte oxidative burst response. *Immunol. Lett.* **1989**, *21*, 177–184. [CrossRef]

147. Behrens, M.M.; Ali, S.S.; Dugan, L.L. Interleukin-6 mediates the increase in NADPH-oxidase in the ketamine model of schizophrenia. *J. Neurosci.* **2008**, *28*, 13957–13966. [CrossRef] [PubMed]

148. Sun, Z.; Klein, A.S.; Radaeva, S.; Hong, F.; El-Assal, O.; Pan, H.N.; Jaruga, B.; Batkai, S.; Hoshino, S.; Tian, Z.; et al. In vitro interleukin-6 treatment prevents mortality associated with fatty liver transplants in rats. *Gastroenterology* **2003**, *125*, 202–215. [CrossRef]

149. Ward, N.S.; Waxman, A.B.; Homer, R.J.; Mantell, L.L.; Einarsson, O.; Du, Y.; Elias, J.A. Interleukin-6-induced protection in hyperoxic acute lung injury. *Am. J. Respir. Cell. Mol. Biol.* **2000**, *22*, 535–542. [CrossRef]

150. Gangemi, S.; Allegra, A.; Alonci, A.; Cristani, M.; Russo, S.; Speciale, A.; Penna, G.; Spatari, G.; Cannavò, A.; Bellomo, G.; et al. Increase of novel biomarkers for oxidative stress in patients with plasma cell disorders and in multiple myeloma patients with bone lesions. *Inflamm. Res.* **2012**, *61*, 1063–1067. [CrossRef]

151. Allegra, A.; Pace, E.; Tartarisco, G.; Innao, V.; DI Salvo, E.; Allegra, A.G.; Ferraro, M.; Musolino, C.; Gangemi, S. Changes in Serum Interleukin-8 and sRAGE Levels in Multiple Myeloma Patients. *Anticancer Res.* **2020**, *40*, 1443–1449. [CrossRef]

152. Ettari, R.; Zappalà, M.; Grasso, S.; Musolino, C.; Innao, V.; Allegra, A. Immunoproteasome-selective and non-selective inhibitors: A promising approach for the treatment of multiple myeloma. *Pharmacol. Ther.* **2018**, *182*, 176–192. [CrossRef]

153. Kuku, I.; Aydogdu, I.; Bayraktar, N.; Kaya, E.; Akyol, O.; Erkurt, M.A. Oxidant/antioxidant parameters and their relationship with medical treatment in multiple myeloma. *Cell. Biochem. Funct.* **2005**, *23*, 47–50. [CrossRef]

154. Sharma, A.; Tripathi, M.; Satyam, A.; Kumar, L. Study of antioxidant levels in patients with multiple myeloma. *Leuk. Lymphoma* **2009**, *50*, 809–815. [CrossRef]

155. Zima, T.; Spicka, I.; Stipek, S.; Crkovska, J.; Platenik, J.; Merta, M.; Nemecek, K.; Tesar, V. Lipid peroxidation and activity of antioxidative enzymes in patients with multiple myeloma. *Cas. Lek. Cesk.* **1996**, *135*, 14–17.

156. Goel, A.; Spitz, D.R.; Weiner, G.J. Manipulation of cellular redox metabolism for improving therapeutic responses in B-cell lymphoma and multiple myeloma. *J. Cell. Biochem.* **2011**, *113*, 419–425. [CrossRef]

157. Brown, C.O.; Salem, K.; Wagner, B.A.; Soumen Bera, S.; Singh, N.; Tiwari, A.; Choudhury, A.; Buettner, G.R.; Goel, A. Interleukin-6 counteracts therapy-induced cellular oxidative stress in multiple myeloma by up-regulating manganese superoxide dismutase. *Biochem. J.* **2012**, *444*, 515–527. [CrossRef]

158. Gougelet, A.; Mansuy, A.; Blay, J.Y.; Alberti, L.; Vermot-Desroches, C. Lymphoma and myeloma cell resistance to cytotoxic agents and ionizing radiations is not affected by exposure to anti-IL-6 antibody. *PLoS ONE* **2009**, *4*, e8026. [CrossRef]

159. Jaramillo, M.C.; Frye, J.B.; Crapo, J.D.; Briehl, M.M.; Tome, M.E. Increased manganese superoxide dismutase expression or treatment with manganese porphyrin potentiates dexamethasone-induced apoptosis in lymphoma cells. *Cancer Res.* **2009**, *69*, 5450–5457. [CrossRef]

160. Yi, J.; Gao, F.; Shi, G.; Li, H.; Wang, Z.; Shi, X.; Tang, X. The inherent cellular level of reactive oxygen species: One of the mechanisms determining apoptotic susceptibility of leukemic cells to arsenic trioxide. *Apoptosis* **2002**, *7*, 209–215. [CrossRef]

161. Bellosillo, B.; Villamor, N.; López-Guillermo, A.; Marcé, S.; Esteve, J.; Campo, E.; Colomer, D.; Montserrat, E. Complement-mediated cell death induced by rituximab in B-cell lymphoproliferative disorders is mediated in vitro by a caspase-independent mechanism involving the generation of reactive oxygen species. *Blood* **2001**, *98*, 2771–2777. [CrossRef] [PubMed]

162. Lee, S.; Lim, M.-J.; Kim, M.-H.; Yu, C.H.; Yun, Y.S.; Ahn, J.; Song, J.Y. An effective strategy for increasing the radiosensitivity of Human lung Cancer cells by blocking Nrf2-dependent antioxidant responses. *Free Radic. Biol. Med.* **2012**, *53*, 807–816. [CrossRef] [PubMed]

163. Allegra, A.; Penna, G.; Alonci, A.; Rizzo, V.; Russo, S.; Musolino, C. Nanoparticles in oncology: The new theragnostic molecules. *Anticancer Agents Med. Chem.* **2011**, *11*, 669–686. [CrossRef] [PubMed]

164. Ma, N.; Liu, P.; He, N.; Gu, N.; Wu, F.G.; Chen, Z. Action of Gold Nanospikes-Based Nanoradiosensitizers: Cellular Internalization, Radiotherapy, and Autophagy. *ACS Appl. Mater. Interfaces* **2017**, *9*, 31526–31542. [CrossRef] [PubMed]

165. Plunkett, W.; Gandhi, V.; Huang, P.; Robertson, L.E.; Yang, L.Y.; Gregoire, V.; Estey, E.; Keating, M.J. Fludarabine: Pharmacokinetics, mechanisms of action, and rationales for combination therapies. *Semin. Oncol.* **1993**, *20*, 2–12.

166. Laurent, D.; Pradier, O.; Schmidberger, H.; Rave-Fränk, M.; Frankenberg, D.; Hess, C.F. Radiation rendered more cytotoxic by fludarabine monophosphate in a human oropharynx carcinoma cell-line than in fetal lung fibroblasts. *J. Cancer Res. Clin. Oncol.* **1998**, *124*, 485–492. [CrossRef]

167. Donehower, R.C. An overview of the clinical experience with hydroxyurea. *Semin. Oncol.* **1992**, *19*, 11–19.

168. Vokes, E.; Panje, W.; Schilsky, R.; Mick, R.; Awan, A.; Moran, W.; Goldman, M.; Tybor, A.; Weichselbaum, R. Hydroxyurea, fluorouracil, and concomitant radiotherapy in poor prognosis head and neck cancer: A phase I-II study. *J. Clin. Oncol.* **1989**, *7*, 761–768. [CrossRef]

169. Allegra, A.; Musolino, C.; Tonacci, A.; Pioggia, G.; Gangemi, S. Interactions between the MicroRNAs and Microbiota in Cancer Development: Roles and Therapeutic opportunities. *Cancers* **2020**, *12*, 805. [CrossRef]

170. Pajic, M.; Froio, D.; Daly, S.; Doculara, L.; Millar, E.; Graham, P.H.; Drury, A.; Steinmann, A.; de Bock, C.E.; Boulghourjian, A.; et al. miR-139-5p Modulates Radiotherapy Resistance in Breast Cancer by Repressing Multiple Gene Networks of DNA Repair and ROS Defense. *Cancer Res.* **2018**, *78*, 501–515. [CrossRef]
171. Mao, Z.; Liu, S.; Cai, J.; Huang, Z.Z.; Lu, S.C. Cloning and functional characterization of the 5′-flanking region of human methionine adenosyltransferase 2A gene. *Biochem. Biophys. Res. Commun.* **1998**, *248*, 479–484. [CrossRef]
172. Innao, V.; Allegra, A.; Pulvirenti, N.; Allegra, A.G.; Musolino, C. Therapeutic potential of antagomiRs in haematological and oncological neoplasms. *Eur. J. Cancer Care* **2020**, *29*, e13208. [CrossRef] [PubMed]
173. Allegra, A.; Innao, V.; Basile, G.; Pugliese, M.; Allegra, A.G.; Pulvirenti, N.; Musolino, C. Post-chemotherapy cognitive impairment in hematological patients: Current understanding of chemobrain in hematology. *Expert Rev. Hematol.* **2020**, *13*, 393–404. [CrossRef] [PubMed]
174. Omuro, A.M.; Ben-Porat, L.S.; Panageas, K.S.; Kim, A.K.; Correa, D.D.; Yahalom, J.; Deangelis, L.M.; Abrey, L.E. Delayed neurotoxicity in primary central nervous system lymphoma. *Arch. Neurol.* **2005**, *62*, 1595–1600. [CrossRef]
175. Weibrich, G.; Kleis, W.K.; Hafner, G.; Hitzler, W.E. Growth factor levels in platelet rich plasma and correlations with donor age, sex, and platelet count. *J. Craniomaxillofac. Surg.* **2002**, *30*, 97–102. [CrossRef] [PubMed]
176. Soliman, A.F.; Saif-Elnasr, M.; Abdel Fattah, S.M. Platelet-rich plasma ameliorates gamma radiation-induced nephrotoxicity via modulating oxidative stress and apoptosis. *Life Sci.* **2019**, *219*, 238–247. [CrossRef]
177. Szumiel, I. From radioresistance to radiosensitivity: In vitro evolution of L5178Y lymphoma. *Int. J. Radiat. Biol.* **2015**, *91*, 465–471. [CrossRef]

Publisher's Note: MDPI stays neutral with regard to jurisdictional claims in published maps and institutional affiliations.

Article

Plasma Treated Water Solutions in Cancer Treatments: The Contrasting Role of RNS

Eloisa Sardella [1,*], Valeria Veronico [2], Roberto Gristina [1], Loris Grossi [3], Savino Cosmai [1], Marinella Striccoli [4], Maura Buttiglione [5], Francesco Fracassi [1,2] and Pietro Favia [1,2]

[1] CNR- Istituto di Nanotecnologia (CNR-NANOTEC) UoS Bari, c/o Dipartimento di Chimica, Università degli Studi di Bari Aldo Moro, via Orabona, 4, 70126 Bari, Italy; roberto.gristina@cnr.it (R.G.); savino.cosmai@cnr.it (S.C.); francesco.fracassi@uniba.it (F.F.); pietro.favia@uniba.it (P.F.)

[2] Dipartimento di Chimica, Università degli Studi di Bari Aldo Moro, via Orabona, 4, 70126 Bari, Italy; valeria.veronico@uniba.it

[3] Dipartimento di "Scienze per la Qualità della Vita" Università di Bologna, Corso d'Augusto, 237, I-47921 Rimini, Italy; loris.grossi@unibo.it

[4] CNR-Istituto per i Processi Chimico-Fisici UoS Bari, via Orabona, 4, 70126 Bari, Italy; m.striccoli@ba.ipcf.cnr.it

[5] Dipartimento di Scienze Biomediche e Oncologia Umana, Scuola di Medicina, Università degli Studi di Bari Aldo Moro, 70124 Bari, Italy; maura.buttiglione@uniba.it

* Correspondence: eloisa.sardella@cnr.it; Tel.: +39-080-544-2295

Abstract: Plasma Treated Water Solutions (PTWS) recently emerged as a novel tool for the generation of Reactive Oxygen and Nitrogen Species (ROS and RNS) in liquids. The presence of ROS with a strong oxidative power, like hydrogen peroxide (H_2O_2), has been proposed as the main effector for the cancer-killing properties of PTWS. A protective role has been postulated for RNS, with nitric oxide (NO) being involved in the activation of antioxidant responses and cell survival. However, recent evidences proved that NO-derivatives in proper mixtures with ROS in PTWS could enhance rather than reduce the selectivity of PTWS-induced cancer cell death through the inhibition of specific antioxidant cancer defenses. In this paper we discuss the formation of RNS in different liquids with a Dielectric Barrier Discharge (DBD), to show that NO is absent in PTWS of complex composition like plasma treated (PT)-cell culture media used for in vitro experiments, as well as its supposed protective role. Nitrite anions (NO_2^-) instead, present in our PTWS, were found to improve the selective death of Saos2 cancer cells compared to EA.hy926 cells by decreasing the cytotoxic threshold of H_2O_2 to non-toxic values for the endothelial cell line.

Keywords: cold atmospheric plasma; reactive oxygen and nitrogen species; oxidative stress; nitrite; cancer treatment

Citation: Sardella, E.; Veronico, V.; Gristina, R.; Grossi, L.; Cosmai, S.; Striccoli, M.; Buttiglione, M.; Fracassi, F.; Favia, P. Plasma Treated Water Solutions in Cancer Treatments: The Contrasting Role of RNS. *Antioxidants* **2021**, *10*, 605. https://doi.org/10.3390/antiox10040605

Academic Editors: Cinzia Domenicotti, Barbara Marengo and Mario Allegra

Received: 3 February 2021
Accepted: 11 April 2021
Published: 14 April 2021

Publisher's Note: MDPI stays neutral with regard to jurisdictional claims in published maps and institutional affiliations.

1. Introduction

Non-Thermal Plasmas (NTP) are partially ionized gases consisting of chemically reactive species such as ions, neutrals, free radicals, electrons, UV-VIS radiation, and electric field. Besides natural events, such as the Aurora Borealis, NTP can be easily generated by applying an electric field to a neutral gas. When nitrogen and oxygen feeds are used, NTP generate ROS and RNS. In this case, NTP are often tested for biomedical purposes through the exposure of biological targets directly to NTP [1] or to liquids or solutions enriched both with ROS and RNS (RONS) by means of NTP [2].

Since it has been established that the efficacy of conventional anti-cancer therapies, chemo- and radiotherapy, actually relies on a release of ROS in cancer cells, PTWS have been widely tested as novel anti-cancer ROS-generating systems, and proved to cause the selective damaging of cancer cells in vitro [3–6], but also a significant inhibition of tumors growth in vivo [7,8]. The mechanism of PTWS-induced cancer inhibition, however, still needs to be fully elucidated [9–11]. So far it is known that the most stable and abundant ROS and RNS in PTWS are H_2O_2, NO_2^-, and nitrate (NO_3^-) ions [12,13]; plasma phase reactions, as well as cross reactions among RONS and liquid components, however, can also produce short-lived molecules such as ozone (O_3), superoxide anion ($O_2^{\bullet-}$), nitric oxide ($^\bullet NO$), and peroxynitrite anion ($ONOO^-$) [14,15].

The presence of ROS with a strong oxidative power like H_2O_2 has been soon argued as one of the main effectors of PTWS-induced cancer cell death [16]; the cytotoxicity of PTWS, though, proved to be generally higher than simple water solutions of H_2O_2 [17].

The biological role postulated for RNS, on the contrary, is controversial. A protective role has been first proposed due to the biological role of nitric oxide. $^\bullet NO$ is a unique diffusible molecular messenger in the vascular and immune system involved in neurotransmission [18], stimulation of the immune system [19], and protection mechanisms against pathogens and cancer cells [20]. Many diseases, including cancer, have been associated to restricted $^\bullet NO$ bioavailability [21] and $^\bullet NO$ donor drugs are nowadays used for many therapeutic purposes [22–24]. In the case of strong oxidative stress, it is quite established that $^\bullet NO$ can play a protective role [25,26] through the upregulation of protective proteins such as heat shock proteins, cyclooxygenase-2, or heme oxygenase-1 as endogenous cytoprotection [27,28], eventually promoting cancer cell survival. The presence of $^\bullet NO$ could thus decrease the potential clinical efficacy of PTWS, due to the antioxidant response induced over cancer cells.

In contrast with this evidence, however, it was also demonstrated that high levels of $^\bullet NO$ may induce direct DNA damage, or lead to the formation of the highly oxidative peroxynitrite $ONOO^-$ [22,29], which, in turn, can damage lipids, DNA, and proteins, addressing cells to necrosis or apoptosis [30]. One of the most important examples of this conversion is the reaction of NO with ROS, especially with $O_2^{\bullet-}$ [31]. Indeed, there is also evidence that nitric oxide synthases (NOX) on tumor cells membranes can produce O_2^- in addition to NO under stress conditions [30], thus generating ideal conditions for the local formation of $ONOO^-$. Under these conditions, therefore, NO was found to strengthen rather than decrease the oxidative stress induced by plasma-generated ROS on cancer cells [29,30]. Regarding this topic, Kamm et al. recently discussed the importance of even exogenous NO-derivatives, nitrites and nitrates, as an alternative way to produce intracellular bioactive $^\bullet NO$, to inhibit cancer proliferation and induce cancer cell death [31]. Furthermore, recent evidence proved the enhancement of selective cancer cell death due to a synergic action of NO-derivatives with oxygenated species also through PTWS [32,33].

Therefore, a double-edge role of RNS as protectors and inducers of oxidative stress emerged. The control of intracellular RNS may thus offer a sophisticated strategy to tune the anticancer effects of RONS-based therapies, including PTWS. To carefully balance the composition of PTWS, however, a deep knowledge of the chemical routes of formation for each species is required.

To date, this knowledge has not been fully achieved. The lack of reliable detection methods of RONS in complex matrices like PTWS, is currently the biggest limitation. Due to the complexity of the study, most of biological effects were reported for PTWS produced by plasma treatment of cell culture media (i.e., liquid media containing carbohydrates, vitamins, growth factors, amino acids, and inorganic salts) to be used in vitro [11,34–37] while, for sake of simplicity, most of the chemical studies investigating RNS formation in PTWS have been performed in simple model liquids like water or similar [38]. With specific regard to $^\bullet NO$, however, chemical routes of formation in PTWS seems to be strongly affected not only by the environment where the plasma is ignited, but also by the chemical composition of the liquid to be treated. In general, the presence of the organic compounds required to improve the biological compatibility of PTWS not only increases the complexity of the study but provides alternative routes for the suppression or promotion of plasma-diffused ROS or RNS in the liquid. Chemical studies, extensively analyzing the strategies for modulation of RNS species even in PTWS of complex composition, are therefore of extreme importance.

The aim of this paper is to investigate the capability of NTP to generate $^\bullet NO$ and its derivatives such as NO_2^- ions in different liquids exposed to a DBD plasma source, with the purpose to better understand the supposed biological role of RNS in PTWS and in promoting anti-cancer effects. For this purpose, Dulbecco Modified Eagle Medium (DMEM), a cell culture medium containing more than 30 organic components (vitamins, carbohydrates, amino acids, antibiotics, and inorganic salts) is here used for the first time to investigate routes of formation of the most important RNS in comparison with water.

Besides the composition of the liquid target, the DBD gas feed was tuned from O_2 to synthetic air, and N_2 in order to demonstrate how plasma composition can be programmed so as to, respectively, exclude or promote the presence of RNS species (i.e., NO_2^-) in PTWS synthesized from DMEM. RNS-enriched and RNS-free PTWS were then used for the incubation of osteosarcoma cancer cell lines (i.e., Saos2) and hybrid endothelial cells (i.e., Ea.hy926 cells), and for ultimately confirming that NO-derivatives increase the extent and selectivity of ROS-induced death of Saos2 cells with respect to the endothelial EA.hy926 ones.

Most tumors are characterized by an inflammatory microenvironment, and correlations between inflammation and cancer progression have been extensively reported in literature [39–41]. It is well established that cancer cells can fuse with endothelial cells to form hybrid cells spontaneously, which facilitates cancer cells traversing the endothelial barrier to form metastases [42]. However, up to now, little is known about the biologic characteristics of hybrid cells. Endothelial cells (EC), as part of the tumor microenvironment, play a crucial role in inflammatory processes, as well as in angiogenesis and could be critical targets of cancer therapy like, as an example irradiation [43], and as in our case, exposure to plasma treated water solutions. For this reason, our work is aimed to assess whether a different behavior is observed for two cell lines used together to represent a study model of cancer, the Saos2, and hybrid endothelial cells, Eahy926, with tumor mass-vessel wall characteristics. Eahy926 cell line was derived from fusion of human umbilical vein endothelial cells with human lung adenocarcinoma cell line A549. A study published in 2009 [44] showed that in this cell line the proliferation ability was similar to that of A549 cells, but the ability in adhesion and migration of Eahy926 cells was higher. In addition, Eahy926 cells had weaker ability of invasion and could not form tumor mass. These characteristics led us to use them as an experimental model to compare with the Saos2 cancer cell line. A recent paper shows that ECs, as part of the inflammatory microenvironment of tumors, are important regulators of the actual tumor response to radiation therapy [43]. Moreover, EAhy926 have been used as model cells to assess whether cordycepin regulates proliferation, migration and angiogenesis in the hepatoma [45]. Following these features we used Saos2 and EAhy926 cell lines as cell model to gain new insights in the role of RNS in combination with ROS against cancer. Our results, therefore, indicate how PTWS could be a powerful and sophisticated tool to balance RNS- and ROS-induced cellular stress in cancer treatment applications, and support the emerging role of NO-derivatives mediated by PTWS as inducers rather than protectors of ROS-induced oxidative stress in cancer cells.

2. Materials and Methods

2.1. Plasma Setup and PTWS Generation

Cold atmospheric plasmas were generated in a DBD PetriPlas$^+$ source, engineered by the Leibniz Institute for Plasma Science and Technologies (INP, Greifswald, GER), and modified by co-authors of this paper. The source consists of a Plexiglas flow unit set with gas connections, sockets, and a discharge unit (Figure 1a), as schematized in Figure 1b.

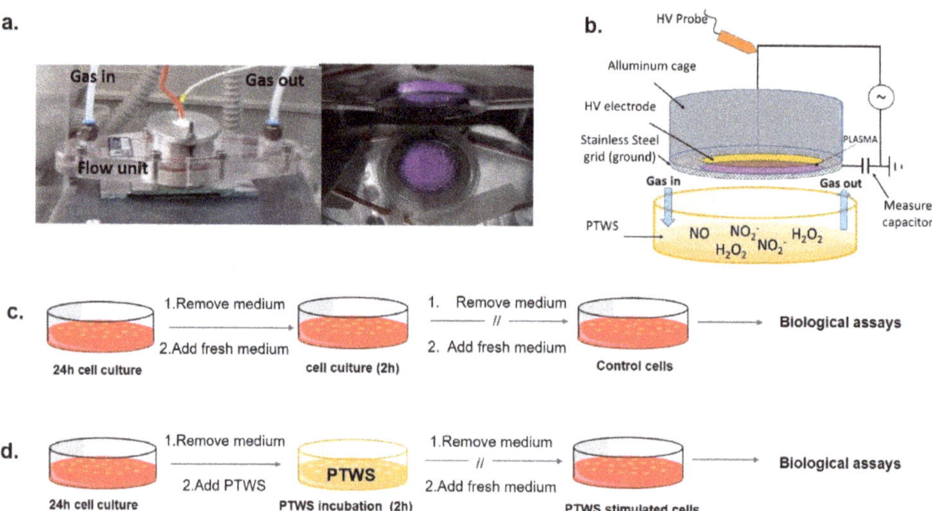

Figure 1. Experimental apparatus and scheme of the biological testing (**a**): left, Plexiglas chamber containing the modified Petri Plas+ plasma source; right, bottom-view of the plasma in air gas feed. (**b**) scheme of the source facing a Petri dish containing the liquid to be treated. After 24 h of cell growth, cells were incubated with 2 mL of (**c**) untreated medium (in case of control samples) and (**d**) Plasma Treated-DMEM (PT-DMEM, in case of PTWS stimulated cells) for 2 h at 37 °C, 5% CO$_2$. After, the liquid was replaced with a cell culture medium, and cell cultures were prolonged until biological assays.

A full description of the experimental apparatus has been provided in our previous papers [5]. An electric field of 6 kHz frequency and peak-to-peak voltage of 13 kV was applied with a power supply connected to a programmable 10 MHz DDS (Direct digital synthesis) function generator (TG1010A, Aim-TTi, Huntingdon, UK) to ignite a volume discharge as wide as the HV electrode (Figure 1a).

Plasma treatments were performed on a commercial TPP® Petri dish (57 mm diameter, Techno Plastic Products (TPP), Trasadingen, CH) containing 2 mL of liquids, positioned beneath the source, at a distance between the liquid and the ground grid of 3 mm. A closed system is set in this way, whose chemical composition can be properly conditioned by purging it with the gas feed before igniting the plasma. High glucose DMEM (catalognumberD1145, Sigma–Aldrich, St. Louis, MO, USA) without phenol red, supplemented with 10% v/v Fetal Bovine Serum (FBS, cat. N. F7524 Sigma–Aldrich), 2mM L-glutamine (cat. N. G7513, Sigma–Aldrich), penicillin and streptomycin solution (20 units mL^{-1}/20 mg mL^{-1}, T4049, Sigma–Aldrich) was used for plasma treatments and, when untreated, as a reference in all biological experiments; for Electron Paramagnetic Resonance (EPR) analyses also serum free DMEM was used to assess the ability of components of DMEM to scavenge •NO; water and water solutions were also used for the EPR experiments. The energy dose delivered during the discharges was calculated with the voltage–charge (i.e., the time integral of the current) Lissajous method described in our previous works [5,6]. Data are summarized in Table 1.

Table 1. Energy doses (J cm^{-2}) of plasma in DBD ignited in different feed gases (O$_2$, N$_2$, and synthetic Air with 0.5 slm of flow rate) and different treatment times. The experimental conditions are 13 kV, 6 kHz, and 25% DC with a period of 100 ms on 2 mL of DMEM.

Treatment Time (s)	Energy Dose (J cm^{-2})		
	O$_2$	N$_2$	Air
30	7.2 ± 0.3	6.12 ± 0.13	6.5 ± 0.7
60	14.5 ± 0.6	12.2 ± 0.3	12.9 ± 1.4
120	28.9 ± 1.3	24.5 ± 0.6	26 ± 3
180	43.4 ± 1.9	36.7 ± 0.8	39 ± 4

2.2. Detection of $^{\bullet}$NO in PTWS

EPR measurements were carried out using an iron(II) N, N-diethyldithiocarbamate, Fe^{2+}(DETC)$_2$ spin trap, synthesized by mixing 25 mL of a water solution containing 2×10^{-2} M of diethyldithiocarbamate (DETC) with 100 mL of FeSO$_4 \cdot$7H$_2$O (10^{-2} M in water) for 60-min. The mixture was stirred for 2 h, and the precipitate Fe^{2+}(DETC)$_2$ was collected by filtration. All processes were run in an N$_2$ atmosphere. Since the spin trap is soluble only in organic solvents, in all EPR experiments a spin trap solution in dichloromethane (CH$_2$Cl$_2$, cat. N. 270997, Sigma–Aldrich), from now on named "trap solution", was used. PTWS or plasma effluents were put in contact with the trap solution by feeding the DBD chamber with N$_2$. The production of $^{\bullet}$NO in plasma phase was assessed by bubbling the gas effluents from the DBD directly in the trap solution for the entire duration of the plasma treatment. The sampling of the plasma effluents was carried out both with and without the liquid into the Petri dish (Figure 2a,b).

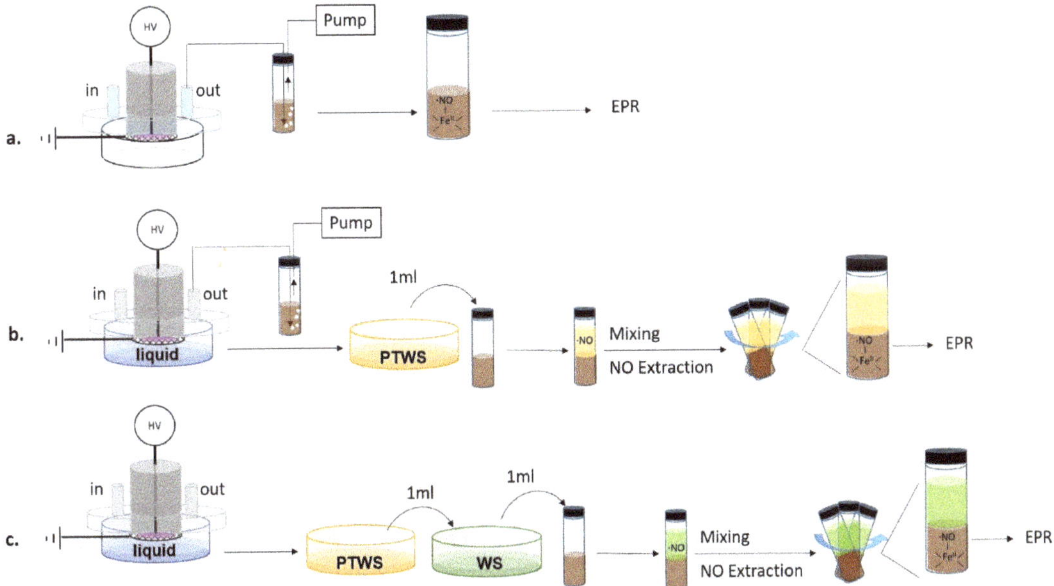

Figure 2. Scheme of the experimental procedures used for the Electron Paramagnetic Resonance (EPR) detection of nitric oxide. (**a**) Detection of nitric oxide ($^{\bullet}$NO) in plasma effluents without the liquid (**b**) Contemporary detection of $^{\bullet}$NO in plasma effluents and PTWS. (**c**) Detection of $^{\bullet}$NO in PTWS after mixture with untreated Water Solutions (WS) before the trapping step.

To analyze the $^\bullet$NO production in the liquid, 1 mL of PTWS was added to 1 mL of trap solution and mixed to promote the extraction of $^\bullet$NO from PTWS (Figure 2c).

$^\bullet$NO was also measured in PTW, NO_2^--water solution (170 μM, prepared by sodium nitrite, Sigma–Aldrich $\geq$ 99.0%); pH = 3), plasma-treated cysteine (PT-Cys) solution (0.006 g L^{-1}, prepared from cat. N. C8755, Sigma–Aldrich), plasma-treated glucose (PT-Glu) solution (45 g L^{-1} solution, prepared from D-(+)-Glucose solution 45% in H_2O, Sigma–Aldrich), and Plasma Treated DMEM (PT-DMEM). The same solutions, untreated, were used as reference (DMEM). In some cases, such liquids were mixed with other PT- or untreated water solutions (WS) as described in the results section. In all mixing experiments, 1 mL of starting liquid was mixed with 1 mL of the WS to be tested, then 1 mL of the mixture (starting liquid+ WS) was mixed with 1 mL of the spin trap.

The samples to be analyzed were transferred to a glass capillary and inserted into the cavity of a 9-GHz Varian E-9 Century Line EPR spectrometer. Instrumental settings were: 12 mT scan width (3000 points, for a spectral resolution of 0.004 mT per point); 100 kHz and 0.16 mT field modulation frequency and amplitude, respectively; 9.5 GHz and 15 mW microwave frequency and power, respectively; 0.064 s time constant; 8 min scan time; room temperature.

As a method alternative to EPR, a spectrofluorometric assay with 4-Amino-5-Methylamino-2$'$,7$'$-Difluorofluorescein Diacetate (DAF-FM-DA, cat. N. D2384, Thermo Fisher Scientific, Waltham, MA, USA) was used to assess the presence of $^\bullet$NO in PTWSs. DAF-FM was obtained by basic hydrolysis of DAF-FM-DA [46]. Spectrofluorometric analyses were performed with a Fluorolog-3 spectrofluorometer (Horiba Jobin-Yvon, Kyoto, JAP) equipped with a 450 W Xe lamp exciting source, double grating excitation and emission monochromators, and a TBX single-photon counter detector. Optical measurements were performed at room temperature; samples were analyzed immediately after incubation with DAF-FM at the excitation wavelength of 450 nm. Positive controls for $^\bullet$NO in water and DMEM were assessed by using the $^\bullet$NO donor Diethylamine NONOate diethyl ammonium salt (DETA-NONOate, cat. N. D185, Sigma–Aldrich). Solutions of DETA-NONOate (100 μM) in DMEM and water were used 2h after the preparation.

2.3. H_2O_2 and NO_2^- Detection in PTWS

The colorimetric detection of H_2O_2 and NO_2^- ions was performed soon after the plasma treatment. H_2O_2 was detected with a copper-phenanthroline assay (Spectroquant$^\circledR$, cat. N. 118789, Merck Millipore, Burlington, MA, USA). Ions NO_2^- were detected with the Griess assay (Spectroquant$^\circledR$, cat. N. 114776, Merck Millipore, Burlington, MA, USA). UV-Vis absorbance measurements were performed with a Cary 60 UV-Vis (Agilent, Santa Clara, CA, USA) spectrophotometer. Details about data processing are described in the supporting information.

2.4. Biological Assays

The biological effects of PT-DMEM with different ROS–RNS ratios were analyzed after cell incubation for 2 h (37 $^\circ$C, 5% CO_2), according to the schemes in Figure 1c,d.

The cytotoxic effect of different PT-DMEM was evaluated with the Saos2 cell line (American Type Culture Collection, ATCC, HTB-85), chosen as a model for cancer cells, and EAhy926 (ATCC, CRL-2922), chosen as a model for endothelial cells. Both cell lines were cultured in high glucose DMEM supplemented with 10% of FBS in a humidified atmosphere containing 5% CO_2 at 37 $^\circ$C. About 20 h before the plasma treatment, cells were seeded in 57mm diameter TPP$^\circledR$ tissue culture Petri dishes with a density of 10^4 cells per dish. There were 3 samples per each PT-DMEM and per each growth time were used. After that, the growing medium was removed and replaced with 2 mL of PT-DMEM. For control cells, the original medium was replaced with 2 mL of untreated fresh medium. After 2 h of incubation (time 0 h), PT-DMEM was removed, cells were rinsed once with untreated DMEM, and left growing for 24 and 72 h.

After 24 and 72 h cells were stained with a Coomassie Brilliant Blue solution, as described in our previous papers [6,37]: cells were fixed with 4% paraformaldehyde in Phosphate Buffer Saline (PBS) for 20 min, rinsed twice with PBS, then a Coomassie Brilliant Blue (R250, Biorad, Hercules, CA, USA) solution in methanol (45% *v/v*), acetic acid (10% *v/v*), and water (45% *v/v*) was added for 3 min. After rinsing with double distilled water (ddH$_2$O), cells were observed using an inverted optic microscope (Nikon Eclipse Ti); at least 7 pictures per time/type/dish were acquired with a Nikon DS Fi2 CCD camera. Image J analysis was performed to calculate the percentage of the substrate area covered by the cells, described as cell density in the Results Section.

2.5. Statistical Analysis

Statistical differences among groups, at least 3 replicates per group, were determined by one-way or two-ways-ANOVA assuming normal distribution of data, followed by Tukey's or Bonferroni's post-hoc tests.

3. Results

3.1. Detection of Plasma Produced Exogenous •NO in Gas and in PTW

EPR analyses were carried out to detect plasma generated •NO through the formation of a mononitrosyl-iron-dithio-carbamate (MNIC), NO-Fe^{2+}(DETC)$_2$ as spin trap for the radical. Spectra were acquired directly from the plasma effluents either during the DBD with no liquid present (Figure 3a), or in the presence of ddH$_2$O underneath the discharge (Figure 3b). To reveal •NO in the plasma, the analysis was carried out by bubbling plasma effluents into a vial filled with the spin trap solution and connected to the pump (Figure 3a).

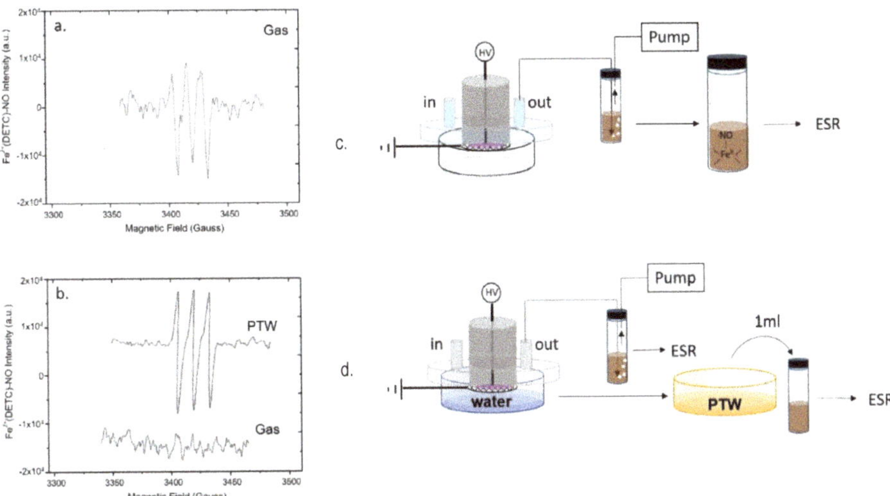

Figure 3. Production of •NO in the plasma phase or in the liquid underneath. EPR spectra of mononitrosyl-iron complex MNIC acquired by (**a**,**c**) bubbling the plasma effluent in the spin trap solution during a DBD with no liquid present, or by (**b**,**d**) simultaneous sampling the plasma effluent and the PTW. Experimental DBD conditions: 15 min, 0.5 (SLM) Air, 13 kV, 6 kHz, 50% DC, 100 ms period. The experimental schemes used for EPR detection are reported on the right of each spectrum.

The EPR spectrum shows a single triplet due to the adduct of •NO to the trap, which was identified as the paramagnetic five-coordinate complex NO-Fe^{2+}(DETC)$_2$, characterized by a hyperfine coupling constant of a$_N$ = 13.10 G, and g-factor = 2.039.

When the same discharge was carried out in presence of ddH$_2$O in the Petri dish the EPR signal of $^\bullet$NO, is revealed only in the liquid (Figure 3b, PTW), but not in the plasma effluents (Figure 3b, gas). The PTW sample was also tested using a spectrofluorometric assay based on the DAF-FM NO probe; the fluorescence intensity emitted from DAF-FM consists of a signal with a maximum at 515 nm attesting the presence of $^\bullet$NO and its derivatives in PTW, in agreement with the EPR, as reported in Figure S1.

3.2. Plasma Generation of Exogenous $^\bullet$NO in PT-DMEM and PTWS Containing Organic Components

PT-DMEM with and without serum was analyzed by EPR similarly to PTW; the resulting spectra are reported in Figure 4. In the same experimental conditions allowing the detection of $^\bullet$NO in PTW, no EPR signal was observed in PT-DMEM (Figure 4a). This was also confirmed with the DAF-FM assay, where almost no changes of the fluorescence intensities were found between PT and untreated DMEM both with and without serum (Figure 4b, Figure S4). To exclude possible interferences of DMEM components, the ability of the fluorescent probe to detect $^\bullet$NO in DMEM was confirmed by using a solution of the $^\bullet$NO donor DETA-NONO in DMEM (100 μM) as a positive control (Figure 4b, Figure S4).

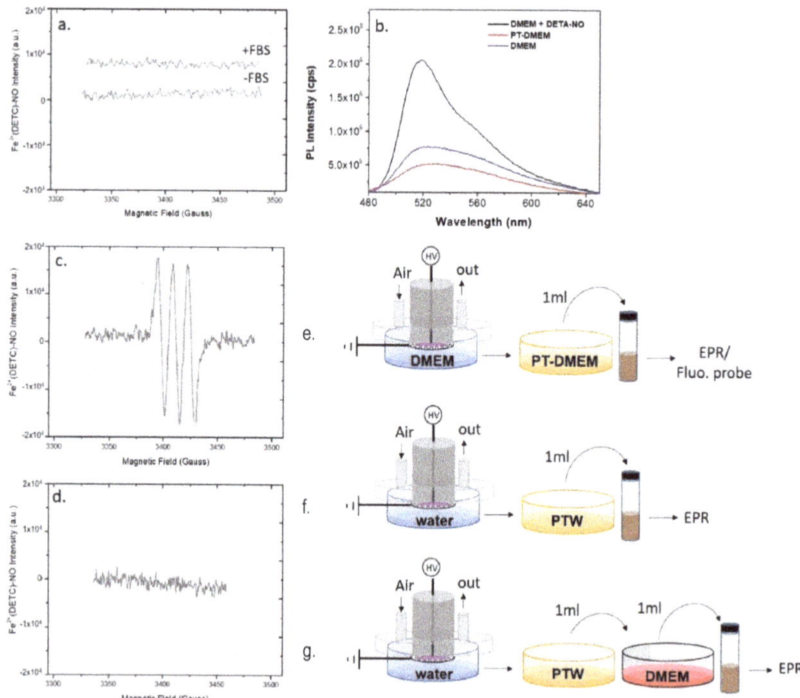

Figure 4. Plasma synthesis of $^\bullet$NO in ddH$_2$O and DMEM. EPR spectra of (**a,e**) PT-DMEM with (+FBS) and without (-FBS) Fetal Bovin Serum (FBS), (**c,f**) PTW, and (**d,g**) PTW mixed with 1 mL of untreated DMEM. (**b,e**) Fluorescent spectra of PT-DMEM, DMEM with the addition of the $^\bullet$NO donor DETA-NONOate (100 μM) and untreated DMEM (NT-DMEM) after the use of the fluorescent probe DAF-FM excited at 450 nm All measurements were acquired 15 min after the DBD. Experimental conditions: 15 min, 0.5 slm Air, 13 kV, 6 kHz, 50% DC, 100 ms period. Schemes of the experimental procedures used for the EPR are reported on the right of each spectrum.

By assuming that $^\bullet$NO is formed in the plasma phase and then diffuses into the liquid, since $^\bullet$NO is not detected in the liquids we assume that some chemical component or conditions (e.g., the buffered pH) in DMEM are responsible for the abatement of the plasma diffused $^\bullet$NO in the treated medium. In Figure 4a is reported the spectrum acquired on DMEM demonstrating the absence of the MNIC signal both in case of DMEM with and without serum. On the other hand, it was found that the addition of untreated DMEM to PTW suppresses the strong EPR signal acquired in PTW (Figure 4c,d), attesting for the presence of components of the serum free medium able to scavenge $^\bullet$NO. Among all DMEM components, L-cysteine and D-glucose were identified as possible scavengers of $^\bullet$NO; L-Cysteine, for its ability to form nitrosothiol groups [47], D-glucose, because it is the most abundant component of DMEM (4.5 g L-1). To ascertain whether these two compounds scavenge or not $^\bullet$NO, their solutions at the same concentrations as in DMEM were plasma treated in the same conditions as for PT-DMEM and PTW, and named PT-L-cysteine and PT-D-glucose, respectively. PTWS were not buffered, to show similar conditions as PTW.

The EPR spectrum of PT-cysteine (Figure 5) shows the MNIC signal; this limits or excludes the role of cysteine as $^\bullet$NO scavenger in our conditions. Moreover, when PT-L-cysteine was mixed with DMEM (Figure 5b), the signal was not suppressed as in the case of DMEM added with PTW (Figure 4d). In particular, by comparing the intensities of the spectra (Figure 5a,b) it can be observed that when PT-cysteine was mixed with untreated DMEM the intensity of the signal decreased but did not disappear, likely due to some of the DMEM components capable of scavenging $^\bullet$NO.

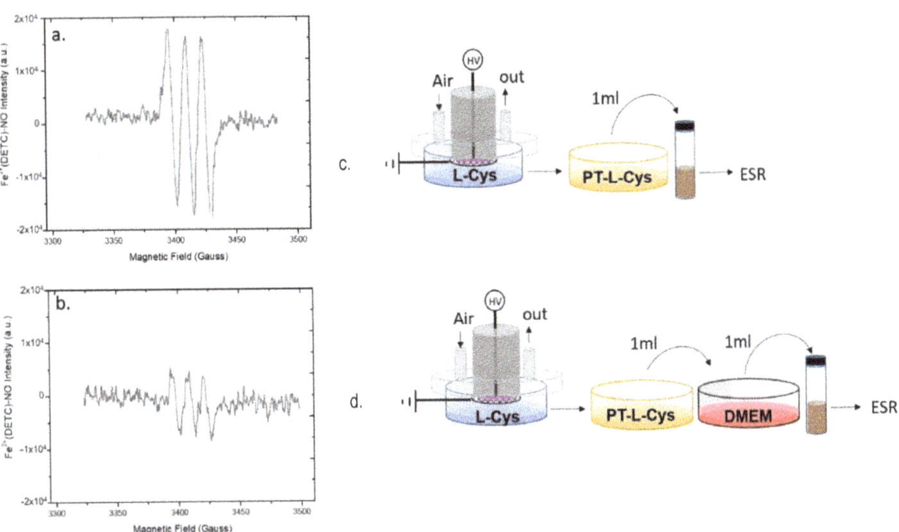

Figure 5. Role of Cysteine in scavenging plasma produced $^\bullet$NO. EPR spectra of (**a,c**) a Plasma Treated L-cysteine (PT-L-cysteine) solution (0.006 g L^{-1}), and (**b,d**) PT-L-cysteine solution mixed with untreated DMEM acquired 15 min after the discharge. Experimental conditions: 0.5 slm Air, 50% DC, 100 ms period, 6 kHz, 13 kV, 2 mL solution, 15 min treatment time. Schemes of the experimental procedures used for EPR analysis are reported on the right of each spectrum.

The EPR spectra acquired on PT-D-glucose are reported in Figure 6. Although the MNIC EPR signal is present in PT-D-glucose (Figure 6a), its intensity was lower than in PTW (Figures 3b and 4c), while in the case of PT-cysteine, the signal was similar to that in PTW (Figure 5a).

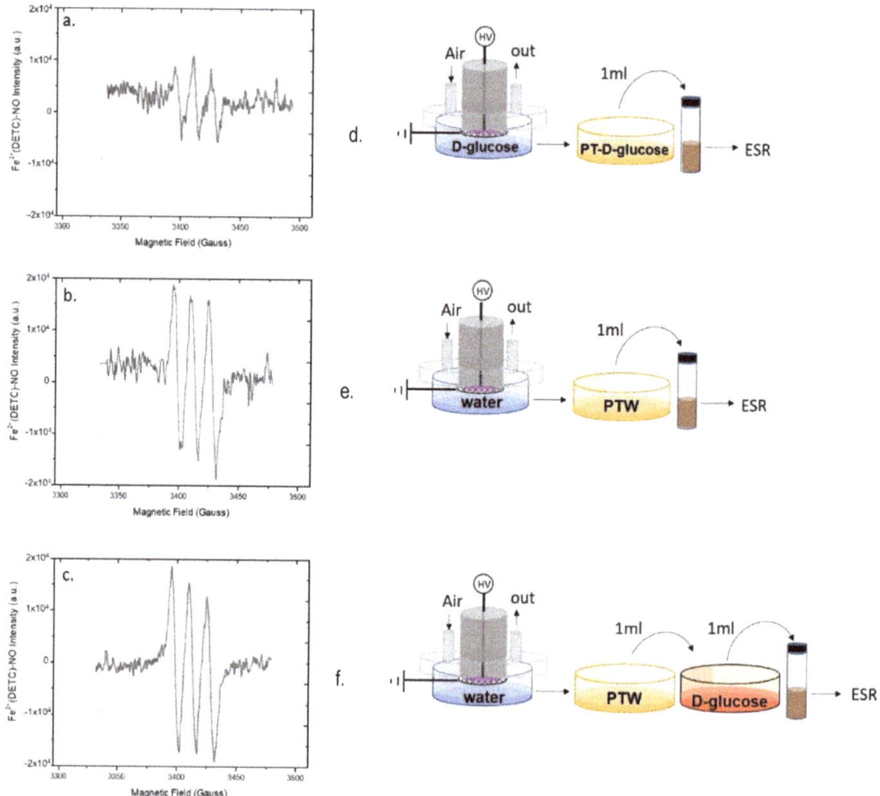

Figure 6. Role of D-Glucose in scavenging plasma produced $^{\bullet}$NO. EPR spectra of (**a,d**) Plasma Tretaed- D-glucose (PT-D-glucose) solution (45 mg L^{-1}), (**b,e**) PTW, and (**c,f**) PT-D-glucose solution mixed with PTW 15 min before spectra acquisition. Experimental conditions: 0.5 slm Air, 50% DC, 100 ms period, 6 kHz, 13kV, 2 mL solution, 15 min treatment time. Schemes of the experimental procedures used for EPR analysis are reported on the right of each spectrum.

This decrease suggests that D-glucose contributes a bit to scavenge $^{\bullet}$NO in PT-DMEM, but the persistence of the signal does not allow to identify D-glucose as the only cause of $^{\bullet}$NO abatement in this liquid. Moreover, when PTW is mixed with the untreated solution of D-glucose, the intensity of the EPR signal was not decreased (Figure 6c). This last result seems to attest that PT-D-Glucose has some $^{\bullet}$NO scavenging ability, likely due to some derivative of D-Glucose after the plasma treatment, while untreated D-Glucose has not.

3.3. Plasma Generation of Exogenous H_2O_2 and NO_2^- in PT-DMEM

From the experiments described so far, it is clear that the plasma-induced generation of $^{\bullet}$NO in DMEM is scavenged by derivatives of D-Glucose generated after the treatment, and by some other compound. The possibility to produce other RNS, such as NO_2^-, in PT-DMEM was therefore tested, by exploring plasma parameters affecting their production. O_2, N_2, and synthetic air were used as gas feed to exclude or promote the presence of RNS in PT-DMEM, respectively, and to relate it to the amount of ROS generated. We decided to analyze the concentration of NO_2^- and H_2O_2 used as markers, respectively, of all RNS and ROS species present in the PT-DMEM due to their estimated long life-time also in complex media. The presence of different species like as an example $ONOO^-$ and $O_2^{\bullet-}$ cannot be in any case excluded [37]. The results of the chemical composition of PT-DMEM in terms of NO_2^- and H_2O_2 concentration after N_2, $O_2^{\bullet-}$ and air-DBDs, performed at different treatment times (30–180 s), are reported in Figure 7. As shown in the graphs, the treatment time affects the amount of the species produced rather than the kind, that is determined only by the composition of the gas feed.

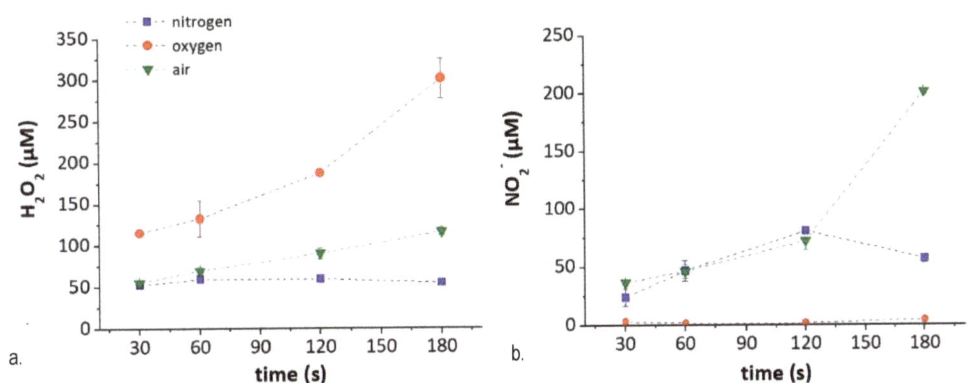

Figure 7. Chemical composition of PTWS as a function of experimental conditions- Content of (a) H_2O_2 and (b) NO_2^- in PT-DMEM according to feed composition and treatment time.

H_2O_2 (i.e., marker of ROS species) in PT-DMEM increases with the percentage of O_2 in the feed, from N_2 (where the source of oxygen in the feed comes from water evaporated from the liquid) to pure O_2. The NO_2^- (i.e., marker of RNS species) concentration in the liquid, instead, is higher when both O_2 and N_2 are fed in the plasma, as in the case of air-treated DMEM, and decreases in liquids treated with N_2-DBD. For the same reason, no NO_2^- is found in PT-DMEM obtained by O_2-DBDs, due to a purging procedure that eliminates any remaining air at the plasma–liquid interface (see Methods).

The possibility to achieve such fine control over the PT-DMEM composition revealed to be decisive to ascertain the role of NO_2^- and other undetected RNS species as stress effector, synergic with ROS, to cells exposed to PTWS. PT-DMEMs with different H_2O_2–NO_2^- ratios, as listed in Table 2, have been thus easily produced for cytotoxicity experiments. We decided to name the different PT-DMEM according to their H_2O_2 and NO_2^- contents, as listed in Table 2.

Table 2. Composition of PT-DMEM (DBDs ignited at 13 kV, 6 kHz, 25% DC, 100 ms period), produced with different feeds (0.5 slm) and treatment times (1 or 3 min) for cell culture experiments.

PTWS	Plasma Condition	H_2O_2 (μM)	NO_2^- (μM)
$(-)H_2O_2(-)NO_2-$	Air 1 min	69 ± 4	46 ± 6
$(+)H_2O_2$	O_2 1 min	130 ± 20	1.60 ± 0.08
$(++)H_2O_2$	O_2 3 min	300 ± 20	4.3 ± 0.4
$(+)H_2O_2(+)NO_2-$	Air 3 min	116 ± 4	201.4 ± 1.7

3.4. Cells Incubation with PT-DMEM Containing Different Exogenous H_2O_2-and-NO_2^- Doses

The modifications of the density of Saos2 and EAhy926 cells after incubation with PT-DMEMs with different H_2O_2–NO_2^- ratios are summarized in Figure 8. Cell density is intended as the area of Petri dish covered by cells, calculated by ImageJ analysis. Some representative images acquired on Coomassie Blue stained cells 72 h after incubation with PT-DMEM at different H_2O_2–NO_2^- ratio are shown in Figure S2 and Figure S3.

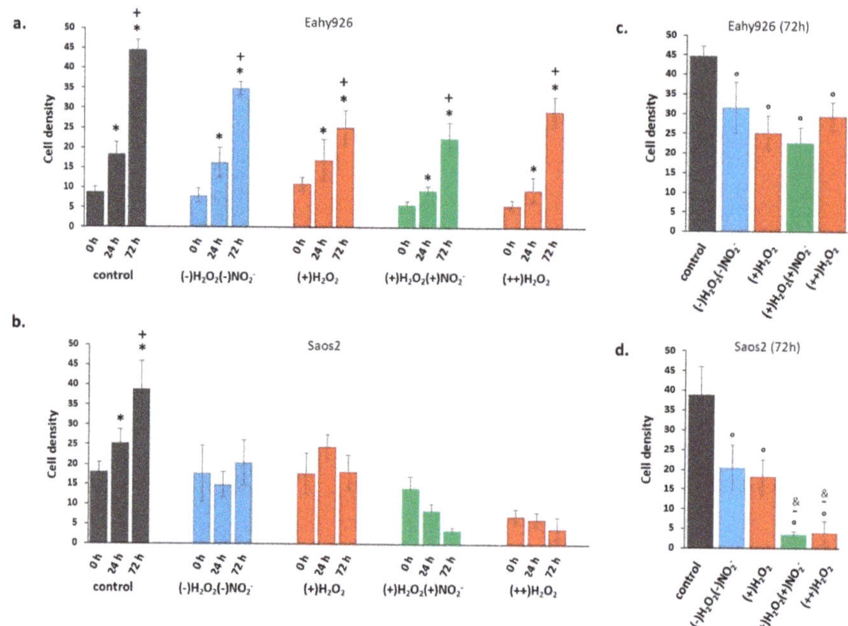

Figure 8. Cell density of cells incubated with PTWS respect to the control. (**a**) Eahy926 and (**b**) Saos2 cells at different time points (0, 24, and 72 h) after incubation in different PTWSs. (**c**) Eahy926 and (**d**) Saos2 cells after 72 h of incubation in different PTWSs. The time 0 h corresponds to cells exposed for 2h to PTWSs, PFA4% fixed, rinsed with PBS and Coomassie blue stained. One-way ANOVA + Tukey's post-test: (*) = $p < 0.01$ vs. 0 h; (+) = $p < 0.01$ vs. 24 h; (°) = $p < 0.01$ vs. control; (-)= $p < 0.01$ vs. (+)H_2O_2; (&) = $p < 0.01$ vs. (-)H_2O_2 (-)NO_2^-.

Cell growth data at different time points, in Figure 8a,b, show that incubation in all PT-DMEM synthesized resulted in no growth for Saos2 cells (Figure 8b). In the case of EAhy926 cells, instead, growth was measured in all considered conditions (Figure 8a). A first selective effect of PT-DMEM as inhibitors of the growth of cancer cell could thus be observed, regardless of the H_2O_2–NO_2^- ratio. However, looking at cell densities after 72 h incubation (Figure 8c,d), a specific response of the Saos2 cells can be clearly observed, depending on the H_2O_2–NO_2^- ratio in the liquid, differently from EAhy926 at the same time point.

Both cell lines show a reduced density after incubation in PT-DMEM in comparison with control cells; however, in the case of EAhy926 cells (Figure 8c), the different H_2O_2–NO_2^- ratio in the liquids did not produce statistically significant differences in cell density. In the case of Saos2 cells, instead, besides a general stronger density decrease in comparison with the control, the specific effect of each liquid changed depending on their specific H_2O_2–NO_2^- ratio. A mild effect was produced after incubation in (-)H_2O_2, (-)NO_2^-, and (+)H_2O_2, while the strongest decrease was caused by (+)H_2O_2, (+)NO_2^-, and (++)H_2O_2 media. The anti-cancer effects of NO_2^- emerged, in particular,: the simultaneous presence of H_2O_2 (about 120 µM) and high doses of nitrite ions (about 200 µM), as in (+)H_2O_2, (+)NO_2^-, decreased Saos2 cell density much more than the same amount of H_2O_2 alone, as in (+)H_2O_2, whose effect, conversely, was quite mild. When Saos2 cells were exposed to (++)H_2O_2 solutions for 2 h (0 h in Figure 8b), before 4%PFA/PBS fixation, round cells (probably dead cells) floating in the liquid were observed attesting for the deleterious effect of such solutions since first hours of observation.

Morphological changes in EAhy926 and Saos2 exposed to PTWS with different ROS/RNS and stained 72h after the exposure can be, respectively, appreciated by looking at Figure 9 that shows pictures of Comassie blu stained cells (higher magnification than the ones shown in Figure S2 and Figure S3). In case of the endothelial cell line, the morphology of the cells is almost retained after the treatment with PTWS. Looking at single cells (Figure 9), it is possible to observe the retention of the typical starry shape of this cell line; while, by analyzing the overall appearance of cells, the PTWS exposed ones do not remarkably differ from control cells for cell density on the substrate and for the capability to form clusters among each other, another important feature of endothelial cell lines (Figure S2). Endothelial cells show only a change on cell density and not in morphology, differently from what has been observed for Saos2 cells. In case of Saos2 cells, in fact, the morphology of PTWS-exposed cells is dramatically impacted. It is evident that a loss in cell clustering is present in all the Saos2 cells incubated with PTWSs. Furthermore, where the cells show the less density also a predominance of spherical cells are present, clear evidences of apoptotic or pre-apoptotic conditions (white arrows of Figure 9 e,f). Then, it is possible to observe the loss of the typical stretched shape of this cell line, in favor of a more condensed form which is generally associated to a stressing environment. These changes gradually occur depending on the specific chemical composition of the tested PTWS: Saos2 cells incubated with $(-)H_2O_2(-)NO_2^-$ show a milder reduction in cell density and only a partial degradation of cell shape, while cells incubated with $(+)H_2O_2(+)NO_2^-$ show a radical change in morphology, with small, spherical and isolated cells being greatly reduced in number.

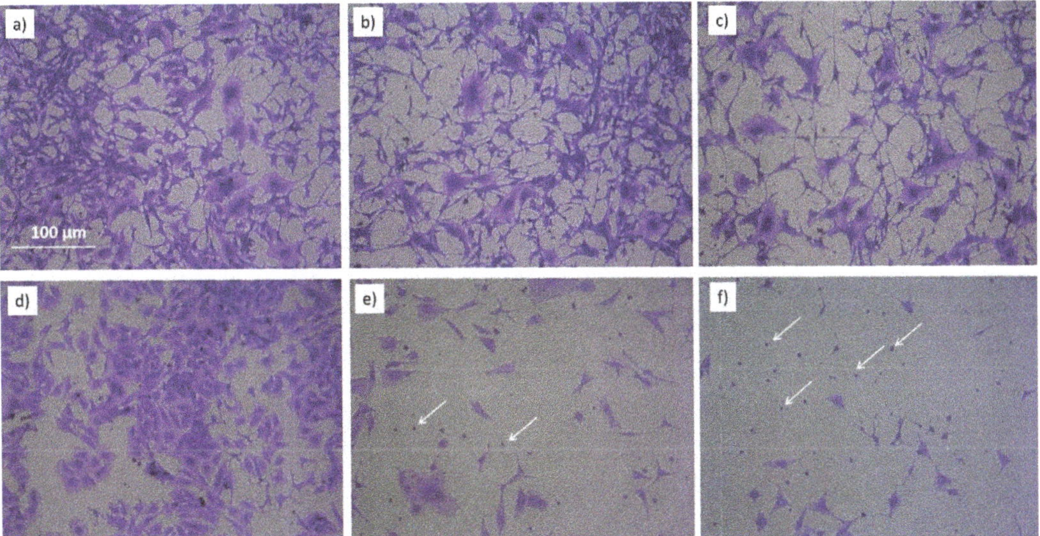

Figure 9. Coomassie Blue images of Eahy926 (**a–c**) and Saos2 (**d–f**) cells incubated with PTWS and let grown for 72h. Cells no incubated with PTWS are shown in (**a,d**). (**b,e**) show cells incubated with $(-)H_2O_2$ $(-)NO_2^-$. (**c,f**) show cells incubated with $(+)H_2O_2$ $(+)NO_2^-$. White arrows in images (**e,f**) show spherical Saos2 cells.

4. Discussion

Understanding the dual role of RNS as both protectors and inducers of oxidative stress is of peculiar importance for controlling intracellular RNS and, ultimately, for tuning anticancer effects of RONS-based approaches and therapies, including those based on PTWS [48]. However, the generation of PTWS with predetermined chemical composition for a specific target requires a deep knowledge of the chemical formation routes of each bioactive specie in the plasma-activated liquid. This knowledge, to date, is highly desired but still far to be achieved, above all in the case of those PTWS generated from bioactive liquids with a complex chemical composition, that would typically be required for in vitro experiments (i.e., cell culture media), or for in vivo administration (i.e., blood, biological fluids). Our study is therefore aimed at clarifying, from one side the routes of formation of RNS in PTWS generated in bioactive liquids like DMEM, and, on the other side, at demonstrating how the modulation of RNS concentrations in PT-DMEM used incubation with cells effectively concurs to tune anticancer effects.

The chemical analysis performed to assess the presence of $^\bullet NO$ in water and in DMEM with (+FBS) and without serum (-FBS) after plasma treatment confirmed that the formation of $^\bullet NO$ strongly depends on the composition of the liquid to be treated. Although $^\bullet NO$ was present in plasma effluents and in PTW (Figure 3a), no trace of $^\bullet NO$ was found in PT-DMEM ($+/-$ FBS), due to its scavenging in part by D-Glucose and by other compounds still to be identified. It is not excluded that the serum could additionally contribute to scavenge $^\bullet NO$, due to its content of organic molecules. However, as well known, the FBS is a not fully defined medium and as such, may vary in composition between batches, thus it is quite difficult to identify the molecule of FBS responsible of such scavenging. In any case, the absence of dissolved free $^\bullet NO$ in PT-DMEM would exclude the involvement of this RNS during incubation with cells, at least in our conditions, although it cannot be excluded that $^\bullet NO$ is first bound by some of the species in the medium, then transferred to cells with no passage through the free solvated form. Also Chauvin et al. [49] could not find trace of $^\bullet NO$ after EPR analysis of PT-PBS and PT-DMEM ($+/-$ FBS), and tentatively explained the lack of $^\bullet NO$, that would have been detected at low pH, with the neutral pH of the buffered media investigated. A possible explanation for this evidence, in fact, emerges from the analysis of the production routes of $^\bullet NO$ in PTWS, schematically summarized in Table 3.

Table 3. Reactions representing the possible chemical formation routes of $^\bullet NO$ in PTWS.

Reaction	Number	References
$N_2{}^*{}_{(g)} + {}^\bullet O_{(g)} \rightarrow {}^\bullet NO_{(g)} + {}^\bullet N_{(g)}$	(1)	[50,51]
${}^\bullet N_{(g)} + O_{2(g)} \rightarrow {}^\bullet NO_{(g)} + {}^\bullet O_{(g)}$	(2)	[50,51]
${}^\bullet N_{(g)} + {}^\bullet OH_{(g)} \rightarrow {}^\bullet NO_{(g)} + {}^\bullet H_{(g)}$	(3)	[50,52]
${}^\bullet NO_{(g)} + O_{3(g)} \rightarrow NO_{2(g)} + O_{2(g)}$	(4)	[52]
${}^\bullet NO_{(g)} + \cdot O_{(g)}{}^\bullet + M_{(g)}{}^\bullet \rightarrow NO_{2(g)} + M_{(g)}$	(5)	[52]
${}^\bullet NO_{(g)} + {}^\bullet OH_{(g)} \rightarrow HNO_{2(g)} \rightarrow HNO_{2(aq)}$	(6)	[52]
$3HNO_{2(aq)} \rightarrow 2{}^\bullet NO_{(aq)} + HNO_{3(aq)} + H_2O$	(7)	[52,53]
${}^\bullet OH_{(aq)} + ONOO^-{}_{(aq)} \rightarrow O_{2(aq)} + OH^-{}_{(aq)} + {}^\bullet NO_{(aq)}$	(8)	[54,55]

$^\bullet NO$ is generally produced in the plasma phase through chemical reactions among N_2, O_2, atomic oxygen (O) and nitrogen (N) through the well-known Zeldovich mechanism (Equations (1) and (2) in Table 3), or by oxidation of atomic N by hydroxyl radicals ($^\bullet OH$) generated from the dissociation of water vapor from the liquid, if is present during the discharge. The diffusion of plasma-generated $^\bullet NO$ into the liquid may be thus supposed as the most likely route for the formation of $^\bullet NO$ in PTWS. This hypothesis would even be supported in our data by the disappearance of MNIC signal in plasma effluents, and by its appearance in PTW registered when the discharge is ignited in presence of water (Figure 3a,b). However, the dissolution of $^\bullet NO$ from the plasma towards the liquid is highly discouraged by the low solubility of $^\bullet NO$ in water media (Henry constant, $K_H = 0.046$ at 25 °C [56]) and by the short diffusion distance reported for $^\bullet NO$ during plasma treatment of liquids (<1 mm) [54], quite lower than the distance between the liquid and the discharge in our conditions (3 mm).

For these reasons, it is rather believed that the presence of $^\bullet NO$ in PTWS may be more likely due to reactions among other plasma-generated RONS that can more easily diffuse in water media, such as NO_2 ($K_H = 0.978$ at 25 °C) [56], HNO_2 ($K_H = 1198$ at 25 °C), and $^\bullet OH$ radicals ($K_H = 733$ at 25 °C) [57]. NO_2 is generally produced in plasma effluents due to $^\bullet NO$ oxidation with oxidants like atomic oxygen (O) or ozone in presence of a third collision body (M) (Equations (4) and (5), Table 3), while HNO_2 could be formed in the plasma from reactions between $^\bullet$NO and $^\bullet$OH (Equation (6), Table 3). The diffusion of HNO_2 in the liquid to form NO_2^- that, in turn, can disproportionate to produce $^\bullet$NO (Equation (7) in Table 3), indeed, is the most cited pathway for the formation of $^\bullet$NO in PTWS in the literature [50,53]. However, besides its very low rate constant (k = 13.4), the conversion of NO_2^- to $^\bullet$NO requires acidic conditions (pH < 3.5, [50]) and could thus be prevented in buffered media like PT-DMEM. In the end, in presence of liquid, e.g., DMEM, underneath the plasma, even the other possible route to produce $^\bullet$NO though reaction between $^\bullet$OH and $ONOO^-$ (Equation (8), Table 3), could be actually prevented since, again, acidic conditions are generally required to form even $ONOO^-$ [58] and, by now, the presence of $ONOO^-$ in PT-cell culture media has never been confirmed. The neutral pH of buffered media like DMEM, that discourage $^\bullet$NO formation routes, can be therefore addressed as the main explanation for the abatement of $^\bullet$NO in PT-DMEM found with EPR and spectrofluorometric analysis (Figure 4a,b).

However, the suppression of the MNIC signal in PTW after the addition of untreated DMEM (Figure 4d) also suggests the possible presence of chemical $^\bullet$NO scavengers in the cell culture medium. In the last case, even the low amounts of plasma-generated $^\bullet$NO that manage to diffuse in the liquid could be rapidly consumed in the medium. The starting composition of DMEM includes several components that can react with plasma-produced RONS, some of them may be particularly keen to react with $^\bullet$NO, such as L-cysteine and D-glucose. L-Cysteine is often reported as an "in vivo $^\bullet$NO modulator" through the reaction of its thiol group with $^\bullet$NO [56,59–61]. Moreover, as recently reported by Lackmann et al. [51], covalent modifications of cysteine treated with different plasmas have been shown. D-glucose is also actively involved in the scavenging of $^\bullet$NO in many natural processes [62]; it has been found that the exposure of human endothelial cells to the same level of glucose as in the plasma of diabetic patients results in a significant blunting of $^\bullet$NO responses [62].

The EPR spectrum acquired from PT-cysteine (Figure 5a), however, shows the MNIC signal as intense as in PTW, thus excluding cysteine as a possible scavenger for $^\bullet$NO, in our experimental conditions. Although the formation of S-nitrous-cysteine derivatives is likely to occur, the amount of $^\bullet$NO produced in the treatment, or immediately after, is probably much higher than that consumed by cysteine. Very interestingly, in contrast with the supposed role of $^\bullet$NO scavenger for such molecule, a role of "$^\bullet$NO protector and modulator" could actually be proposed, since, when PT-L-cysteine was mixed with DMEM (Figure 5b), the signal was not suppressed, differently from the case of PTW (Figure 4d). This evidence can be explained by considering that ROS and RNS in PTWS form nitroso-thiol (S-NO) functionalities on cysteine [63]. Since S-NO bond is very unstable in physiological conditions, a $^\bullet$NO release from S-nitrous-cysteine is often reported [63] and could thus contribute to sustain the MNIC signal when PT-cysteine is mixed with untreated DMEM.

On the opposite, a role of "$^\bullet$NO scavenger" emerged for D-glucose, albeit quite low. The intensity of the EPR signal in PT-D-glucose solution (Figure 6a) was much lower than that obtained in PTW (Figure 6b), suggesting that D-glucose contributes to suppress the signal in PT-DMEM. Nevertheless, when PTW was mixed with an untreated D-glucose solution, the intensity of the EPR signal of the resulting solution was not decreased at all (Figure 6c). This evidence seems to be in contrast with the results reported in Figure 6a unless structural modifications of the molecule due to plasma treatment are proposed. Indeed, plasma-induced modifications could probably enhance the reactivity of glucose towards $^\bullet$NO and, as a consequence, its capability to scavenge it. D-glucose, however, cannot be indicated as the unique cause for the MNIC abatement in DMEM, since the signal was still present even at PT-D-glucose with a concentration (45 g L^{-1}) 10 times higher than that in DMEM (4.5 g L^{-1}), as shown in Figure 6.

We, therefore, can conclude that the absence of $^\bullet$NO in PT-DMEM may be due at least to two concurrent causes, namely, the neutral pH preventing the main formation pathways for $^\bullet$NO, and compounds, such as glucose, able to sequester $^\bullet$NO possibly diffused by the plasma. These evidences undermine the role hypothesized so far for plasma-produced $^\bullet$NO as a protector or inducer of nitrosative stress on cells incubated in PTWS, which must therefore be sought among other RNS such as NO_2^-, at least when PTWS are generated in cell culture media. As already mentioned, though, it cannot be excluded that $^\bullet$NO is first bound by some of the species in the medium (e.g., cysteine or others), and then transferred to cells with no passage through the free solvated form.

The presence of NO_2^- anions in PT- culture media, in fact, could act as a source of nitrosative stress as well, due to the biological conversion, internal to cells, in the more reactive species of $^\bullet$NO and $ONOO^-$ [6]. Indeed, nitrite ions are considered the largest intravascular and tissue source of $^\bullet$NO in the human body [64]. In turn, nitrite-derived $^\bullet$NO inside cells can be converted to $ONOO^-$ as shown above; and, in presence of peroxides such as H_2O_2, nitrites can be even directly converted to $ONOO^-$, thus providing a second route for the formation of this molecule [65] in biological environments.

Recently Bauer [66,67] supposed that anti-cancer effects of PTWS can be due by local reactions of $^\bullet$NO and NO_2^- with H_2O_2 near cell membranes These species are implicated in the formation of $ONOO^-$ and singlet oxygen $^1O_2$ near the membrane of cancer cells, leading to local inactivation of the redox enzyme catalase. Tumor cells contribute to their death through further membrane-associated catalase inactivation and reactivation of intercellular apoptosis-inducing ROS and RNS signaling [65].

NO_2^- remain stable many weeks in PTWS generated from culture media, despite the presence of many organic compounds [59,60]. These features thus promote NO_2^- as the ideal reservoir for nitrosative stress in PTWS of complex composition. Moreover, the production of NO_2^- during DBD treatments can be finely tuned by changing conditions, such as treatment time and gas feed. As shown in this paper, the first is important for the amount, while the second for the kind of plasma-generated species in PT-DMEM. Ranging from O_2-DBD to N_2- to air-DBD, the presence of NO_2^- in PT-DMEM can be, respectively, excluded or promoted (Figure 7b). In our conditions, H_2O_2 formation in PT-DMEM can be promoted by increasing the percentage of O_2 in the feed, in agreement with the formation mechanism of H_2O_2 from OH recombination at the plasma–liquid interface, as discussed in our previous studies [5]. In turn, the NO_2^- concentration in the liquid can be enhanced if both O_2 and N_2 are fed in the plasma. The trends obtained are similar to those obtained with a 2D fluid-dynamic model proposed by Verlackt et al. [61], that describes the interaction between a plasma jet and water buffered at pH around 7, as a function of treatment time. The possibility to achieve fine control over ROS and RNS composition in PTWS by our system was revealed to be fundamental for assessing the role of each specific species as effectors in the PTWS-induced stress on cancer cells.

Within the tumor microenvironment, low to moderate levels of $^\bullet$NO derived from cancer and endothelial cells can activate angiogenesis and epithelial-to-mesenchymal transition, promoting an aggressive phenotype [68]. For this reason, to assess what happens when cancer and hybrid ECs are exposed to NOx species, excluding $^\bullet$NO, in combination with ROS (i.e., H_2O_2) is fundamental for further application of such liquids in cancer therapy and in general to gain new insights in tumor progression. The higher sensitivity of Saos2 cancer cells to PT-DMEMs with controlled H_2O_2–NO_2^- ratio, in comparison with the endothelial EAhy926 cells confirmed that RNS, mainly nitrites, can induce a cell-dependent anti-cancer effect (Figure 8). The key at the base of such selectivity seems to be the decrease in the cytotoxic threshold for H_2O_2 in cancer cells, when PTWS with high doses of NO_2^- are used.

From previous experiments it is known that 2 h of incubation in a 300 μM untreated solution of H_2O_2 in DMEM (i.e., unexposed to any plasma) is enough to kill Saos2 cells after a few minutes, with this concentration representing a sort of "tolerance threshold" for this cell line (Figure S5). As a consequence, the response of Saos2 cells to PT-DMEM (++)H_2O_2 containing about 300 μM of H_2O_2 was a strong decrease in cell density, as expected, despite the absence of nitrites. Moreover, it is well known that high doses of H_2O_2 can trigger many apoptotic pathways in cancer cells, most of them involving the passage of exogenous

H_2O_2 through aquaporin (AQP) channels and the direct intracellular damage via Fenton reaction, as discussed by us [6], and by Yusupov et al. [69]. It has also been suggested by Yan et al. [70] that the increased concentration of AQP channels on tumor cells could explain the anti-tumor effects of PTWS.

Besides a higher number of AQP channels, however, cancer cells also specifically exhibit higher amounts of antioxidant defenses that could seize most of the H_2O_2 exogenously delivered through PTWS, resulting in the increase in the cytotoxic threshold for most cancer cell lines. This could be the reason why, indeed, the direct H_2O_2-apoptosis induction via PTWS is often reported as non-specific for cancer cells [66,71,72], since the high doses of H_2O_2 required to overcome membrane-associated catalase on tumor cells could largely exceed the tolerance of endothelial cells too, which lack this protection.

In all these cases, to selectively interact with a specific-redox chemistry on the surface of tumor cells, a mechanism in which NO_2^- cooperates with lower doses of H_2O_2 (i.e., lower than the tolerance threshold for endothelial cells) could be more effective to provide specific apoptotic pathways for cancer cells. Indeed, our results are in agreement with those reported by Bauer [65,71,73], where a synergic effect of NO_2^- and H_2O_2 was proposed on human KKN-45 gastric carcinoma cells when the H_2O_2 concentration was below 160 μM, a threshold for H_2O_2-dependent apoptosis for that cancer line. Similar results on the synergic anti-cancer effect of NO_2^- and H_2O_2 in PTWS were also reported by Girard et al. on HCT116 human colon cancer cells and Lu1205 human melanoma cells [33], and by Kurake et al. on glioblastoma tumor cells [32], as discussed by Von Woetdtke and Jablonowski [9]. In agreement with these findings by other authors, in this papers it is shown that the selectivity of PTWS-induced cancer cell death relies on a window of H_2O_2 concentration where the synergic action of H_2O_2 and NO_2^- can induce death in tumor cells but does not affect hybrid endothelial cells. Endothelial cells (EC), as part of the tumor microenvironment, play a crucial role in inflammatory processes, as well as in angiogenesis, and could be critical targets of cancer therapy. In order to promote a deleterious effect contemporary on cancer cells and hybrid endothelial cells as part of the tumor microenvironment, it is necessary to overcome the threshold toxic level for H_2O_2. In this way the apoptosis is promoted also in hybrid ECs as attested by the presence of round shaped cells in Figure S3d. In our experience, we have observed that the changes in morphology (reduced numbers of clusters, spherical shape of cells) are important markers of cell reaction to PTWS, while the changes in cell density are representative of the reduction, retention, or increase in cells number (and, indirectly, of their proliferation). Thus we can conclude that ECs have a capability to overcome the oxidative stress differently from that of osteosarcoma cells. This cell susceptibility threshold and the concurrent action of H_2O_2 and NO_2^- was supposed by Bauer et al. [65], and the results shown in our paper attest that an exogenous delivery of RNS in proper combination with ROS actually enhances the selectivity of oxidative stress induced in cancer cells. However, much more research has to be done before a clinical application of PTWS will be possible.

5. Conclusions

The routes of formation of RNS such as $^\bullet NO$ and NO_2^- after DBD plasma treatment of DMEM have been investigated and their involvement in anti-cancer effects on malignant cells was evaluated. A double detection method for $^\bullet NO$ was used for the first time to demonstrate that no trace of $^\bullet NO$ could be found in PT-DMEM, where, instead, high amounts of NO_2^- are formed. NO_2^- concentration in PT-DMEM revealed to be finely tunable with the plasma operating conditions such as treatment time, feed composition, and composition of the starting liquid. The addition of PT-DMEM loaded with different $H_2O_2 - NO_2^-$ ratio to cultures of Saos2 and EAhy926 cells allowed to define the contribution of NO_2^- as supporter rather than protector of oxidative stress induced on cancer cells by the H_2O_2 ROS present in PTWS. In particular, high doses of NO_2^- were found to increase the selectivity of anti-cancer effects by decreasing the cytotoxic threshold of H_2O_2 for cancer cells to unharmful values for hybrid endothelial cell line.

Supplementary Materials: The following are available online at https://www.mdpi.com/article/10.3390/antiox10040605/s1. Figure S1: Fluorescence spectra of untreated water (blue line), and PTW (red line) 15 min after the discharge; Figure S2: Coomassie blue stained Saos2 cells, fixed after 72 h of incubation with PT-DMEMs at different H_2O_2–NO_2^- ratio; Figure S3: Coomassie blue stained Eahy926 cells, fixed after 72 h of incubation in PT-DMEMs at different H_2O_2–NO_2^- ratio; Figure S4: EPR spectra and fluorescence spectra of PT-DMEM with and without the serum; Figure S5: Coomassie blue stained Saos2 cells 48 h after incubation in 300 μM H_2O_2 solution in DMEM compared with the control.

Author Contributions: Conceptualization, E.S.; methodology, E.S., V.V., L.G., R.G.; software, E.S., V.V., R.G., S.C.; validation, R.G., E.S.; formal analysis, E.S., R.G.; investigation, E.S., R.G., V.V., M.S., L.G.; resources, M.B., P.F., F.F.; data curation, E.S., V.V., R.G.; writing—original draft preparation, E.S., V.V.; writing—review and editing, all authors; visualization, R.G.; supervision, E.S.; project administration, E.S.; funding acquisition, P.F., F.F. All authors have read and agreed to the published version of the manuscript.

Funding: This research was funded by financial support of Regione Puglia, under grants "LIPP" (grant no. 51, within the Framework Programme Agreement APQ "Ricerca Scientifica", II atto integrative—Reti di Laboratori Pubblici di Ricerca) and of Regione Puglia MIUR PON Ricerca e Competitività 2007–2013 Avviso 254/Ric. del 18/05/2011, Project PONa3 00369 "Laboratorio SISTEMA".

Institutional Review Board Statement: Not applicable.

Informed Consent Statement: Not applicable.

Data Availability Statement: Not applicable.

Acknowledgments: We thank K-D Weltmann and M. Schmidt (Leibniz Institute for Plasma Science and Technology, INP, Greifswald, GER) for the cooperation offered with the development of the Petriplus+ plasma source. D. Benedetti (University Bari) is gratefully acknowledged for his support in the laboratory and F. Ciminale (Univ. Bari) for laboratory support and constructive discussions on EPR.

Conflicts of Interest: The authors declare no conflict of interest.

References

1. Bekeschus, S.; Favia, P.; Robert, E.; Von Woedtke, T. White paper on plasma for medicine and hygiene: Future in plasma health sciences. *Plasma Process. Polym.* **2019**, *16*, 1800033. [CrossRef]
2. Adamovich, I.; Baalrud, S.D.; Bogaerts, A.; Bruggeman, P.J.; Cappelli, M.; Colombo, V.; Czarnetzki, U.; Ebert, U.; Eden, J.G.; Favia, P.; et al. The 2017 Plasma Roadmap: Low temperature plasma science and technology. *J. Phys. D Appl. Phys.* **2017**, *50*, 323001. [CrossRef]
3. Tanaka, H.; Mizuno, M.; Ishikawa, K.; Nakamura, K.; Utsumi, F.; Kajiyama, H.; Kano, H.; Maruyama, S.; Kikkawa, F.; Hori, M. Cell survival and proliferation signaling pathways are downregulated by plasma-activated medium in glioblastoma brain tumor cells. *Plasma Med.* **2012**, *2*, 207–220. [CrossRef]
4. Liedtke, K.R.; Freund, E.; Hackbarth, C.; Heidecke, C.-D.; Partecke, L.-I.; Bekeschus, S. A myeloid and lymphoid infiltrate in murine pancreatic tumors exposed to plasma-treated medium. *Clin. Plasma Med.* **2018**, *11*, 10–17. [CrossRef]
5. Azzariti, A.; Iacobazzi, R.M.; Di Fonte, R.; Porcelli, L.; Gristina, R.; Favia, P.; Fracassi, F.; Trizio, I.; Silvestris, N.; Guida, G.; et al. Plasma-activated medium triggers cell death and the presentation of immune activating danger signals in melanoma and pancreatic cancer cells. *Sci. Rep.* **2019**, *9*, 4099. [CrossRef] [PubMed]
6. Sardella, E.; Mola, M.G.; Gristina, R.; Piccione, M.; Veronico, V.; De Bellis, M.; Cibelli, A.; Buttiglione, M.; Armenise, V.; Favia, P.; et al. A Synergistic Effect of Reactive Oxygen and Reactive Nitrogen Species in Plasma Activated Liquid Media Triggers Astrocyte Wound Healing. *Int. J. Mol. Sci.* **2020**, *21*, 3343. [CrossRef] [PubMed]
7. Utsumi, F.; Kajiyama, H.; Nakamura, K.; Tanaka, H.; Mizuno, M.; Ishikawa, K.; Kondo, H.; Kano, H.; Hori, M.; Kikkawa, F. Effect of Indirect Nonequilibrium Atmospheric Pressure Plasma on Anti-Proliferative Activity against Chronic Chemo-Resistant Ovarian Cancer Cells In Vitro and In Vivo. *PLoS ONE* **2013**, *8*, e81576. [CrossRef] [PubMed]
8. Freund, E.; Bekeschus, S. Gas plasma-oxidized liquids for cancer treatment: Pre-clinical relevance, immuno-oncology, and clinical obstacles. *IEEE Trans. Radiat. Plasma Med. Sci.* **2020**, 1. [CrossRef]
9. Jablonowski, H.; von Woedtke, T. Research on plasma medicine-relevant plasma-liquid interaction: What happened in the past five years? *Clin. Plasma Med.* **2015**, *3*, 42–52. [CrossRef]
10. Bekeschus, S.; Schmidt, A.; Niessner, F.; Gerling, T.; Weltmann, K.-D.; Wende, K. Basic Research in Plasma Medicine—A Throughput Approach from Liquids to Cells. *J. Vis. Exp.* **2017**, e56331. [CrossRef]

11. Yan, D.; Talbot, A.; Nourmohammadi, N.; Cheng, X.; Canady, J.; Sherman, J.H.; Keidar, M. Principles of using Cold Atmospheric Plasma Stimulated Media for Cancer Treatment. *Sci. Rep.* **2015**, *5*, 18339. [CrossRef] [PubMed]

12. Ramli, N.A.H.; Zaaba, S.K.; Mustaffa, M.T.; Zakaria, A.; Shahriman, A.B. Review on the development of plasma discharge in liquid solution. *AIP Conf. Proc.* **2017**, *1824*, 030015. [CrossRef]

13. Anderson, C.E.; Cha, N.R.; Lindsay, A.D.; Clark, D.S.; Graves, D.B. The Role of Interfacial Reactions in Determining Plasma–Liquid Chemistry. *Plasma Chem. Plasma Process.* **2016**, *36*, 1393–1415. [CrossRef]

14. Kurake, N.; Tanaka, H.; Ishikawa, K.; Takeda, K.; Hashizume, H.; Nakamura, K.; Kajiyama, H.; Kondo, T.; Kikkawa, F.; Mizuno, M.; et al. Effects of •OH and •NO radicals in the aqueous phase on H_2O_2 and NO_2^- generated in plasma-activated medium. *J. Phys. D Appl. Phys.* **2017**, *50*, 155202. [CrossRef]

15. Lin, A.G.; Xiang, B.; Merlino, D.J.; Baybutt, T.R.; Sahu, J.; Fridman, A.; Snook, A.E.; Miller, V. Non-thermal plasma induces immunogenic cell death in vivo in murine CT26 colorectal tumors. *OncoImmunology* **2018**, *7*, e1484978. [CrossRef] [PubMed]

16. Mitra, S.; Nguyen, L.N.; Akter, M.; Park, G.; Choi, E.H.; Kaushik, N.K. Impact of ROS Generated by Chemical, Physical, and Plasma Techniques on Cancer Attenuation. *Cancers* **2019**, *11*, 1030. [CrossRef]

17. Mateu-Sanz, M.; Tornín, J.; Brulin, B.; Khlyustova, A.; Ginebra, M.-P.; Layrolle, P.; Canal, C. Cold Plasma-Treated Ringer's Saline: A Weapon to Target Osteosarcoma. *Cancers* **2020**, *12*, 227. [CrossRef] [PubMed]

18. Bredt, D.S.; Hwang, P.M.; Snyder, S.H. Localization of nitric oxide synthase indicating a neural role for nitric oxide. *Nat. Cell Biol.* **1990**, *347*, 768–770. [CrossRef] [PubMed]

19. Granger, D.L.; Hibbs, J.B.; Perfect, J.R.; Durack, D.T. Metabolic fate of L-arginine in relation to microbiostatic capability of murine macrophages. *J. Clin. Investig.* **1990**, *85*, 264–273. [CrossRef] [PubMed]

20. Stuehr, D.J.; Nathan, C.F. Nitric oxide. A macrophage product responsible for cytostasis and respiratory inhibition in tumor target cells. *J. Exp. Med.* **1989**, *169*, 1543–1555. [CrossRef] [PubMed]

21. David, G.H. Nitrosative stress in cancer therapy. *Front. Biosci.* **2007**, *12*, 3406–3418. [CrossRef]

22. Pacher, P.; Beckman, J.S.; Liaudet, L. Nitric Oxide and Peroxynitrite in Health and Disease. *Physiol. Rev.* **2007**, *87*, 315–424. [CrossRef]

23. Malik, M.A. Water Purification by Plasmas: Which Reactors are Most Energy Efficient? *Plasma Chem. Plasma Process.* **2009**, *30*, 21–31. [CrossRef]

24. Qian, L.; Wenlu, Z.; Hong, W.; Juan, D.; Xinli, T.; Linlin, W.; Hong, S. Design and study of nitric oxide portable producing device using continuous discharging arc plasma reaction keeping low energy efficiency for viral pneumonia emergency therapy. *PLoS ONE* **2020**, *15*, e0237604. [CrossRef]

25. Motterlini, R.; Foresti, R.; Intaglietta, M.; Winslow, R.M. NO-mediated activation of heme oxygenase: Endogenous cytoprotection against oxidative stress to endothelium. *Am. J. Physiol. Circ. Physiol.* **1996**, *270*, H107–H114. [CrossRef] [PubMed]

26. Li, Q.-Y.; Niu, H.-B.; Yin, J.; Wang, M.-B.; Shao, H.-B.; Deng, D.-Z.; Chen, X.-X.; Ren, J.-P.; Li, Y.-C. Protective role of exogenous nitric oxide against oxidative-stress induced by salt stress in barley (*Hordeum vulgare*). *Colloids Surfaces B Biointerfaces* **2008**, *65*, 220–225. [CrossRef]

27. Iwata, M.; Inoue, T.; Asai, Y.; Hori, K.; Fujiwara, M.; Matsuo, S.; Tsuchida, W.; Suzuki, S. The protective role of localized nitric oxide production during inflammation may be mediated by the heme oxygenase-1/carbon monoxide pathway. *Biochem. Biophys. Rep.* **2020**, *23*, 100790. [CrossRef] [PubMed]

28. Brüne, B.; von Knethen, A.; Sandau, K.B. Nitric oxide and its role in apoptosis. *Eur. J. Pharmacol.* **1998**, *351*, 261–272. [CrossRef]

29. Radi, R. Oxygen radicals, nitric oxide, and peroxynitrite: Redox pathways in molecular medicine. *PNAS* **2018**, *115*, 23–5839. [CrossRef] [PubMed]

30. Graves, D.B. The emerging role of reactive oxygen and nitrogen species in redox biology and some implications for plasma applications to medicine and biology. *J. Phys. D Appl. Phys.* **2012**, *45*, 263001. [CrossRef]

31. Kamm, A.; Przychodzen, P.; Kuban-Jankowska, A.; Jacewicz, D.; Dabrowska, A.M.; Nussberger, S.; Wozniak, M.; Gorska-Ponikowska, M. Nitric oxide and its derivatives in the cancer battlefield. *Nitric Oxide* **2019**, *93*, 102–114. [CrossRef]

32. Kurake, N.; Tanaka, H.; Ishikawa, K.; Kondo, T.; Sekine, M.; Nakamura, K.; Kajiyama, H.; Kikkawa, F.; Mizuno, M.; Hori, M. Cell survival of glioblastoma grown in medium containing hydrogen peroxide and/or nitrite, or in plasma-activated medium. *Arch. Biochem. Biophys.* **2016**, *605*, 102–108. [CrossRef] [PubMed]

33. Girard, P.-M.; Arbabian, A.; Fleury, M.; Bauville, G.; Puech, V.; Dutreix, M.; Sousa, J.S. Synergistic Effect of H_2O_2 and NO_2 in Cell Death Induced by Cold Atmospheric He Plasma. *Sci. Rep.* **2016**, *6*, 29098. [CrossRef]

34. Biscop, E.; Lin, A.; Van Boxem, W.; Van Loenhout, J.; De Backer, J.; Deben, C.; Dewilde, S.; Smits, E.; Bogaerts, A.A. Influence of Cell Type and Culture Medium on Determining Cancer Selectivity of Cold Atmospheric Plasma Treatment. *Cancers* **2019**, *11*, 1287. [CrossRef]

35. Mohades, S.; Barekzi, N.; Razavi, H.; Maruthamuthu, V.; Laroussi, M. Temporal evaluation of the anti-tumor efficiency of plasma-activated media. *Plasma Process. Polym.* **2016**, *13*, 1206–1211. [CrossRef]

36. Koensgen, D.; Besic, I.; Gümbel, D.; Kaul, A.; Weiss, M.; Diesing, K.; Kramer, A.; Bekeschus, S.; Mustea, A.; Stope, M.B. Cold Atmospheric Plasma (CAP) and CAP-Stimulated Cell Culture Media Suppress Ovarian Cancer Cell Growth—A Putative Treatment Option in Ovarian Cancer Therapy. *Anticancer. Res.* **2017**, *37*. [CrossRef]

37. Trizio, I.; Rizzi, V.; Gristina, R.; Sardella, E.; Cosma, P.; Francioso, E.; Von Woedtke, T.; Favia, P. Plasma generated RONS in cell culture medium for in vitro studies of eukaryotic cells on Tissue Engineering scaffolds. *Plasma Process. Polym.* **2017**, *14*. [CrossRef]

38. Lukes, P.; Dolezalova, E.; Sisrova, I.; Clupek, M. Aqueous-phase chemistry and bactericidal effects from an air discharge plasma in contact with water: Evidence for the formation of peroxynitrite through a pseudo-second-order post-discharge reaction of H_2O_2 and HNO_2. *Plasma Sources Sci. Technol.* **2014**, *23*. [CrossRef]

39. Maishi, N.; Hida, K. Tumor endothelial cells accelerate tumor metastasis. *Cancer Sci.* **2017**, *108*, 1921–1926. [CrossRef] [PubMed]

40. Lau, A.N.; Heiden, M.G.V. Metabolism in the Tumor Microenvironment. *Annu. Rev. Cancer Biol.* **2020**, *4*, 17–40. [CrossRef]

41. Da Cunha, B.R.; Domingos, C.; Stefanini, A.C.B.; Henrique, T.; Polachini, G.M.; Castelo-Branco, P.; Tajara, E.H. Cellular Interactions in the Tumor Microenvironment: The Role of Secretome. *J. Cancer* **2019**, *10*, 4574–4587. [CrossRef]

42. Shenoy, A.K.; Lu, J. Cancer cells remodel themselves and vasculature to overcome the endothelial barrier. *Cancer Lett.* **2016**, *380*, 534–544. [CrossRef] [PubMed]

43. Schröder, S.; Broese, S.; Baake, J.; Juerß, D.; Kriesen, S.; Hildebrandt, G.; Manda, K. Effect of Ionizing Radiation on Human EA.hy926 Endothelial Cells under Inflammatory Conditions and Their Interactions with A549 Tumour Cells. *J. Immunol. Res.* **2019**, *2019*, 1–14. [CrossRef]

44. Lu, Z.J.; Ren, Y.Q.; Wang, G.P.; Song, Q.; Li, M.; Jiang, S.S.; Ning, T.; Guan, Y.S.; Yang, J.L.; Luo, F. Biological behaviors and proteomics analysis of hybrid cell line EAhy926 and its parent cell line A549. *J. Exp. Clin. Cancer Res.* **2009**, *28*, 1–10. [CrossRef]

45. Lu, H.; Li, X.; Zhang, J.; Shi, H.; Zhu, X.; He, X. Effects of cordycepin on HepG2 and EA.hy926 cells: Potential antiproliferative, antimetastatic and anti-angiogenic effects on hepatocellular carcinoma. *Oncol. Lett.* **2014**, *7*, 1556–1562. [CrossRef]

46. Alcaide, M.; Serrano, M.-C.; Pagani, R.; Sánchez-Salcedo, S.; Vallet-Regí, M.; Portolés, M.-T. Biocompatibility markers for the study of interactions between osteoblasts and composite biomaterials. *Biomaterials* **2009**, *30*, 45–51. [CrossRef]

47. Kogelheide, F.; Kartaschew, K.; Strack, M.; Baldus, S.; Metzler-Nolte, N.; Havenith, M.; Awakowicz, P.; Stapelmann, K.; Lackmann, J.-W. FTIR spectroscopy of cysteine as a ready-to-use method for the investigation of plasma-induced chemical modifications of macromolecules. *J. Phys. D Appl. Phys.* **2016**, *49*, 084004. [CrossRef]

48. Sies, H.; Jones, D.P. Reactive oxygen species (ROS) as pleiotropic physiological signalling agents. *Nat. Rev. Mol. Cell Biol.* **2020**, *21*, 363–383. [CrossRef] [PubMed]

49. Chauvin, J.; Judée, F.; Yousfi, M.; Vicendo, P.; Merbahi, N. Analysis of reactive oxygen and nitrogen species generated in three liquid media by low temperature helium plasma jet. *Sci. Rep.* **2017**, *7*, 1–15. [CrossRef] [PubMed]

50. Jablonowski, H.; Schmidt-Bleker, A.; Weltmann, K.-D.; Von Woedtke, T.; Wende, K. Non-touching plasma–liquid interaction— Where is aqueous nitric oxide generated? *Phys. Chem. Chem. Phys.* **2018**, *20*, 25387–25398. [CrossRef] [PubMed]

51. Lackmann, J.-W.; Wende, K.; Verlackt, C.; Golda, J.; Volzke, J.; Kogelheide, F.; Held, J.; Bekeschus, S.; Bogaerts, A.; Der Gathen, V.S.-V.; et al. Chemical fingerprints of cold physical plasmas—An experimental and computational study using cysteine as tracer compound. *Sci. Rep.* **2018**, *8*, 1–14. [CrossRef] [PubMed]

52. Girard, F.; Peret, M.; Dumont, N.; Badets, V.; Blanc, S.; Gazeli, K.; Noël, C.; Belmonte, T.; Marlin, L.; Cambus, J.-P.; et al. Correlations between gaseous and liquid phase chemistries induced by cold atmospheric plasmas in a physiological buffer. *Phys. Chem. Chem. Phys.* **2018**, *20*, 9198–9210. [CrossRef] [PubMed]

53. Williams, D. *Nitrosation Reactions and the Chemistry of Nitric Oxide*; Elsevier: Amsterdam, The Netherlands, 2004.

54. Liu, Z.C.; Liu, D.X.; Chen, C.; Li, D.; Yang, A.J.; Rong, M.Z.; Chen, H.L.; Kong, M.G. Physicochemical processes in the indirect interaction between surface air plasma and deionized water. *J. Phys. D Appl. Phys.* **2015**, *48*, 495201. [CrossRef]

55. Tarabová, B.; Lukeš, P.; Hammer, M.U.; Jablonowski, H.; Von Woedtke, T.; Reuter, S.; Machala, Z. Fluorescence measurements of peroxynitrite/peroxynitrous acid in cold air plasma treated aqueous solutions. *Phys. Chem. Chem. Phys.* **2019**, *21*, 8883–8896. [CrossRef] [PubMed]

56. Lee, Y.N.; Schwartz, S.E. Reaction kinetics of nitrogen dioxide with liquid water at low partial pressure. *J. Phys. Chem.* **1981**, *85*, 840–848. [CrossRef]

57. Hanson, D.R.; Burkholder, J.B.; Howard, C.J.; Ravishankara, A.R. Measurement of hydroxyl and hydroperoxy radical uptake coefficients on water and sulfuric acid surfaces. *J. Phys. Chem.* **1992**, *96*, 4979–4985. [CrossRef]

58. Machala, Z.; Tarabova, B.; Hensel, K.; Spetlikova, E.; Sikurova, L.; Lukes, P. Formation of ROS and RNS in Water Electro-Sprayed through Transient Spark Discharge in Air and their Bactericidal Effects. *Plasma Process. Polym.* **2013**, *10*, 649–659. [CrossRef]

59. Khlyustova, A.; Labay, C.; Machala, Z.; Ginebra, M.-P.; Canal, C. Important parameters in plasma jets for the production of RONS in liquids for plasma medicine: A brief review. *Front. Chem. Sci. Eng.* **2019**, *13*, 238–252. [CrossRef]

60. Kaushik, N.K.; Ghimire, B.; Li, Y.; Adhikari, M.; Veerana, M.; Kaushik, N.; Jha, N.; Adhikari, B.; Lee, S.-J.; Masur, K.; et al. Biological and medical applications of plasma-activated media, water and solutions. *Biol. Chem.* **2018**, *400*, 39–62. [CrossRef] [PubMed]

61. Verlackt, C.C.W.; Van Boxem, W.; Bogaerts, A. Transport and accumulation of plasma generated species in aqueous solution. *Phys. Chem. Chem. Phys.* **2018**, *20*, 6845–6859. [CrossRef]

62. Brodsky, S.V.; Morrishow, A.M.; Dharia, N.; Gross, S.S.; Goligorsky, M.S. Glucose scavenging of nitric oxide. *Am. J. Physiol. Physiol.* **2001**, *280*, F480–F486. [CrossRef]

63. Villa, L.M.; Salas, E.; Darley-Usmar, V.M.; Radomski, M.W.; Moncada, S. Peroxynitrite induces both vasodilatation and impaired vascular relaxation in the isolated perfused rat heart. *Proc. Natl. Acad. Sci. USA* **1994**, *91*, 12383–12387. [CrossRef] [PubMed]

64. Dejam, A.; Hunter, C.J.; Schechter, A.N.; Gladwin, M.T. Emerging role of nitrite in human biology. *Blood Cells Mol. Dis.* **2004**, *32*, 423–429. [CrossRef] [PubMed]

65. Bauer, G. Intercellular singlet oxygen-mediated bystander signaling triggered by long-lived species of cold atmospheric plasma and plasma-activated medium. *Redox Biol.* **2019**, *26*, 101301. [CrossRef] [PubMed]
66. Bauer, G.; Sersenová, D.; Graves, D.B.; Machala, Z. Cold Atmospheric Plasma and Plasma-Activated Medium Trigger RONS-Based Tumor Cell Apoptosis. *Sci. Rep.* **2019**, *9*, 1–28. [CrossRef]
67. Bauer, G. Autoamplificatory singlet oxygen generation sensitizes tumor cells for intercellular apoptosis-inducing signaling. *Mech. Ageing Dev.* **2018**, *172*, 59–77. [CrossRef]
68. Khan, F.H.; Dervan, E.; Bhattacharyya, D.D.; McAuliffe, J.D.; Miranda, K.M.; Glynn, S.A. The Role of Nitric Oxide in Cancer: Master Regulator or Not? *Int. J. Mol. Sci.* **2020**, *21*, 9393. [CrossRef] [PubMed]
69. Yusupov, M.; Yan, D.; Cordeiro, R.M.; Bogaerts, A. Atomic scale simulation of H_2O_2 permeation through aquaporin: Toward the understanding of plasma cancer treatment. *J. Phys. D Appl. Phys.* **2018**, *51*, 125401. [CrossRef]
70. Yan, D.; Xiao, H.; Zhu, W.; Nourmohammadi, N.; Zhang, L.G.; Bian, K.; Keidar, M. The role of aquaporins in the anti-glioblastoma capacity of the cold plasma-stimulated medium. *J. Phys. D Appl. Phys.* **2017**, *50*, 055401. [CrossRef]
71. Bauer, G.; Graves, D.B. Mechanisms of Selective Antitumor Action of Cold Atmospheric Plasma-Derived Reactive Oxygen and Nitrogen Species. *Plasma Process. Polym.* **2016**, *13*, 1157–1178. [CrossRef]
72. Gorbanev, Y.; Stehling, N.; O'Connell, D.; Chechik, V. Reactions of nitroxide radicals in aqueous solutions exposed to non-thermal plasma: Limitations of spin trapping of the plasma induced species. *Plasma Sources Sci. Technol.* **2016**, *25*, 55017. [CrossRef]
73. Bauer, G. Cold Atmospheric Plasma and Plasma-Activated Medium: Antitumor Cell Effects with Inherent Synergistic Potential. *Plasma Med.* **2019**, *9*, 57–88. [CrossRef]

Article

Using C-doped TiO$_2$ Nanoparticles as a Novel Sonosensitizer for Cancer Treatment

Chun-Chen Yang [1], Chong-Xuan Wang [1], Che-Yung Kuan [2,3], Chih-Ying Chi [2,3], Ching-Yun Chen [3,4], Yu-Ying Lin [2,3], Gin-Shin Chen [3], Chun-Han Hou [5,*] and Feng-Huei Lin [3,6,*]

[1]   Department of Materials Science and Engineering, National Taiwan University, Taipei 10617, Taiwan; d03527007@ntu.edu.tw (C.-C.Y.); r05527013@ntu.edu.tw (C.-X.W.)

[2]   PhD Program in Tissue Engineering and Regenerative Medicine, National Chung Hsing University, Taichung 40227, Taiwan; 090115@nhri.edu.tw (C.-Y.K.); d103001707@mail.nchu.edu.tw (C.-Y.C.); 024016@nhri.edu.tw (Y.-Y.L.)

[3]   Institute of Biomedical Engineering and Nanomedicine, National Health Research Institutes, Miaoli County 35053, Taiwan; chingyun523@nhri.edu.tw (C.-Y.C.); gschen@nhri.edu.tw (G.-S.C.)

[4]   Department of Biomedical Sciences & Engineering, National Central University, Taoyuan City 32001, Taiwan

[5]   Department of Orthopedic Surgery, National Taiwan University, Taipei 10617, Taiwan

[6]   Institute of Biomedical Engineering, National Taiwan University, Taipei 10617, Taiwan

*   Correspondence: chhou@ntu.edu.tw (C.-H.H.); double@ntu.edu.tw (F.-H.L.); Tel.: +886-2-23123456#65274 (C.-H.H.); +886-2-27327474 (F.-H.L.)

Received: 18 August 2020; Accepted: 15 September 2020; Published: 17 September 2020

Abstract: Sonodynamic therapy is an effective treatment for eliminating tumor cells by irradiating sonosentitizer in a patient's body with higher penetration ultrasound and inducing the free radicals. Titanium dioxide has attracted the most attention due to its properties among many nanosensitizers. Hence, in this study, carbon doped titanium dioxide, one of inorganic materials, is applied to avoid the foregoing, and furthermore, carbon doped titanium dioxide is used to generate ROS under ultrasound irradiation to eliminate tumor cells. Spherical carbon doped titanium dioxide nanoparticles are synthesized by the sol-gel process. The forming of C-Ti-O bond may also induce defects in lattice which would be beneficial for the phenomenon of sonoluminescence to improve the effectiveness of sonodynamic therapy. By dint of DCFDA, WST-1, LDH and the Live/Dead test, carbon doped titanium dioxide nanoparticles are shown to be a biocompatible material which may induce ROS radicals to suppress the proliferation of 4T1 breast cancer cells under ultrasound treatment. From in vivo study, carbon doped titanium dioxide nanoparticles activated by ultrasound may inhibit the growth of the 4T1 tumor, and it showed a significant difference between sonodynamic therapy (SDT) and the other groups on the seventh day of the treatment.

Keywords: sonodynamic therapy; carbon doped titanium dioxide; sonosensitizers; ultrasound; cancer treatment; breast cancer treatment

---

1. Introduction

Cancer has been the leading cause of death in the US for 40 consecutive years. In 2019, there were 606,880 deaths from cancer projected to occur in the US [1]. Surgery, radiotherapy and chemotherapy are the most fundamental and effective cancer treatments. Nevertheless, it is hard to remove tumor cells comprehensively via surgery; radiotherapy and chemotherapy may not only kill the cancer cells but cause harm to healthy cells nearby and make patients feel fatigue. Immunotherapy is likely to interfere with the immune system and cause autoimmune disease. Hyperthermia is probably resisted by cancer cells after several heat treatments. Photodynamic therapy (PDT) is limited due to the shallow penetration depth of light sources into tumor tissue. In previous studies, we used X-ray as an alternative light source which provided a novel therapeutic approach for deep-seated tumor/cancer treatment [2–4]; however, the annual radiation dose limit was another issue. Thus, an alternative therapy with fewer side effects was proposed by Umemura and Yumita, called "Sonodynamic therapy (SDT)" [5]. SDT can focus the ultrasound energy on the deeply located tumor site, which overcomes the shortcoming of PDT. SDT is considered to be a safer and more acceptable therapy for patients compared to radiotherapy and chemotherapy [6]. It is noninvasive, and the apparatus is relatively inexpensive [7,8].

SDT consists of three basic elements: ultrasound, sonosensitizer and oxygen molecules. The mechanism of SDT is that the nonthermal effect of acoustic cavitation generated by sonoluminescence, and the sonoluminescent light activates the sonosensitizer, leading to the electronic excitement of the sonosensitizer [7]. When the excited sonosensitizer decays back to the ground state, the released energy transfers to oxygen to generate the highly reactive singlet oxygen ($^1O_2$) [9]. Meanwhile, the energy may lead to pyrolysis reaction of the water near the exposed site of ultrasound and generate hydroxyl radicals ($\bullet$OH). These reactive oxygen species (ROS) may cause the death of the tumor cells afterwards [10]. ROS plays an important role in cellular signaling pathways, such as metabolism, growth, differentiation and death signaling, and react with molecules by reversible oxidative modifications. Excess generation of ROS may cause cell senescence and death to intracellular biomacromolecules, such as protein, lipid, RNA and DNA, via oxidative damage [11].

Ultrasound is a mechanical wave with periodic vibrations in a continuous medium at frequencies greater than 20 kHz [12]. Ultrasound is able to penetrate tissue with less attenuation of energy. Therefore, it can be applied to medical diagnosis and therapeutic use. For a medical diagnosis purpose, the ultrasound is irradiated at a frequency of 2.0 to 28.0 MHz with low-energy irradiation to prevent tissue from damaging. For therapeutic use, the ultrasound is irradiated at a frequency of 0.5 to 3.0 MHz with higher doses of energy to generate the desired biological results [13]. For SDT, low-intensity ultrasound is used to induce the non-thermal and sono-chemical effects to activate sonosensitizer to cause the damage and even the death of tumor cells [14]. The non-thermal effect of ultrasound in SDT is cavitation that involves formation, growth and collapse of cavitation bubbles [7]. Under ultrasound irradiation, the static pressure of the aqueous solution decreases below the vapor pressure, and water may evaporate into gas bubbles. The cavitation bubbles nucleate in the presence of impurities or pre-existing bubbles in solution and oscillate in the phase under irradiation [15]. During the ultrasound irradiation, bubbles grow increasingly larger and stop growing when the static pressure equals the vapor pressure. They may start to break down from its weakest spot when the static pressure exceeds the vapor pressure, and then collapse (known as inertial cavitation) led to a highly concentrated energy release [16,17]. The released energy leads to the pyrolysis reaction of the water, which generate ($\bullet$OH) and short light pulses (known as sonoluminescence) [18]. Sonoluminescence involves intense ultraviolet-visible light, which can excite sonosensitizer to generate ROS [19].

Sonosensitizers play a critical role in SDT that can enhance the effect of ultrasound. The development of sonosensitizers had grown swiftly in recent decades due to the known mechanisms of cell apoptosis for SDT [20]. The porphyrin-based sonosensitizers, such as photofrin, hematoporphyrin, 5-ALA (5-aminolevulinic acid) and chlorin-e6, are the most often used sonosensitizers in SDT research [7]. However, porphyrin-based sonosensitizers have phototoxicity on the skin that may affect both tumor cells and normal cells under a certain wavelength of light or energy irradiation in PDT studies, which means that this issue may also take place in SDT [20]. On the other hand, most sonosensitizers were hydrophobic and easy to aggregate in physiological condition, leading to a reduction in their ROS production [21]. Nonetheless, the development of nanoparticles shows a promising potential to solve these problems. Among many nanosensitizers, titanium dioxide (TiO_2) has attracted the most attention due to its properties [21]. TiO_2 is widely used in many territories based on low toxicity, high stability, high photocatalytic activity and low cost [22,23]. Compared to porphyrin-based sonosensitizers which are quickly degraded under oxidizing conditions, TiO_2 exhibits high stability because it is highly resistant to degradation by ROS. TiO_2 exhibits three kinds of crystal structures, namely anatase, rutile and brookite.

Anatase and rutile are the most common in the utilization of crystal structures, and brookite is less used in industrial application. Even though anatase (E_g = 3.2 eV) has a wider bandgap than rutile (E_g = 3.0 eV), anatase shows higher photoactivity due to its larger specific surface area that anatase is more suitable to be used as a photocatalyst [24]. In previous studies, the anatase structure of TiO_2 has been utilized as a sonocatalyst to generate ROS under ultrasound irradiation [10,14]. Nonetheless, the wide bandgap of anatase requires a greater energy to trigger. Carbon has previously been doped in the semiconductors to form a new valence band, thus narrowing the bandgap [25]. The addition of carbon may give TiO_2 an excess of conducting electrons or holes which is important for lowering the bandgap [23,24].

Hence, in this study, the sonosensitizer C-doped TiO_2 was synthesized through doping carbon into the anatase structure of TiO_2 to diminish the bandgap. A square wave of the ultrasound at a resonant frequency of 1.0 MHz, intensity of 0.33 MPa and duty cycle of 50% was used to induce the inertial cavitation and generate sonoluminescent light to activate the synthesized C-doped TiO_2.

2. Materials and Methods

2.1. Preparation of C-Doped TiO_2

C-doped TiO_2 was synthesized by the sol-gel method [23]. First, 2 g of Pluronic® F127 (Sigma-Aldrich, St. Louis, MO, USA) was dissolved in 40 mL 95% ethanol completely with vigorous stirring. Then, 5 mL titanium(IV) isopropoxide (TIP, Ti(OCH(CH$_3$)$_2$)$_4$, purity > 97%, Sigma-Aldrich) was added into the solution with magnetic stirring of 600 rpm. An amount of 3 g of D-(+)-glucose ($C_6H_{12}O_6$, Sigma-Aldrich), which was used as the carbon source, was dissolved in 6 mL ddH$_2$O. Then, the glucose solution was dropped into the TIP/Pluronic® F127 solution. The mixed solution was kept vigorously stirring at room temperature for 30 min. The precipitate was collected by centrifugation at 5000 rpm and washed with 95% ethanol for three times. The precipitate was then calcined at 400 °C for 2 h to obtain particles.

2.2. Material Characterization

The crystal structure of the synthesized C-doped TiO_2 was determined by X-ray diffraction (XRD; Rigaku TTRAX 3) with Cu Kα radiation at a speed of 2° per minute at 40 kV and 30 mA from 20° to 60°.The surface morphology and particle size of C-doped TiO_2 were characterized by a scanning electron microscope (SEM; Nova Nano SEM 450). The structure and diffraction pattern of the material were analyzed by a transmission electron microscope (TEM; JEOL 2010F). The particle size of the material was reconfirmed via Zetasizer (Malvern Nano ZS). The chemical composition of the material was analyzed by an energy dispersive spectrometer (EDS; Nova Nano SEM 450) SEM attachment. To confirm that carbon had been doped in titanium dioxide and formed the bonding, C-doped TiO_2 was measured by auger electron spectroscopy (AES; JEOL JAMP 9510F) and X-ray photoelectron spectroscopy (XPS; Theta Probe).

2.3. Ultrasound Apparatus

Ultrasound irradiation was conducted with a function generator (Agilent 33521A) at a resonant frequency of 1.0 MHz and a duty cycle of 50% and amplified by a power amplifier (E&I 1040L) to generate a square wave with a negative pressure of 0.33 MPa and intensity of 1.8 W/cm^2 for 90 s in the whole experiment. Figure 1 shows the entire apparatus.

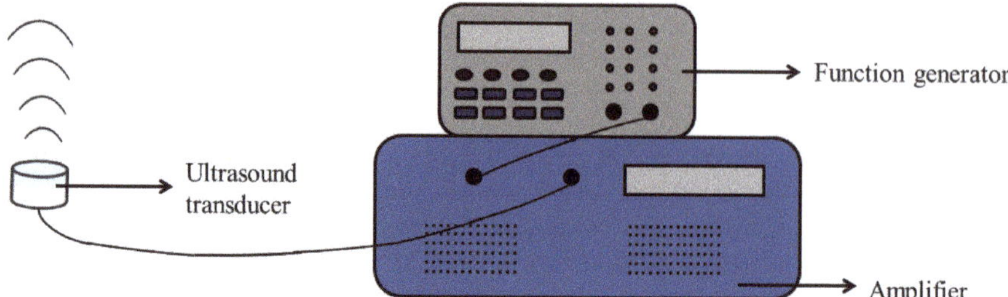

Figure 1. Ultrasound apparatus. The ultrasound transducer with a diameter of 2.0 cm was fixed with aluminum block by a rubber band to keep the transducer facing upward in the degassed water tank.

2.4. Preparation of Terephthalic Acid Solution and Evaluation of Ultrasound Parameter

First, 2 mM terephthalic acid (TA; Sigma-Aldrich) was dissolved in 800 mL ddH$_2$O. During ultrasound treatment, the experiments should be carried out under alkaline condition. Hence, 5 mL 1 M of NaOH was added in the solution. TA solution was stirred for 1 h under 4 °C and dark conditions to avoid photochemical reaction [26].

The experiment was divided into two groups. No US group was named as the control group: the US group was the group irradiated by ultrasound. The fluorescence signal intensity was measured by a multi-label plate reader with excitation and emission wavelengths at 310 and 420 nm, respectively.

2.5. Cell Culture and Animal Model

Briefly, L929 cells obtained from National Health Research Institutes were cultured in MEM medium (Minimum Essential Media, Sigma-Aldrich) with 10% Fetal bovine serum (FBS, Gibco) and 1% Antibiotic-Antimycotic (Gibco) and maintained in a 5% CO$_2$ atmosphere at 37 °C.

Breast cancer 4T1 cells obtained from National Health Research Institutes were used as model cancer cells in this study. The 4T1 cells were cultured in RPMI medium (RPMI-1640 Media (Sigma-Aldrich) with 10% Fetal bovine serum (FBS, Gibco) and 1% Antibiotic-Antimycotic (Gibco) and maintained in 5% CO$_2$ atmosphere at 37 °C.

6-week-old male BALB/c nude mice were used as a xenograft animal model in this study. The nude mice were injected subcutaneously with 2×10^6 4T1 cells suspended in 200 μL 1X PBS into the right thigh. All the animal experiments were approved by National Health Research Institutes (NHRI-IACUC-107013).

2.6. Evaluation of Biocompatibility for the Synthesized C-Doped TiO$_2$

The evaluation of biocompatibility in this study was based on International Standard ISO 10993. The cell viability induced by the synthesized C-doped TiO$_2$ was determined by the WST-1 test (TAKARA). L929 cells were seeded on a 96-well cell culture plate with a cell density of 10^4 cells per well and incubated in 5% CO$_2$ atmosphere at 37 °C. Briefly, 0.2 g/mL of the materials was immersed in MEM medium for 16 to 24 h in 5% CO$_2$ atmosphere at 37 °C as material extracts. The control group included L929 cells without treatment. Zinc diethyldithiocarbamate (ZDEC, Sigma-Aldrich) and aluminum oxide (Sigma-Aldrich) were indicated as the "Positive Control" and "Negative Control" group, respectively. The C-doped TiO$_2$ group included cells cultured with C-doped TiO$_2$. After one-day incubation, the previous solution in each well was removed, and 100 μL of the supernatant of the material extracts was added into each well. After one-day incubation, the previous solution in each well was removed, and 100 μL fresh MEM medium with 10% WST-1 reagent was added. After 2 h incubation in 5% CO$_2$ atmosphere at 37 °C kept in the dark, the solution in each well was collected to measure absorbance at 490 nm by a microplate reader.

2.7. Detection of Cellular ROS Generation

ROS generation induced by the synthesized C-doped TiO_2 in SDT was measured by a DCFDA-cellular ROS detection assay kit (abcam). 4T1 breast cancer cells were seeded on 12-well cell culture plates with a cell density of 6×10^4 cells per well and incubated in 5% CO_2 atmosphere at 37 °C. After one-day incubation, the previous solution in each well was removed, and 1 mL 15 mg/mL synthesized C-doped TiO_2 in RPMI medium was added into each well. After 6 h incubation, ultrasound was then irradiated from the bottom of the cell culture plates in degassed water. The distance between ultrasound transducer and the bottom of the cell culture plate was around 5 mm. After US treatment, the cell culture plates were incubated for 2 h. The previous solution in each well was then removed and washed once with 1× PBS. An amount of 1 mL 20 μM 2′,7′ –dichlorofluorescin diacetate (DCFDA, abcam) in 1× buffer was added into each well for 30 min in 5% CO_2 atmosphere at 37 °C in the dark. The previous solution in each well was removed, and cells were washed with 1× PBS once. An amount of 1 mL 1× PBS was added into each well. The fluorescence signal intensity was measured by a multilabel plate reader with excitation and emission wavelengths at 485 and 535 nm, respectively [27]. The control group included 4T1 cells without treatment. The C-doped TiO_2 group included cells cultured with C-doped TiO_2. The US group included cells subjected to ultrasound irradiation. The SDT group included cells cultured with C-doped TiO_2 and subjected to ultrasound irradiation.

2.8. Evaluation of the Synthesized C-Doped TiO_2 in SDT In Vitro

WST-1 and LDH tests were used to evaluate the cell viability and cytotoxicity of the synthesized C-doped TiO_2 in SDT. In vitro, 6×10^4 4T1 cells suspended in 1 mL RPMI medium were seeded on 12-well cell culture plates and incubated in 5% CO_2 atmosphere at 37 °C. After one-day incubation, the previous solution in each well was removed, and 1 mL 15 mg/mL synthesized C-doped TiO_2 in RPMI medium was added into each well. After 6 h incubation, ultrasound was then irradiated from the bottom of the cell culture plates in degassed water. The distance between the ultrasound transducer and the bottom of the cell culture plate was around 5 mm. After ultrasound treatment, the cell culture plates were incubated for one day. For the WST-1 test, the previous solution in each well was removed, and 400 μL fresh RPMI medium with 10% WST-1 reagent was added. After 1 h incubation in 5% CO_2 atmosphere at 37 °C kept in the dark, the solution in each well was collected to measure absorbance at 490 nm by an Elisa reader. The control group included 4T1 cells without treatment. Zinc diethyldithiocarbamate (ZDEC) and aluminum oxide were indicated as the "Positive Control" and "Negative Control" group, respectively. The C-doped TiO_2 group included cells cultured with C-doped TiO_2. The US group included cells subjected to ultrasound irradiation. The SDT group included cells cultured with C-doped TiO_2 and subjected to ultrasound irradiation.

For the LDH test, 50 μL lysis solution was then added into each well of the Lysis group and then incubated in 5% CO_2 atmosphere at 37 °C for 30 min. Fifty μL of supernatant from each well was then transferred to a 96-well cell culture plate, and 50 μL LDH reagent was added into each well. After 15 min of reaction at room temperature in the dark, the cytotoxicity was measured by a microplate reader with absorbance at 490 nm. The C-doped TiO_2 group included cells cultured with C-doped TiO_2. The US group included cells subjected to ultrasound irradiation. The SDT group included cells cultured with C-doped TiO_2 and subjected to ultrasound irradiation.

2.9. Evaluation of SDT in In Vivo Tumor Growth

There were 4 different treatments exerted on nude mice: (1) the control group was injected with 200 μL 1× PBS at day 0 and day 7 without ultrasound irradiation; (2) the C-doped TiO_2 group was injected with 200 μL synthesized C-doped TiO_2 (150 mg/mL) in 1× PBS at day 0 and day 7 without ultrasound irradiation; (3) the US group was injected with 200 μL of 1× PBS at day 0 and day 7 with US irradiation; (4) the SDT group was injected with 200 μL synthesized C-doped TiO_2 (150 mg/mL) in 1× PBS at day 0 and day 7 with US irradiation. There were 5 nude mice in each group.

The tumor volume was measured by an electronic slide caliper using Equation (1), where L is the longest dimension and W is the shortest dimension, parallel to the mouse body [28].

$$V = 0.5 \times L \times W^2 \tag{1}$$

2.10. Histopathology

During sacrifice, cardiac puncture was used to obtain the blood samples. The samples were collected into tubes with 7.5% EDTA solution. Tumor and internal organs, including heart, liver, spleen, lung and kidney were also harvested and then fixed with 4% formaldehyde. These organs were embedded into the paraffin and further stained with hematoxylin and eosin (H&E) for histologic examination.

2.11. Statistics

All data are indicated as the mean ± standard deviation. Statistical analysis was performed using one-way ANOVA followed by multiple comparisons with the Dunnett test, and the difference between the groups was deemed to be statistically significant when $p < 0.05$.

3. Results

The XRD pattern of the synthesized C-doped TiO_2 is shown in Figure 2a. It is indicated that the 2θ at 25.2°, 38.3°, 48°, 54.3° and 55.1° corresponded to the crystal form of (101), (004), (200), (105) and (211), which was completely matched with the standard pattern of the anatase structure of TiO_2 (JCPDS Card No. 21-1272). The surface morphology of C-doped TiO_2 was characterized by SEM as shown in Figure 2b,c. Figure 2d shows the chemical composition of the material analyzed by EDS, Ti, O and C elements with an atomic percent (At%) of 62.8, 32.8 and 4.4, respectively. The particles were spherical with a rough surface. The structure and diffraction pattern of the material were examined by TEM as shown in Figure 2e,f, respectively. C-doped TiO_2 formed in the aggregation of nanograins, and its particle size was in the range of 100 to 200 nm. The selected area diffraction pattern (SADP) of C-doped TiO_2 showed a ring pattern with the concentric rings from interior to outside, representing the crystal planes of (101), (004) and (105) which corresponded to the anatase phase. The particle size was precisely measured by Zetasizer as shown in Figure 2g. The precise particle size was 156.9 nm with a polydispersity index (PDI) of 0.137, which was in agreement with the analysis of SEM particle size (148.9 ± 26.3 nm) as shown in Figure 2h. Figure 3a shows the detection of chemical composition of the material by AES. In addition to the peaks related to Ti and O, C was detected at the energy of 263 eV. Figure 3b shows the XPS spectra of C-doped TiO_2 was at a binding energy of 282, 462 and 562 eV, representing O 1s, Ti 2p and C 1s, respectively. Figure 3c shows the XPS spectra of the C 1s scan: the peak of TiC was detected at binding energy around 282 eV [29]. Figure 3d shows the XPS spectra of the Ti 2p scan, the peak of Ti^{4+} was detected at a binding energy of 462 (Ti $2p_{1/2}$) and 456 eV (Ti $2p_{3/2}$). Figure 3e shows the XPS spectra of the O 1s scan: the peak of (Ti^{4+}/Ti^{3+})-O was detected at a binding energy of 527 eV.

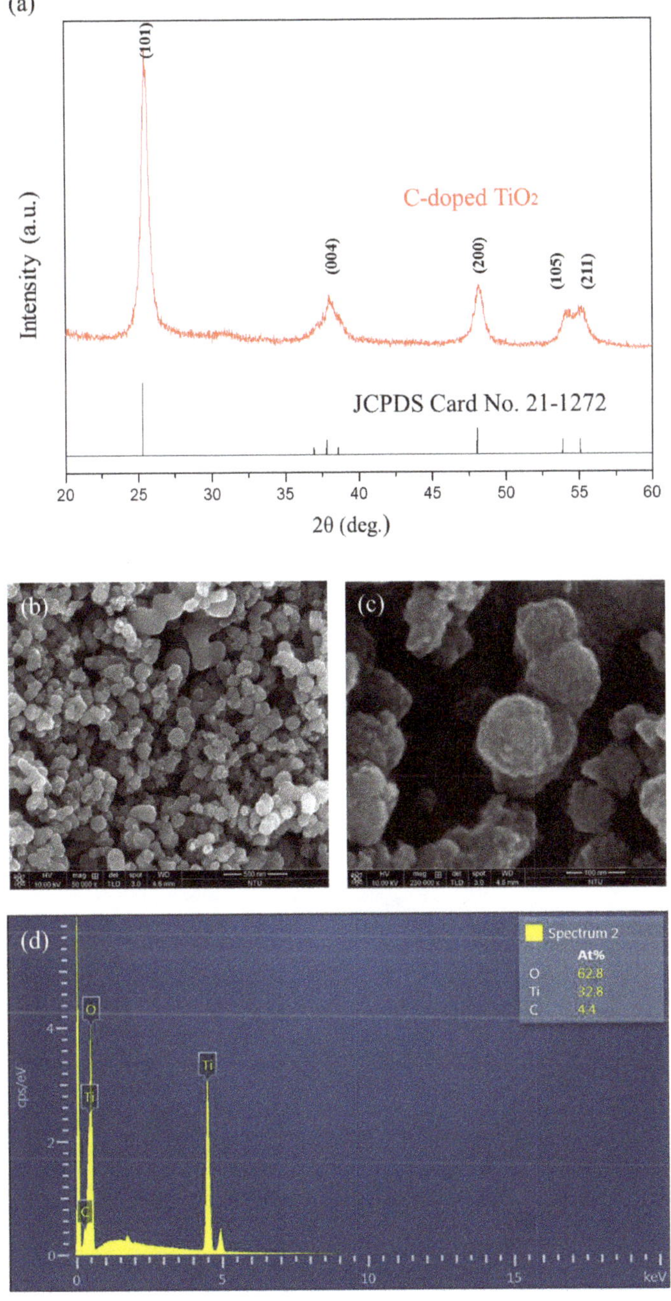

Figure 2. *Cont.*

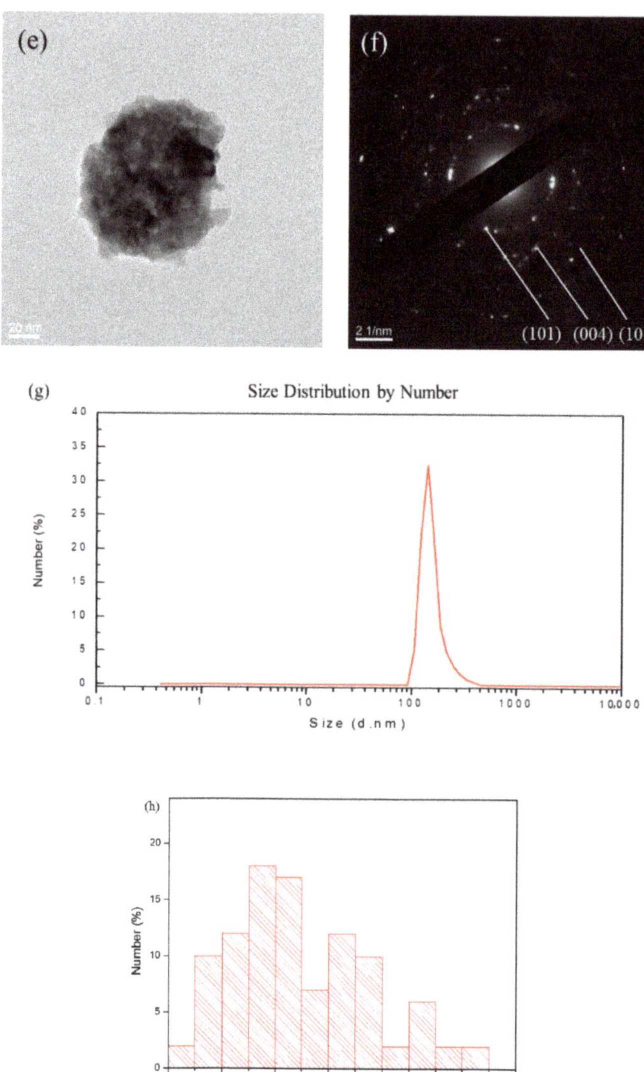

Figure 2. (**a**) XRD pattern of C-doped TiO_2, (**b**) low-magnification SEM images of C-doped TiO_2, (**c**) high-magnification SEM images of C-doped TiO_2, (**d**) EDS of C-doped TiO_2, (**e**) TEM images of C-doped TiO_2, (**f**) Diffraction pattern of C-doped TiO_2, (**g**) Size distribution of C-doped TiO_2 and (**h**) SEM particle size distribution of C-doped TiO_2.

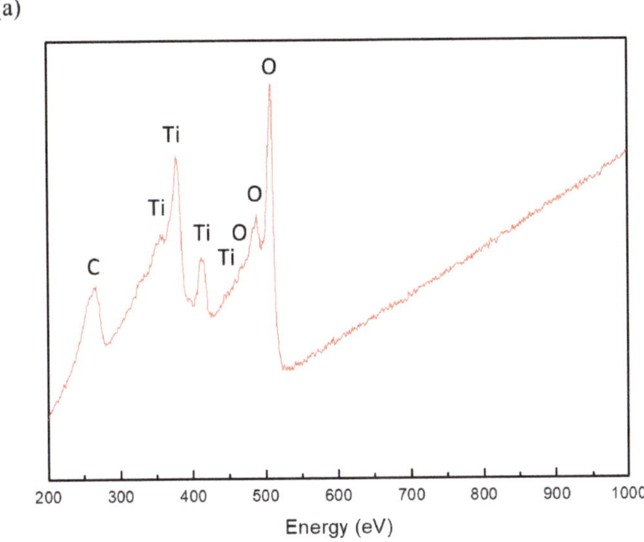

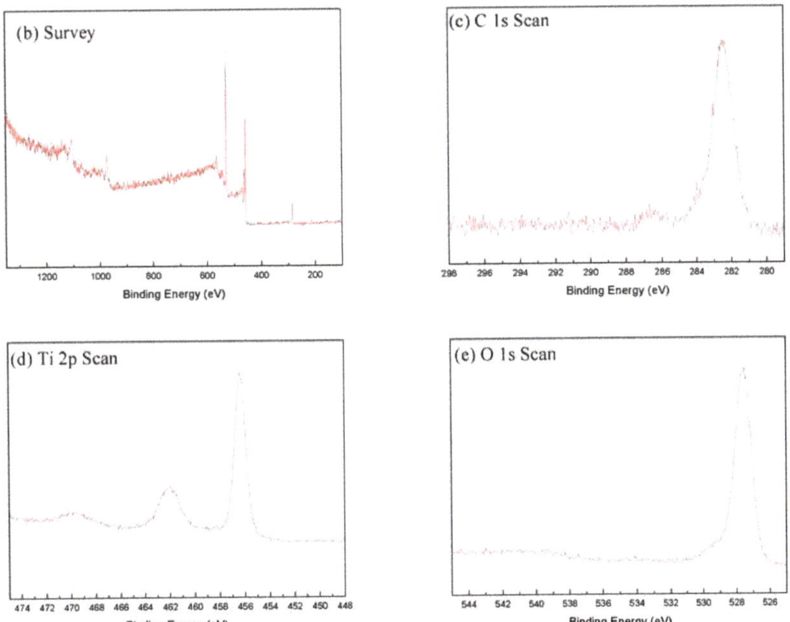

Figure 3. (**a**) AES spectra of C-doped TiO_2, (**b**) XPS spectra of C-doped TiO_2, (**c**) C 1s scan, (**d**) Ti 2p scan and (**e**) O 1s scan.

The 2-hydroxyterephthalic acid (HTA) fluorescence emission of the TA solution with or without US irradiation is shown in Figure 4. TA is a non-fluorescent compound and may further react with hydroxyl radicals to produce a highly fluorescent HTA. The fluorescence was 106 ± 4% (US group) and 100 ± 2%

(control group), respectively. The fluorescence emission of the control group was 100% as the baseline. Each experiment was repeated 6 times. Based on the result, we believe that the US parameter in this study is able to induce inertial cavitation.

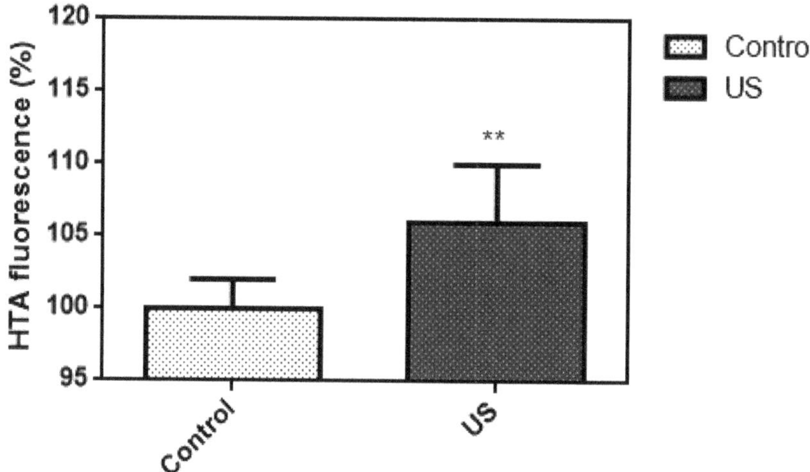

Figure 4. Oxidation of terephthalic acid (TA) to 2-hydroxyterephthalic acid (HTA) in the presence of hydroxyl radicals induced by inertial cavitation (one-way ANOVA, mean $\pm$ SD, $n = 6$, **: $p < 0.01$).

The biocompatibility test of C-doped TiO_2 on cell viability by the WST-1 test is shown in Figure 5. After treatment with C-doped TiO_2, the cell viability was $0.25 \pm 0.24\%$ (positive control group), $99.54 \pm 10\%$ (negative control group), $100 \pm 24\%$ (control group) and $81.49 \pm 20\%$ (C-doped TiO_2 group), respectively. The cell viability of the control group was 100% as the baseline. Each experiment was repeated 5 times. The biocompatibility test followed by the regulations of ISO 10,993 indicated that there would be no potential toxicity when the cell viability of the experiment group was over 75%. Based on the WST-1 test, C-doped TiO_2 showed no potential toxicity to normal cells.

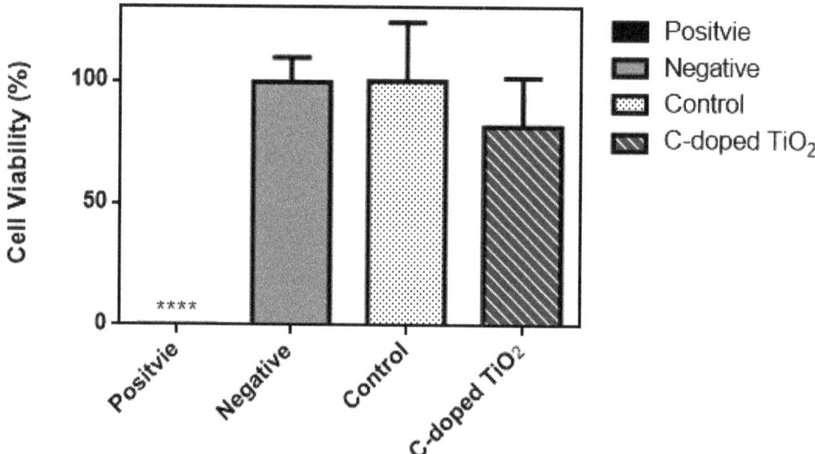

Figure 5. WST-1 test for the biocompatibility of C-doped TiO_2 (one-way ANOVA, mean $\pm$ SD, $n = 5$, ****: $p < 0.0001$).

ROS generation of C-doped TiO$_2$ under US irradiation by the DCFDA test is shown in Figure 6. DCFDA was able to permeate into 4T1 cells and be deacetylated by cellular esterases to a non-fluorescent DCFH. DCFH was then oxidized to a highly fluorescent 2′, 7′–dichlorofluorescein (DCF) in the presence of ROS. After the treatments, the fluorescence was 100 ± 7% (control group), 97 ± 12% (C-doped TiO$_2$ group), 119 ± 13% (US group) and 141 ± 11% (SDT group), respectively. The fluorescence of the control group was 100% as the baseline. Each experiment was repeated 6 times.

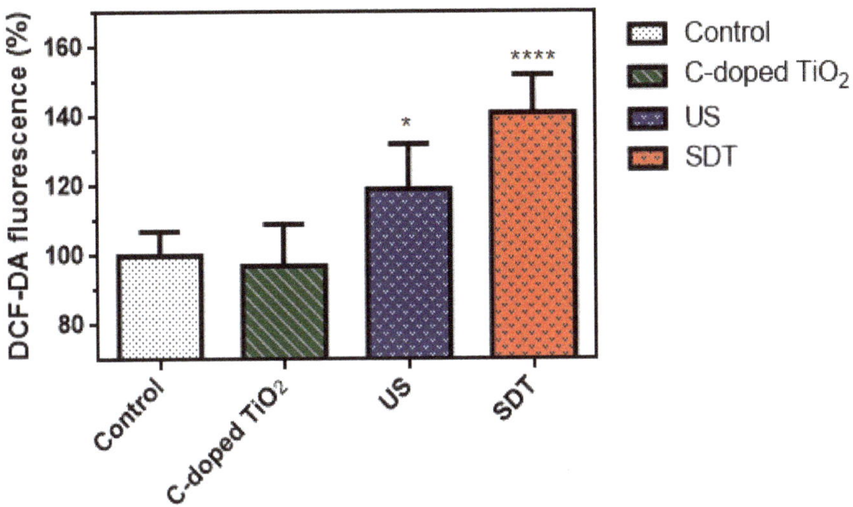

Figure 6. DCFDA test for relative ROS concentration induced by C-doped TiO$_2$ activated by US (one-way ANOVA, mean ± SD, $n = 6$, *: $p < 0.05$, ****: $p < 0.0001$).

The effect of C-doped TiO$_2$ in SDT on cell viability and cytotoxicity of 4T1 cells was examined by WST-1 and LDH tests. The results of live/dead assay are shown in Figure S1 to evaluate cell viability in C-doped TiO$_2$ during ultrasound irradiation. After the treatments, the cell viability was 0 ± 1% (positive control group), 90 ± 13% (negative control group), 100 ± 14% (control group), 84 ± 3% (C-doped TiO$_2$ group), 77 ± 2% (US group) and 67 ± 7% (SDT group), respectively, as shown in Figure 7a. The cell viability of the control group was 100% as the baseline. Each experiment was repeated 6 times. After the treatments, the cytotoxicity was 100 ± 18% (total lysis group), 0 ± 3% (C-doped TiO$_2$ group), 7 ± 13% (US group) and 23 ± 13% (SDT group), respectively, as shown in Figure 7b. The cytotoxicity of the control group was 0% as the baseline.

The antitumor efficacy of C-doped TiO$_2$ under ultrasound irradiation was evaluated using BALB/c nude mice subcutaneously injected with 4T1 cells, where C-doped TiO$_2$ particles were delivered into mice during the entire experiment and then treated with ultrasound irradiation twice—on day 0 and day 7 (Figures S2 and S3). The variation of relative tumor volume is shown in Figure 8, which was used to evaluate the antitumor effect of C-doped TiO$_2$ in SDT for 4T1 cells. At 14 days, the relative tumor volume was 367 ± 176% (control group), 407 ± 73% (C-doped TiO$_2$ group), 335 ± 82% (US group) and 165 ± 22% (SDT group), respectively. The H&E-stained images of organs are shown in Figure S4. The result showed that the C-doped TiO$_2$ group and US group could not suppress the growth of the tumor. On the other hand, the SDT group could retard the tumor growth significantly ($p < 0.05$). The histologic staining of the tumor site is shown in Figure 9. H&E-stained tumor tissue images confirmed that SDT enhanced the ability to cause the death of the 4T1 cells compared to the other groups.

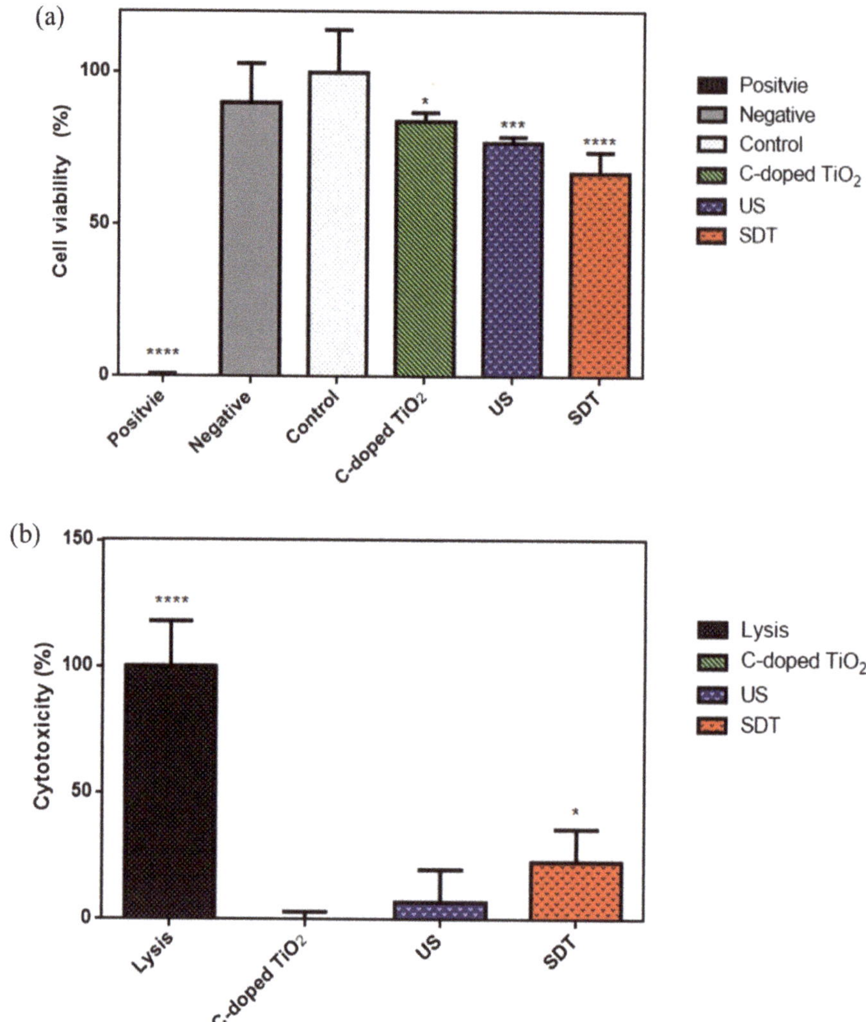

Figure 7. (**a**) Cell viability induced by C-doped TiO$_2$ activated by US evaluated by WST-1 assay (one-way ANOVA, mean ± SD, n = 6), (**b**) cytotoxicity induced by C-doped TiO$_2$ activated by US evaluated by LDH assay (one-way ANOVA, mean ± SD, n = 6, *: $p < 0.05$, ***: $p < 0.001$, ****: $p < 0.0001$).

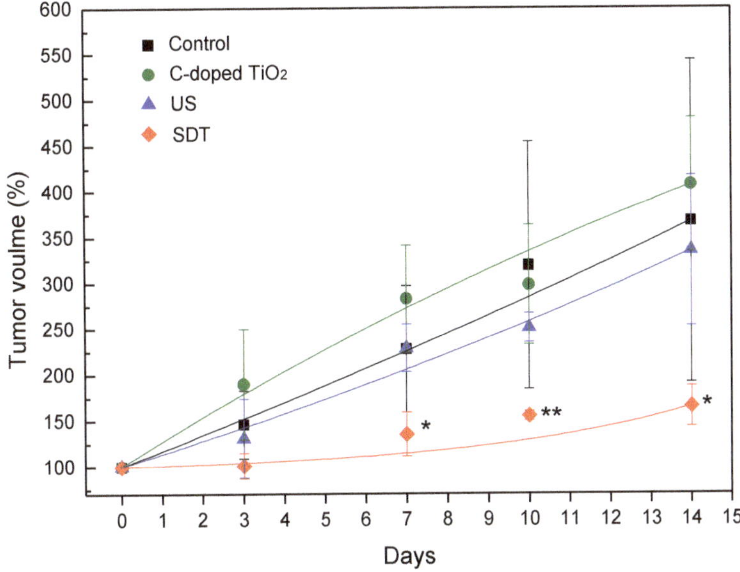

Figure 8. The variation of relative tumor volume in BALB/c nude mice induced by C-doped TiO_2 activated by US (one-way ANOVA, mean $\pm$ SD, $n = 5$, *: $p < 0.05$, **: $p < 0.01$).

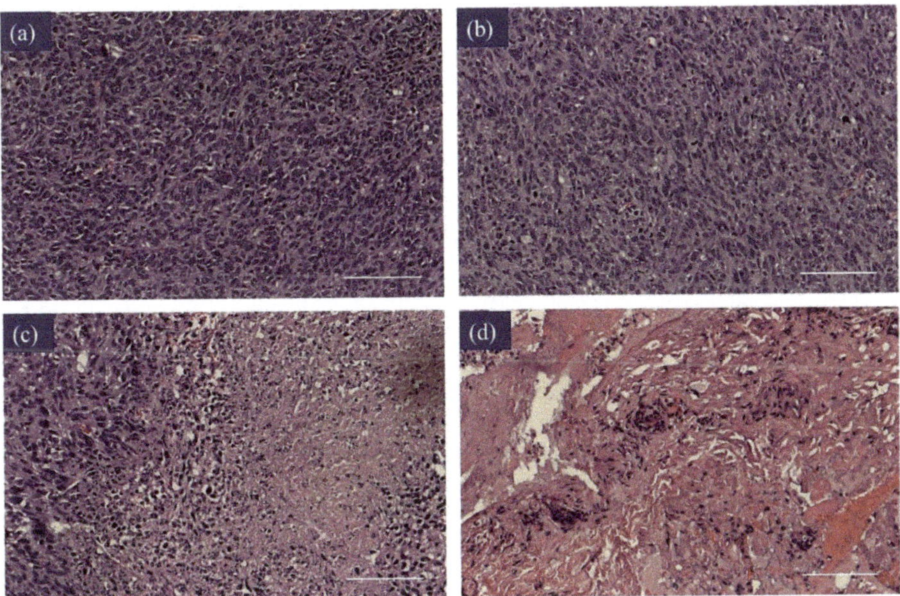

Figure 9. H&E-stained images of tumor tissue. (**a**) Control group, (**b**) C-doped TiO_2 group, (**c**) US group and (**d**) sonodynamic therapy (SDT) group.

4. Discussion

TiO_2 is one of most representative material studied in inorganic sonosensitizers. Nonetheless, the low quantum yield of ROS limits the effectiveness of TiO_2 as a sonosensitizer by rapid recombination of the free electron and electron hole. The addition of noble metals, such as Pt and Au, have been reported to retard carrier combination [30,31]. Pt-doped or Au-doped TiO_2 has been confirmed to show therapeutic efficacy and suppress the growth of tumors significantly. However, the price of novel metals would increase the cost of material preparation dramatically. Carbon shows highly promising dopant to narrow the bandgap, reaching similar therapeutic efficacy to a novel metal-based system in a more economical way. The comparative table is listed in Table 1.

Table 1. The sonosensitizers and the matched ultrasound parameters.

Tumor/Source	Host	Sonosensitizers	Ultrasound Parameter			Reference
			MHz (f)	W/cm^2 (I)	Duration (s)	
Heptaic/human	mouse	TiO_2	0.5/1.0	0.8/0.4	60	[32]
Skin/mouse	mouse	TiO_2	1	1.0	120	[33]
Breast/human	mouse	TiO_2	1	0.1	30	[34]
Lung/mouse	mouse	Au-doped TiO_2	1.5	30	30	[31]
Breast/human	mouse	Pt-doped TiO_2	1	1.5	300	[30]

In Figure 2a, the crystal structure of synthesized C-doped TiO_2 was matched with the standard pattern of the anatase structure without significant structure change when carbon atoms played the substitutional foreign atoms, replacing some of the oxygen atoms. The results corresponded with the previous study [35]. In Figure 3c, the peak at around 282 eV was ascribed to Ti-C bonds. Thus, the Kröger–Vink notation of possible substitution of O^{2-} by C was proposed and shown in Equation (2). The creation of oxygen vacancies may form the impurity band that contributes a band tailing effect, thus diminishing the bandgap [36,37]. In the previous study [38], comparing the typical Ti-O binding energy, the clear blue-shift was shown, which indicated the presence of oxygen vacancy after the reduction.

$$TiC \xrightarrow{TiO_2} Ti_{Ti}^{\times} + Co'' + Vo^{\bullet\bullet} \tag{2}$$

In Figure 2c,g, the particle size of C-doped TiO_2 was 156.9 nm which made the nanoparticles possible to accumulate in tumor tissue without entering the normal tissue [39].

Mechanical index (MI) was one of the indicators for evaluating the occurrence of the inertial cavitation. MI was calculated using Equation (3), where P^- is the peak-negative acoustic pressure and fc is the center frequency of the ultrasound transducer. It was shown that 0.3 MPa of the peak-negative acoustic pressure was the least pressure to induce the inertial cavitation in aqueous solution [40]. In Figure 4, under US irradiation, the production of HTA increased significantly compared to the control group. In other words, US irradiation is able to elevate the OH generation in TA solution. Hence, the US parameter in this study was confirmed to be effective to induce the inertial cavitation.

$$MI = \frac{P}{\sqrt{f_c}} \tag{3}$$

In Figure 6, compared to the control group, 4T1 cells treated with C-doped TiO_2 did not increase the level of ROS; 4T1 cells treated with US could be effective to increase the level of ROS, ascribed to the pyrolysis reaction of the water to produce hydroxyl radicals; 4T1 cells treated with C-doped TiO_2 plus US could increase the most significant level of ROS than the other groups, ascribed to the highly reactive singlet oxygen and hydroxyl radicals. This result corroborated that, under US irradiation, ROS generation could be increased in the presence of C-doped TiO_2. However, the amount of extracellular generation of ROS may affect the cell toxicity of tumor cells. Figure 7 shows that SDT induced higher toxicity in 4T1 cells than the other three groups. The results show that only US treatment seemed to cause insufficient ability to damage the 4T1 cells with the production of hydroxyl radicals. Nonetheless, US treatment in the presence of C-doped TiO_2 seemed to significantly suppress the proliferation of 4T1 cells. The cellular uptake of C-doped TiO_2 to 4T1 cells was supposed to occur via endocytosis or pinocytosis [41]. The possible pathway of cell damage induced by SDT may be speculated as: Under US irradiation, C-doped TiO_2 was activated by the sonoluminescent light, induced by inertial cavitation, to an excited state. Carbon was doped in the anatase TiO_2 to form a new valence band, thus narrowing the bandgap. The addition of carbon may give TiO_2 an excess of conducting electrons or holes, which is important for lowering the bandgap [23,24]. The unstableness of the excited state may cause decay back to the ground state, leading to energy release. The released energy may transfer to oxygen to generate the highly reactive singlet oxygen and water to generate hydroxyl radicals (Figure 10). With the increasing concentration of singlet oxygen and hydroxyl radicals, 4T1 cells would be gradually damaged and lead to cell senescence and death due to oxidative damaging effects [11,42,43]. Due to the enhanced permeability and retention (EPR) effect, C-doped TiO_2 preferentially accumulates in tumor cells eliciting efficient ROS generation [44] and further increasing the effectiveness of the SDT treatment. It was also revealed that the suppression of the tumor is due to the elevated level of ROS which results in both direct tumor cell death and blood vessel stasis. ROS can induce blood stasis via platelet aggregation or vessel constriction by destroying the endothelial layer [45]. Furthermore, SDT may elevate the level of inflammatory-associated cytokine (including TNF-alpha, IL-6, IL-1) production which is known to stimulate the maturation and function of granulocytes and macrophages [46].

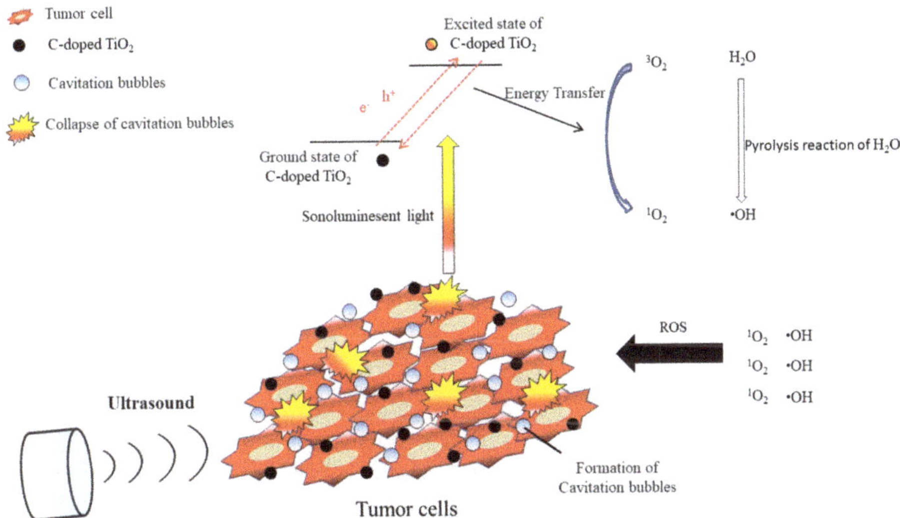

Figure 10. The possible pathway of 4T1 tumor cell damage induced by SDT.

To further investigate the cell death of 4T1 cells induce by SDT, in-vivo study was implemented in this study. Figure 8 shows that SDT could be more effective in suppressing the growth of the 4T1 tumor with a significant difference between SDT and the other groups on day 7. The relative tumor volume of SDT ended with almost half compared to the other groups at the 14th day of the treatment. H&E-stained tumor tissue is shown in Figure 9. In the tumors treated with 1× PBS or C-doped TiO_2 individually, integral tumor cellularity with spindle shape nuclei was visible. In the tumors treated with US, attributed to the inertial cavitation effect of US, a small area of nuclear fragmentation with little necrosis was visible. However, in the tumors treated with SDT, with the synergistic effect of hydroxyl radicals and singlet oxygen, a huge necrotic area with shrinkage and fragments of nuclei was seen. These results corroborated that SDT could inhibit the proliferation of 4T1 tumor cells.

5. Conclusions

The availability of C-doped TiO_2 nanoparticles as sonosensitizers activated by low-intensity ultrasound to generate ROS for cancer treatment was investigated in this study. Spherical C-doped TiO_2 nanoparticles with an average particle size of around 157 nm were synthesized by the sol-gel process. From the WST-1 test, C-doped TiO_2 showed no potential toxicity to normal cells. From in vitro study, it showed more extracellular generation of ROS in 4T1 breast cancer cells under SDT (C-doped TiO_2/US treatment). Moreover, the effect of C-doped TiO_2 in SDT could reduce the cell viability and increase the cytotoxicity of 4T1 cells. From in vivo study, the antitumor effect of C-doped TiO_2 in SDT may inhibit the proliferation of the 4T1 tumor and lead to a huge necrotic area with shrinkage and fragments of nuclei in the tumor site. We believe that C-doped TiO_2 nanoparticles could be viewed as sonosensitizers in SDT activated by low-intensity ultrasound for ROS generation as an effective strategy for alternative cancer treatments in the future.

Supplementary Materials: The following are available online at http://www.mdpi.com/2076-3921/9/9/880/s1, Figure S1: Live/Dead assay; Figure S2: survival rate of BALB/c nude mice; Figure S3: average weight of BALB/c nude mice; Figure S4: H&E-stained images of organs.

Author Contributions: Conceptualization, C.-C.Y. and F.-H.L.; methodology, G.-S.C. and C.-H.H.; software, C.-X.W.; validation, C.-Y.K., C.-Y.C. (Chih-Ying Chi), C.-Y.C. (Ching-Yun Chen) and Y.-Y.L.; formal analysis, C.-X.W.; investigation, C.-Y.K., C.-Y.C. (Chih-Ying Chi), C.-Y.C. (Ching-Yun Chen) and Y.-Y.L.; resources, C.-H.H. and F.-H.L.; data curation, C.-C.Y.; writing—original draft preparation, C.-X.W.; writing—review and editing, C.-C.Y.; visualization, C.-X.W.; supervision, C.-H.H., G.-S.C., and F.-H.L.; project administration, F.-H.L. All authors have read and agreed to the published version of the manuscript.

Funding: This research received no external funding.

Conflicts of Interest: The authors declare no conflict of interest.

References

1. Siegel, R.L.; Miller, K.D.; Jemal, A. Cancer statistics, 2019. *CA Cancer J. Clin.* **2019**, *69*, 7–34. [CrossRef] [PubMed]
2. Yang, C.-C.; Wang, W.-Y.; Lin, F.-H.; Hou, C.-H. Rare-Earth-Doped Calcium Carbonate Exposed to X-ray Irradiation to Induce Reactive Oxygen Species for Tumor Treatment. *Int. J. Mol. Sci.* **2019**, *20*, 1148. [CrossRef] [PubMed]
3. Yang, C.-C.; Tsai, M.-H.; Li, K.-Y.; Hou, C.-H.; Lin, F.-H. Carbon-Doped TiO2 Activated by X-Ray Irradiation for the Generation of Reactive Oxygen Species to Enhance Photodynamic Therapy in Tumor Treatment. *Int. J. Mol. Sci.* **2019**, *20*, 2072. [CrossRef] [PubMed]
4. Yang, C.-C.; Sun, Y.-J.; Chung, P.-H.; Chen, W.-Y.; Swieszkowski, W.; Tian, W.; Lin, F.-H. Development of Ce-doped TiO2 activated by X-ray irradiation for alternative cancer treatment. *Ceram. Int.* **2017**, *43*, 12675–12683. [CrossRef]
5. Umemura, S.; Yumita, N.; Nishigaki, R.; Umemura, K. Mechanism of cell damage by ultrasound in combination with hematoporphyrin. *Jpn. J. Cancer Res. Gann* **1990**, *81*, 962–966. [CrossRef]
6. Serpe, L.; Foglietta, F.; Canaparo, R. Nanosonotechnology: The next challenge in cancer sonodynamic therapy. *Nanotechnol. Rev.* **2012**, *1*. [CrossRef]
7. Shibaguchi, H.; Tsuru, H.; Kuroki, M.; Kuroki, M. Sonodynamic cancer therapy: A non-invasive and repeatable approach using low-intensity ultrasound with a sonosensitizer. *Anticancer Res.* **2011**, *31*, 2425–2430.
8. Wood, A.K.; Sehgal, C.M. A review of low-intensity ultrasound for cancer therapy. *Ultrasound Med. Biol.* **2015**, *41*, 905–928. [CrossRef] [PubMed]
9. Rosenthal, I.; Sostaric, J.Z.; Riesz, P. Sonodynamic therapy–a review of the synergistic effects of drugs and ultrasound. *Ultrason. Sonochem.* **2004**, *11*, 349–363. [CrossRef]
10. Qian, X.; Zheng, Y.; Chen, Y. Micro/Nanoparticle-Augmented Sonodynamic Therapy (SDT): Breaking the Depth Shallow of Photoactivation. *Adv. Mater.* **2016**, *28*, 8097–8129. [CrossRef]

11. Zou, Z.; Chang, H.; Li, H.; Wang, S. Induction of reactive oxygen species: An emerging approach for cancer therapy. *Apoptosis* **2017**, *22*, 1321–1335. [CrossRef]

12. Sadanala, K.C.; Chaturvedi, P.K.; Seo, Y.M.; Kim, J.M.; Jo, Y.S.; Lee, Y.K.; Ahn, W.S. Sono-photodynamic combination therapy: A review on sensitizers. *Anticancer Res.* **2014**, *34*, 4657–4664. [PubMed]

13. Serpe, L.; Giuntini, F. Sonodynamic antimicrobial chemotherapy: First steps towards a sound approach for microbe inactivation. *J. Photochem. Photobiol. B* **2015**, *150*, 44–49. [CrossRef] [PubMed]

14. Ninomiya, K.; Ogino, C.; Oshima, S.; Sonoke, S.; Kuroda, S.; Shimizu, N. Targeted sonodynamic therapy using protein-modified TiO_2 nanoparticles. *Ultrason. Sonochem.* **2012**, *19*, 607–614. [CrossRef] [PubMed]

15. Zhu, W.; Alkhazal, M.; Cho, M.; Xiao, S. Microbubble generation by piezotransducer for biological studies. *Rev. Sci. Instrum.* **2015**, *86*, 124901. [CrossRef] [PubMed]

16. Bhangu, S.K.; Ashokkumar, M. Theory of Sonochemistry. *Top. Curr. Chem.* **2016**, *374*, 56. [CrossRef]

17. McHale, A.P.; Callan, J.F.; Nomikou, N.; Fowley, C.; Callan, B. Sonodynamic therapy: Concept, mechanism and application to cancer treatment. In *Advances in Experimental Medicine and Biology*; Springer: Cham, Switzerland, 2016; Volume 880, pp. 429–450.

18. Paliwal, S.; Mitragotri, S. Ultrasound-induced cavitation: Applications in drug and gene delivery. *Expert Opin. Drug Deliv.* **2006**, *3*, 713–726. [CrossRef]

19. Khataee, A.; Kayan, B.; Gholami, P.; Kalderis, D.; Akay, S. Sonocatalytic degradation of an anthraquinone dye using TiO_2-biochar nanocomposite. *Ultrason Sonochem.* **2017**, *39*, 120–128. [CrossRef]

20. Wan, G.Y.; Liu, Y.; Chen, B.W.; Liu, Y.Y.; Wang, Y.S.; Zhang, N. Recent advances of sonodynamic therapy in cancer treatment. *Cancer Biol. Med.* **2016**, *13*, 325–338. [CrossRef]

21. Xu, H.; Zhang, X.; Han, R.; Yang, P.; Ma, H.; Song, Y.; Lu, Z.; Yin, W.; Wu, X.; Wang, H. Nanoparticles in sonodynamic therapy: State of the art review. *RSC Adv.* **2016**, *6*, 50697–50705. [CrossRef]

22. Tong, H.; Enomoto, N.; Inada, M.; Tanaka, Y.; Hojo, J. Synthesis of mesoporous TiO_2 spheres and aggregates by sol–gel method for dye-sensitized solar cells. *Mater. Lett.* **2015**, *141*, 259–262. [CrossRef]

23. Lin, Y.-T.; Weng, C.-H.; Lin, Y.-H.; Shiesh, C.-C.; Chen, F.-Y. Effect of C content and calcination temperature on the photocatalytic activity of C-doped TiO_2 catalyst. *Sep. Purif. Technol.* **2013**, *116*, 114–123. [CrossRef]

24. Palanivelu, K.; Im, J.-S.; Lee, Y.-S. Carbon Doping of TiO_2 for Visible Light Photo Catalysis—A review. *Carbon Lett.* **2007**, *8*, 214–224. [CrossRef]

25. Friehs, E.; AlSalka, Y.; Jonczyk, R.; Lavrentieva, A.; Jochums, A.; Walter, J.-G.; Stahl, F.; Scheper, T.; Bahnemann, D. Toxicity, phototoxicity and biocidal activity of nanoparticles employed in photocatalysis. *J. Photochem. Photobiol. C Photochem. Rev.* **2016**, *29*, 1–28. [CrossRef]

26. Sazgarnia, A.; Shanei, A. Evaluation of Acoustic Cavitation in Terephthalic Acid Solutions Containing Gold Nanoparticles by the Spectrofluorometry Method. *Int. J. Photoenergy* **2012**, *2012*, 376047. [CrossRef]

27. Kuan, C.-Y.; Lin, Y.-Y.; Chen, C.-Y.; Yang, C.-C.; Chi, C.-Y.; Li, C.-H.; Dong, G.-C.; Lin, F.-H. The preparation of oxidized methylcellulose crosslinked by adipic acid dihydrazide loaded with vitamin C for traumatic brain injury. *J. Mater. Chem. B* **2019**, *7*, 4499–4508. [CrossRef]

28. Tomayko, M.M.; Reynolds, C.P. Determination of subcutaneous tumor size in athymic (nude) mice*. *Cancer Chemother Pharm.* **1989**, *24*, 148–154. [CrossRef] [PubMed]

29. Huang, W.H.; Lee, S.; Zhang, L.; Li, G.; Yu, J.C. Effect of Carbon Doping on the Mesoporous Structure of Nanocrystalline Titanium Dioxide and Its Solar-Light-Driven Photocatalytic Degradation of NOx. *Langmuir* **2008**, *24*, 3510–3516. [CrossRef] [PubMed]

30. Liang, S.; Deng, X.; Xu, G.; Xiao, X.; Wang, M.; Guo, X.; Ma, P.A.; Cheng, Z.; Zhang, D.; Lin, J. A Novel Pt–TiO_2 Heterostructure with Oxygen-Deficient Layer as Bilaterally Enhanced Sonosensitizer for Synergistic Chemo-Sonodynamic Cancer Therapy. *Adv. Funct. Mater.* **2020**, *30*, 1908598. [CrossRef]

31. Deepagan, V.G.; You, D.G.; Um, W.; Ko, H.; Kwon, S.; Choi, K.Y.; Yi, G.-R.; Lee, J.Y.; Lee, D.S.; Kim, K.; et al. Long-Circulating Au-TiO_2 Nanocomposite as a Sonosensitizer for ROS-Mediated Eradication of Cancer. *Nano Lett.* **2016**, *16*, 6257–6264. [CrossRef]

32. Ninomiya, K.; Noda, K.; Ogino, C.; Kuroda, S.-I.; Shimizu, N. Enhanced OH radical generation by dual-frequency ultrasound with TiO_2 nanoparticles: Its application to targeted sonodynamic therapy. *Ultrason. Sonochem.* **2014**, *21*, 289–294. [CrossRef]

33. Harada, Y.; Ogawa, K.; Irie, Y.; Endo, H.; Feril, L.B.; Uemura, T.; Tachibana, K. Ultrasound activation of TiO_2 in melanoma tumors. *J. Control. Release* **2011**, *149*, 190–195. [CrossRef]

34. Ninomiya, K.; Fukuda, A.; Ogino, C.; Shimizu, N. Targeted sonocatalytic cancer cell injury using avidin-conjugated titanium dioxide nanoparticles. *Ultrason. Sonochem.* **2014**, *21*, 1624–1628. [CrossRef]

35. Kamisaka, H.; Adachi, T.; Yamashita, K. Theoretical study of the structure and optical properties of carbon-doped rutile and anatase titanium oxides. *J. Chem. Phys.* **2005**, *123*, 084704. [CrossRef] [PubMed]

36. Noorimotlagh, Z.; Kazeminezhad, I.; Jaafarzadeh, N.; Ahmadi, M.; Ramezani, Z. Improved performance of immobilized TiO_2 under visible light for the commercial surfactant degradation: Role of carbon doped TiO_2 and anatase/rutile ratio. *Catal. Today* **2020**, *348*, 277–289. [CrossRef]

37. Saiful Amran, S.N.B.; Wongso, V.; Abdul Halim, N.S.; Husni, M.K.; Sambudi, N.S.; Wirzal, M.D.H. Immobilized carbon-doped TiO_2 in polyamide fibers for the degradation of methylene blue. *J. Asian Ceram. Soc.* **2019**, *7*, 321–330. [CrossRef]

38. Han, X.; Huang, J.; Jing, X.; Yang, D.; Lin, H.; Wang, Z.; Li, P.; Chen, Y. Oxygen-Deficient Black Titania for Synergistic/Enhanced Sonodynamic and Photoinduced Cancer Therapy at Near Infrared-II Biowindow. *ACS Nano* **2018**, *12*, 4545–4555. [CrossRef]

39. Zhou, Y.; Kopeček, J. Biological rationale for the design of polymeric anti-cancer nanomedicines. *J. Drug Target.* **2013**, *21*, 1–26. [CrossRef]

40. Church, C.C. Frequency, pulse length, and the mechanical index. *Acoust. Res. Lett. Online* **2005**, *6*, 162–168. [CrossRef]

41. Xie, J.; Pan, X.; Wang, M.; Yao, L.; Liang, X.; Ma, J.; Fei, Y.; Wang, P.-N.; Mi, L. Targeting and Photodynamic Killing of Cancer Cell by Nitrogen-Doped Titanium Dioxide Coupled with Folic Acid. *Nanomaterials* **2016**, *6*, 113. [CrossRef] [PubMed]

42. Wang, L.; Mao, J.; Zhang, G.-H.; Tu, M.-J. Nano-cerium-element-doped titanium dioxide induces apoptosis of Bel 7402 human hepatoma cells in the presence of visible light. *World J. Gastroenterol. WJG* **2007**, *13*, 4011–4014. [CrossRef]

43. Li, Y.; Zhou, Q.; Deng, Z.; Pan, M.; Liu, X.; Wu, J.; Yan, F.; Zheng, H. IR-780 Dye as a Sonosensitizer for Sonodynamic Therapy of Breast Tumor. *Sci. Rep.* **2016**, *6*, 25968. [CrossRef]

44. Hasanzadeh Kafshgari, M.; Goldmann, W.H. Insights into Theranostic Properties of Titanium Dioxide for Nanomedicine. *Nano-Micro Lett.* **2020**, *12*, 22. [CrossRef]

45. Volanti, C.; Gloire, G.; Vanderplasschen, A.; Jacobs, N.; Habraken, Y.; Piette, J. Downregulation of ICAM-1 and VCAM-1 expression in endothelial cells treated by photodynamic therapy. *Oncogene* **2004**, *23*, 8649–8658. [CrossRef]

46. Shi, Y.; Liu, C.H.; Roberts, A.I.; Das, J.; Xu, G.; Ren, G.; Zhang, Y.; Zhang, L.; Yuan, Z.R.; Tan, H.S.W.; et al. Granulocyte-macrophage colony-stimulating factor (GM-CSF) and T-cell responses: What we do and don't know. *Cell Res.* **2006**, *16*, 126–133. [CrossRef]

 antioxidants

Article

Water Soluble Iron-Based Coordination Trimers as Synergistic Adjuvants for Pancreatic Cancer

Marco Cordani [1], Esther Resines-Urien [1], Arturo Gamonal [1], Paula Milán-Rois [1], Lionel Salmon [2], Azzedine Bousseksou [2], Jose Sanchez Costa [1,*] and Álvaro Somoza [1,*]

[1]   IMDEA Nanociencia, Faraday 9, Ciudad Universitaria de Cantoblanco, 28049 Madrid, Spain; marco.cordani@imdea.org (M.C.); esther.resines@imdea.org (E.R.-U.); arturo.gamonal@imdea.org (A.G.); paula.milan@imdea.org (P.M.-R.)

[2]   Laboratoire de Chimie de Coordination, UPR8241, 205 Route de Narbonne, CEDEX 4, 31077 Toulouse, France; lionel.salmon@lcc-toulouse.fr (L.S.); azzedine.bousseksou@lcc-toulouse.fr (A.B.)

*   Correspondence: jose.sanchezcosta@imdea.org (J.S.C.); alvaro.somoza@imdea.org (Á.S.); Tel.: +34-912-998-848 (J.S.C.); +34-912-998-856 (Á.S.)

Abstract: Pancreatic cancer is a usually fatal disease that needs innovative therapeutic approaches since the current treatments are poorly effective. In this study, based on cell lines, triazole-based coordination trimers made with soluble Fe(II) in an aqueous media were explored for the first time as adjuvant agents for the treatment of this condition. These coordination complexes were effective at relatively high concentrations and led to an increase in reactive oxygen species (ROS) in two pancreatic cancer cell lines, PANC-1 and BXPC-3, and this effect was accompanied by a significant reduction in cell viability in the presence of gemcitabine (GEM). Importantly, the tested compounds enhanced the effect of GEM, an approved drug for pancreatic cancer, through apoptosis induction and downregulation of the mTOR pathway. Although further evaluation in animal-based models of pancreatic cancer is needed, these results open novel avenues for exploring these iron-based materials in biomedicine in general and in pancreatic cancer treatment.

Keywords: pancreatic cancer; antitumor agents; coordination polymers; bioinorganic chemistry; reactive oxygen species

Citation: Cordani, M.; Resines-Urien, E.; Gamonal, A.; Milán-Rois, P.; Salmon, L.; Bousseksou, A.; Costa, J.S.; Somoza, Á. Water Soluble Iron-Based Coordination Trimers as Synergistic Adjuvants for Pancreatic Cancer. *Antioxidants* **2021**, *10*, 66. https://doi.org/10.3390/antiox10010066

Received: 16 December 2020
Accepted: 3 January 2021
Published: 7 January 2021

Publisher's Note: MDPI stays neutral with regard to jurisdictional claims in published maps and institutional affiliations.

1. Introduction

Pancreatic ductal adenocarcinoma (PDAC) is an orphan disease with a terrible prognosis, even when diagnosed early. The survival rate after five years remains below 5%, despite tremendous efforts at the preclinical and clinical stages. Worldwide, in 2018 there were 460,000 new cases of pancreatic cancer and 430,000 associated deaths [1]. The early symptoms of the disease are extremely rare, and, therefore, most patients present with locally advanced disease and/or metastasis at the time of diagnosis, limiting their options of being treated by surgery [2]. The standard treatment for the advanced disease includes gemcitabine (2′,2′-difluoro-2′-deoxycytidine; GEM), with a response rate of less than 20% [3]. Therefore, the identification of novel therapeutic strategies to overcome the limitations of current therapies against PDAC is urgent [4].

In this regard, we are exploring the use of coordination polymers (CPs) as adjuvants for the treatment of pancreatic cancer. These materials are built from metal ions linked by ligands, forming different topologies ranging from 1D to 3D structures. A great variety of CPs have been prepared with different properties and functionalities, such as porous frameworks for gas sorption and separation, catalysis, molecular and ion-exchange, molecular magnets, luminescence, among others [5]. The biological applications of these materials have been limited mainly due to their low solubility in biological media. Nevertheless, they are extremely appealing due to the versatility that confers the use of different biocompatible metals or ligands [6]. Thus, CPs have been employed as nanocarriers [7–11], to control the release of therapeutics [12–15], as contrast agents in MRIs [16,17] or optical imaging [18], and even in theranostic approaches [6].

A particularly interesting family of iron CPs is the chain-like polymeric Fe(II) system, with general formula {[Fe(Rtrz)3]X2}n, (Rtrz = substituted triazole; X = monovalent anion), which have been the focus of remarkable studies in a solid state [19]. These Fe(II) systems present the physical phenomenon of spin-crossover, which is accompanied by a strong thermochromism and a magnetic transition between diamagnetic and paramagnetic behavior in the solid state either as powder [20,21] or as nanoparticles [22]. However, less attention has been paid to the phenomenon in an aqueous solution [23], mainly due to their poor solubility [24]. To overcome this limitation, shorter polymeric chains (more precisely, trinuclear oligomers) are used in this report, providing water-soluble derivatives.

Reactive oxygen species (ROS) are highly reactive molecules containing oxygen produced by cellular metabolism, and they participate in several biological processes, including signal transduction, enzyme activation, gene expression, and protein post-translational modifications [25]. However, although the role of ROS in cellular physiology is well established, when produced in excess, they may cause irreversible cellular damage through the oxidation of biomolecules such as lipid membranes, enzymes, or DNA, which generally leads to cellular death [26]. ROS can also lead to carcinogenesis and tumor progression by inducing DNA mutations, genomic instability, and pro-oncogenic signaling pathways [27]. In this regard, cancer cells frequently exhibit increased ROS, mainly resulting from their high metabolic activity and oncogene activation [27]. However, this increased oxidative stress can promote cell proliferation without leading to cell death, but may make cancer cells vulnerable to exogenous oxidizing agents that generate additional ROS, which may increase oxidative stress levels above the cytotoxic threshold [28]. For this reason, the overstimulation of ROS by anticancer agents has been largely exploited and represents a main mechanism through which chemotherapy drugs kill cancer cells [29,30].

In this regard, once the oligomers were synthesized and characterized, their effect on cell viability and ROS induction, in the presence or absence of GEM, were assessed in two pancreatic cancer cell lines (PANC-1 and BXPC-3) as well as in a non-tumoral cell line (HaCaT). Furthermore, the oligomer's (or CP's) effect on apoptosis and the mTOR pathway within pancreatic cancer cells was evaluated to dissect the molecular and functional aspects underlying the mechanism of action of such structures. Finally, the involvement of ROS in the regulation of cell viability and the mTOR pathway was analyzed. Overall, this study provides new insight into the molecular mechanisms of action of CPs, which could be exploited as adjuvants for the treatment of pancreatic cancer.

2. Materials and Methods

2.1. Chemicals

Roswell Park Memorial Institute (RPMI) medium, Dulbecco's Modified Eagle's (DMEM) medium, streptomycin–penicillin (100×), fetal bovine serum (FBS), l-glutamine (100×), trypsin (10×), phosphate-buffered saline (PBS), and cell culture plasticware were purchased from VWR. Gemcitabine was purchased from Fluorochem. *N*-acetyl cysteine was purchased from Sigma Aldrich, Saint Louis, MO, USA. Chemicals and reagents were purchased from commercial suppliers and used as received following the indications reported.

2.2. Cell Lines and Culture Conditions

PANC-1 and HaCaT cells were purchased from American Type Culture Collection (ATCC, Rockville, MD, USA) and cultured in DMEM medium with 10% FBS, 1% streptomycin–penicillin, and 1% l-glutamine at 37 °C in a Binder CB210 incubator (5% CO_2). BXPC-3 was a gift from Ibane Abasolo Olaortua (CIBBIM-Nanomedicine Institut de Recerca Hospital Universitari Vall d'Hebron) and was cultured in RPMI medium with 10% FBS, 1% streptomycin–penicillin, and 1% l-glutamine at 37 °C in a Binder CB210 incubator (5% CO_2). All the procedures were performed inside a laminar flow hood Telstar CV-30/70 (Telstar, Terrassa, Spain).

2.3. Analysis of ROS

The non-fluorescent diacetylated 2′,7′-dichlorofluorescein (DCF-DA) probe (Sigma Aldrich, Saint Louis, MO, USA), which becomes highly fluorescent upon oxidation, was used to evaluate ROS production.

The ability of trimers to generate ROS in free-cell water was assessed after their incubation at three different concentrations (1 μM, 100 nM, or 1 nM) with 1 μM DCF-DA for 48 h. Salt and hydrogen peroxide were incubated with the probe as control. DCF-DA fluorescence was measured by using a multimode plate reader (λexc = 485 nm and λem = 535 nm) (Synergy H4 Hybrid reader (BioTEK, Winooski, VT, USA).

To evaluate ROS activity of C1, C2, and C3 in cellular models, the cell lines were plated in 96-well plates (5×10^3 cells/well) and 24 h later were treated with various compounds and with 5 μM DCF-DA for 4 min at 37 °C. After incubation, cells were washed with $1\times$ PBS (phosphate buffered saline), pH 7.4 (VWR), and DCF-DA fluorescence was measured by using a multimode plate reader (λexc = 485 nm and λem = 535 nm) (Synergy H4 Hybrid reader (BioTEK)). Values were normalized on cell proliferation by an alamarBlue viability assay. Representative images of cells were obtained using a Leica DMI3000 M inverted microscope (Leica, Wetzlar, Germany) at $20\times$ magnification. Images were analyzed using ImageJ software (NIH Image, Bethesda, MD, USA).

2.4. Alamar Blue Viability Assay

Cells were seeded in 96-well plates and the day after were incubated with various coordination compounds at the indicated conditions (see figure legends). At the end of the treatments, a stock solution of resazurin sodium salt (Sigma-Aldrich, St. Louis, MO, USA) (1 mg/mL) in PBS was diluted 1% (*v/v*) in complete DMEM or RPMI medium and added to the cells. After 3 h in the incubator (37 °C), the fluorescence was measured at 25 °C in a plate reader Synergy H4 Hybrid reader (BioTEK), λex = 550 nm, λem = 590 nm.

The fluorescent intensity measurements were processed using the following Equation:

$$\% \text{ Cell viability} = ((\text{Sample data} - \text{Negative control})/(\text{Positive control} - \text{Negative control})) \times 100$$

The positive control corresponded with untreated cells. A resazurin solution without cells was used as a negative control.

2.5. Western Blot Analysis

Cells were harvested, washed in PBS, and re-suspended in RIPA buffer (Tebu-BIO #AR0105) in the presence of a protease inhibitor cocktail (Thermo Scientific™, Madrid, Spain #A32955). After incubation on ice for 30 min, the lysates were centrifuged at $14,000\times g$ for 10 min at 4 °C and the supernatant fractions were used for Western blot analysis. Protein concentration was measured by Bradford reagent (Bio-Rad protein assay) using bovine serum albumin as a standard. Protein extracts (30 μg/lane) were resolved on a 10% SDS-polyacrylamide gel and electro-blotted onto PVDF membranes (AmershamTM ProtranTM 0.45 μm NC). Membranes were blocked in 5% low-fat milk in TBST or 5% BSA (50 mM Tris pH 7.5, 0.9% NaCl, 0.1% Tween 20) for 1 h at room temperature and probed overnight at 4 °C with a rabbit polyclonal anti-Bcl-2 (1:1000) (Cell Signaling, #2872), rabbit polyclonal anti-p70 S6 kinase (1:1000) (Cell Signaling, #9202), rabbit polyclonal anti-phospho-p70 S6 kinase (Ser371) (1:1000) (Cell Signaling, #9208), or mouse monoclonal anti-GAPDH (1:1000) (Santa Cruz, sc-47724). Horseradish peroxidase conjugated anti-mouse or anti-rabbit IgGs (1:5000 in blocking solution) (Santa Cruz, Spain) were used as secondary antibodies. Immunodetection was carried out using Bio-Rad chemiluminescent substrates and recorded using a Gel Documentation System (Syngene).

2.6. Acquisition of NMR Spectra

NMR spectra were recorded on a Bruker Advance 300 (^{1}H: 400 MHz) spectrometer at 298 K using partially deuterated solvents as internal standards. Chemical shifts (δ) are denoted in ppm. Multiplicities are denoted as follows: s = singlet, d = doublet, t = triplet.

2.7. Acquisition of FT-IR Spectra

FT-IT spectra were recorded as neat samples in the range 400–4000 cm^{-1} on a Bruker Tensor 27 (ATR device) Spectrometer.

2.8. Powder X-ray Diffraction Collection

PXRD data were collected in a Rigaku Smartlab SE diffractometer with a Bragg-Brentano configuration, using Cu-Kα radiation ($\lambda = 0.1541$ nm). Samples were measured between 5° and 50° with a speed of 1.8° min^{-1} under an X-ray fluorescence reduction mode, at room temperature.

2.9. Analysis of Synergy/Antagonism from Combination Studies

To determine possible additive and synergistic effects when using combinations of **C-1**, **C-2** and **C-3** with GEM, the data from cell viability assays were first analyzed using the freely available software Combenefit [31], which simultaneously assesses synergy/antagonism using three published models (Highest single agent (HSA), Bliss, and Loewe).

Three concentrations were employed for **C-1**, **C-2**, and **C-3** (25, 500, and 1000 μM), while six concentrations were employed for GEM (0.5, 4.5, 20, 40, 60, 80 μM). All data were normalized to untreated controls, and imported into Combenefit software, where the Loewe Additivity model was employed to identify synergistic, additive, or antagonistic drug combinations. The software calculates the combination index (CI) for each drug combination, where a CI value < 1 indicates synergy, CI = 1 is additive and CI > 1 indicates antagonism.

2.10. Annexin-V Assay

Cells were seeded in 6-well plates and the day after were incubated with various coordination compounds at the indicated conditions (see figure legends). At the end of the treatments, the dead cells were collected and the attached ones were trypsinized and collected. The cells were washed with PBS 1× in suspension by centrifugation at 177× *g* for 5 min. Cells were resuspended in 100 μL binding buffer 1× and then 10 μL annexin V 1:10 was added and the cells were incubated for 15 min at 4 °C in darkness. After that, 380 μL binding buffer 1× were added to the samples and then 10 μL propidium iodide. The acquisition was performed in a Beckman Coulter Cytomics 500 Flow Cytometer using 20,000 cells in the Flow Cytometry Service at the CNB-CSIC.

2.11. Statistical Analysis

Comparisons among groups were analyzed via the independent-samples one-factor ANOVA test using Prism GraphPad software. All statistical data were obtained using a two-tailed student's *t*-test and homogeneity of variance tests (*p* values < 0.05 were considered significant).

3. Results

3.1. Synthesis and Characterization of Three Iron-Based Trimers

As mentioned previously, an elegant alternative to employ insoluble CPs in biological media is the use of the triazole-based trimer system [Fe$_3$(NH$_2$-trz)$_6$]X$_6$ and its derivatives [Fe$_3$(RN-trz)$_6$]X$_6$. These systems are characterized by a backbone of three linearly arranged Fe(II) ions connected by three triazole ligands (Figure 1) [32–34]. The reduced length contributes to their solubility in contrast to that of the larger CPs. In addition, the steric hindrance of the ligands and the strong Fe-*N* coordination bond contribute to the stabilization of the bivalent iron under physiological conditions. The synthetic approach to obtain these water-soluble trinuclear systems resides in the use of p-toluenesulfonate (OTs) as counterions in highly diluted conditions to prevent or limit the formation of larger "Fe(II)/1,2,4-triazole" polymeric chains.

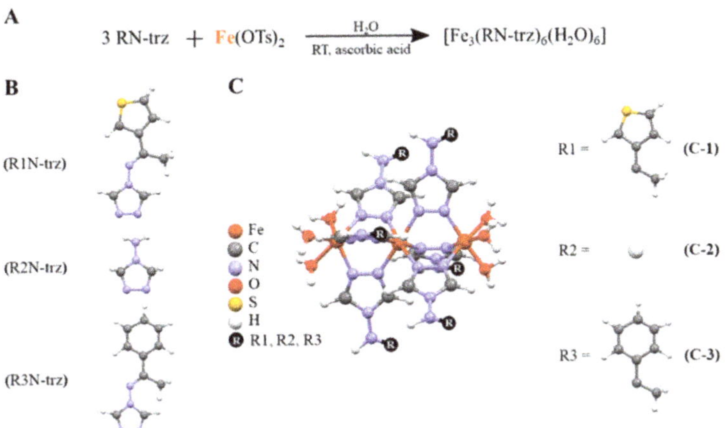

Figure 1. (**A**). Reaction required to obtain the trimers. (**B**). Representation of the different triazole derivatives employed (**C**). Schematic representation of the iron triazole-based trimers, where Fe is represented in orange, C in grey, N in blue, O in red, S in yellow, H in white, and the R group in black.

Three different Fe(II) trinuclear triazole-based CPs were synthesized in air conditions using ascorbic acid as the antioxidant, with general structure [Fe$_3$(RN-trz)$_6$(H$_2$O)$_6$], using iron(II) tosylate salt and triazole-based ligands: [(E)-*N*-(1-(thiophen-2-yl)ethylidene)-4H-1,2,4-triazol-4-amine (**C-1**), 4-amino-1,2,4-triazole (**C-2**), and (E)-*N*-(1-phenylethylidene)-4H-1,2,4-triazol-4-amine] (**C-3**) (see Figures S1–S5). The stability of these in situ generated complexes in water and phosphate-buffered saline (PBS) was monitored through ESI-MS and specifically with UV-Vis spectra at the incubation conditions (i.e., at 10^{-5} mol·L^{-1}, 37 °C and the spectra were taken at 0 h, 24 h, and 48 h) (Figures S6 and S7). It is noteworthy that the enhancement of the intensity of the band around 250 nm takes place without any shift, and also there is an absence of additional bands, which ratifies the stability of the CPs under these conditions. To further characterized these CPs, the solution was evaporated until precipitation and the resulting powders were characterized by FTIR and PXRD (See Figures S8–S11).

3.2. Antitumoral Activity of Coordination Polymers

Once the trimers were obtained, their effect on cell viability was evaluated in pancreatic cancer cell lines, based on previous reports where similar structures were employed in other tumor systems [29,30]. Intriguingly, after 48 h from the treatment, **C-1** and **C-2** were the most active in PANC-1, whereas **C-3** exerted a cytotoxic effect only in BXPC-3 cancer cells (Figure 2A,B). However, as illustrated in Figure S12, lower concentrations of trimers did not significantly affect cell viability.

Then, the different responses of pancreatic ductal adenocarcinoma cells to GEM were studied. First, as reported in previous studies [35], we observed that PANC-1 showed intrinsic chemoresistance to chemotherapy (Figure S13). To evaluate the potential effect as chemotherapy adjuvants of such derivates, the cell viability was studied when the trimers were combined with GEM. In the case of BXPC-3, the cell viability obtained in the presence of GEM did not change when it was combined with the trimers. Interestingly, in PANC-1, the combination of the trimers with GEM was able to overcome their intrinsic chemoresistance to this drug, especially when the chemotherapeutic was used at lower concentrations (0.5 and 4.5 μM) (Figure 2C and Figure S14). In particular, the combined treatment led to a synergistic enhancement of the cytotoxic effect, as shown by heat maps obtained with the Loewe Additivity model (Figure 2E) [31]. The synergistic effect is particularly relevant when the CPs were used at the concentration of 1000 μM and GEM at concentrations ranging from 0.5 to 20 μM. Then, to assess the specificity of these materials for cancer cells, similar experiments were also carried out in HaCaT cells, a non-cancerous cellular model. Interestingly, none of the three compounds exhibited toxicity in this cell line (Figure S15A) and did not enhance the cytotoxic activity of GEM (Figure S15B). These data suggest that these derivates are safe in non-tumor models and have therapeutic potential in the selective treatment of pancreatic cancer cells resistant to GEM.

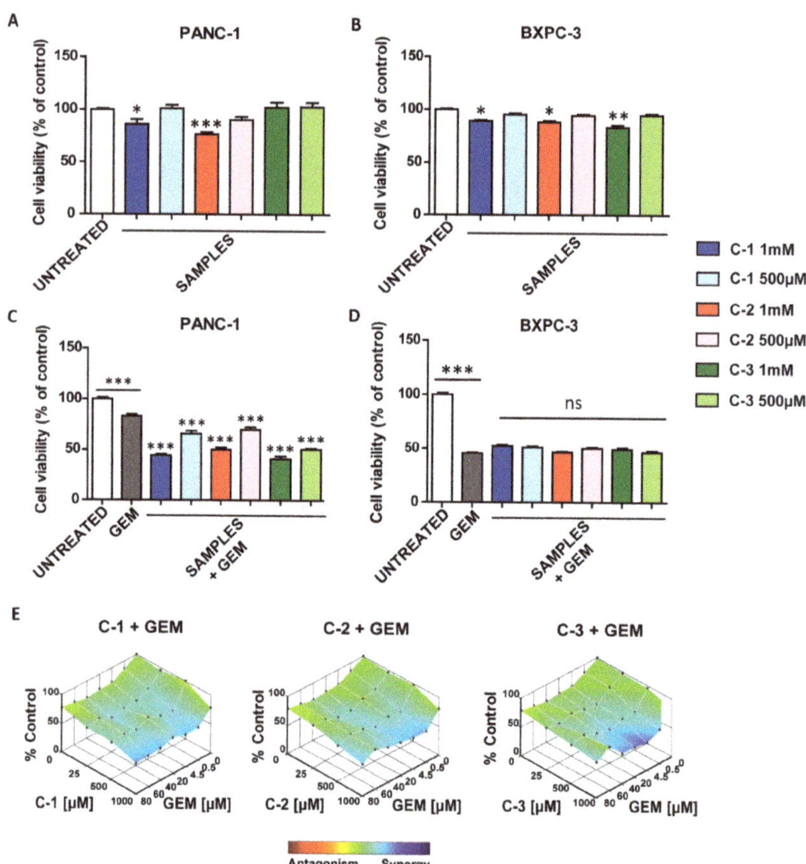

Figure 2. Antitumoral activity of coordination polymers (CPs). (**A,B**) Cell viability studies in pancreatic cancer cell lines treated with **C-1, C-2**, and **C-3** for 48 h. (**C,D**) Cell viability studies in pancreatic cancer cell lines treated with 4.5 μM gemcitabine (GEM) and **C-1, C-2**, and **C-3** for 48 h. The values of treated cells were normalized to the untreated controls and reported as mean ± SEM. Statistical analysis was performed using one-way ANO VA (each group vs. control). (*** $p < 0.001$, ** $p < 0.01$, * $p < 0.05$). (**E**) Combenefit-mapped surface output for the drug combinations involving **C-1, C-2**, and **C-3** with GEM in PANC-1 cells. The concentrations of each drug are plotted along the horizontal axes, while the percentages of cells remaining relative to untreated controls are plotted on the vertical axes. A heat map is used to represent the level of synergy (blue color) at each concentration. A heat map is used to represent the level of synergy (blue color) at each concentration. All experiments were conducted at least three times.

3.3. The Combined Treatment Increased Apoptosis and Inhibited the mTOR Pathway in Pancreatic Cancer Cells

To better understand the mechanism behind the antitumoral activity, some molecular markers related to apoptosis and cell growth in PANC-1 cancer cells were evaluated by Western blots (e.g., Bcl-2, p70S6K) and by Annexin-V assay. First, the effect of the trimers in Bcl-2 was assessed. This protein contributes to cancer formation and progression by promoting the survival of cancer cells and represents a canonical target for cancer therapy [36]. Interestingly, CPs did not change the levels of Bcl-2 when administered alone (Figure S16), however, a reduction of Bcl-2 was observed when PANC-1 cells were treated with **C-1** and GEM. However, no significant reductions in this protein were observed after the treatment with **C-2** or **C-3** in combination with GEM (Figure 3A,B).

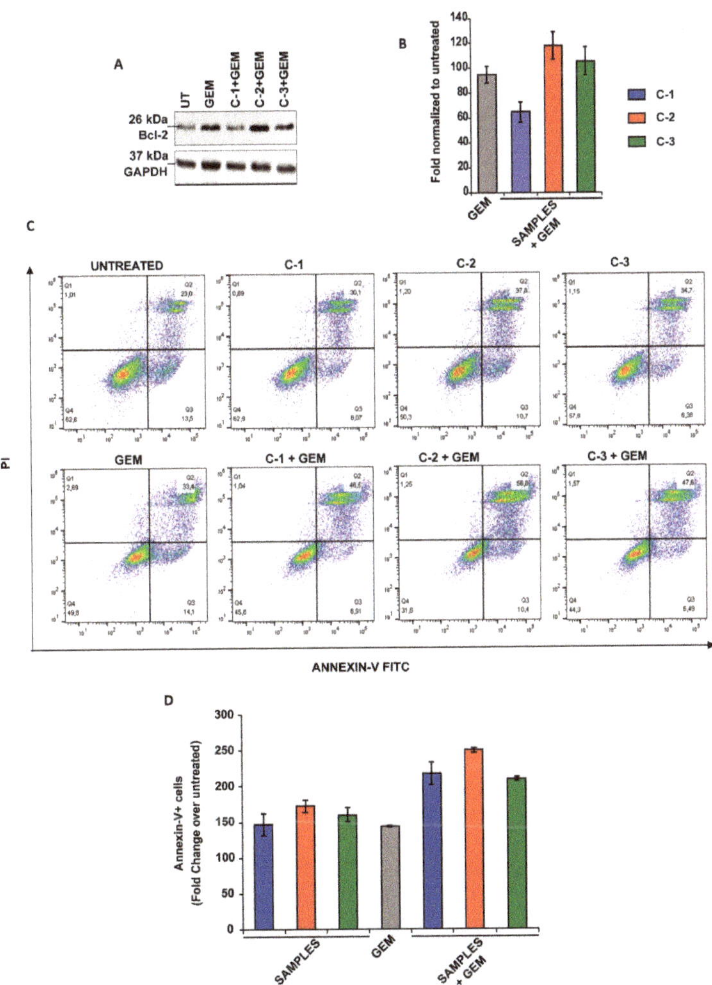

Figure 3. Apoptotic studies in pancreatic cancer cell lines. (**A**) Western blot analysis of Bcl-2 in PANC-1 cells after treatment with 1 mM **C-1, C-2, C-3**, and 4.5 µM GEM for 48 h. GAPDH protein level in the same extract was used as a control loading. (**B**) Densitometry of bands was performed using NIH Image J software and reported as fold change with respect to the untreated condition. The values reported are the mean of three independent experiments and were reported as mean ± SEM. (**C**) Cells were stained with Annexin V and propidium iodide (PI) and analyzed by flow cytometry. The percentage of cells in each group within the gated areas is indicated; the upper right panel represents cells undergoing late apoptosis, and the lower right panel represents cells undergoing early apoptosis. (**D**) Fold change in apoptosis. The values reported are the mean of two independent experiments and were reported as mean ±SEM.

Then, the cells were stained with fluorescein isothiocyanate (Annexin-V FITC) and propidium iodide (PI) and were analyzed by flow cytometry (Figure 3C,D). Interestingly, we observed an increase in Annexin-V fluorescence when the cells were treated with all three derivates and GEM, suggesting that the activation of apoptotic cell death may underlie the chemosensitivity observed.

Moreover, the activation of mTOR signaling after combined treatment was evaluated. It is worth mentioning that the inhibition in the mTOR signaling represents one of the mechanisms exploited by chemotherapy drugs to exert their antitumoral action [37]. Thus, when only GEM was administrated, chemoresistance in pancreatic cancer cells was induced. Remarkably, the administration of the therapeutic mixtures containing **C-1** or **C-2** or **C-3** and GEM, strongly abolished the chemoresistance mentioned before. Particularly, as discussed in below, they reduced the phosphorylation at Ser371 of p70S6 protein), indicating the downregulation of the mTOR pathway.

3.4. Reactive Oxygen Species (ROS) Generated by Coordination Complexes

Once the trimers were evaluated for their capabilities to synergistically increase the effect of GEM in PANC-1, the amount of ROS was evaluated [38], since they are usually generated in a variety of antitumoral systems. Thus, **C-1** to **C-3** were incubated with a standard ROS probe (DCFDA) at three different concentrations (1 μM, 100 nM, 1 nM) for 48 h. In this case, the fluorescent intensity increased in a concentration-dependent manner (Figure S17), highlighting the new trimers' potential use as ROS generators. Interestingly, the ROS production by addition of $Fe(OTs)_2$ was significantly lower compared to the results obtained with CPs at the concentration of 100 nM (Figure S18). Then, the new trimers' activity was evaluated in two pancreatic cancer cell lines, PANC-1 and BXPC-3. Interestingly, **C-1** and **C-3** produced an enhancement on the ROS levels, whereas **C-2** did not provide any relevant effect compared to that of untreated samples (Figure 4A,B). This result suggests that **C-2** might be processed inside the cell differently, compared to **C-1** and **C-3**. Interestingly, the ROS production by addition of $Fe(OTs)_2$ was significantly lower than in the case of **C-1** and **C-3** (Figure S19). These results suggest that iron-based CPs, such as **C-1** and **C-3**, can be employed to modify the oxidative status in pancreatic cancer cells, opening avenues to be exploited for therapeutic purposes.

Then, the activity of CPs was studied in combination with GEM, which is a standard agent used for the treatment of pancreatic cancer. It is important to point out that the cell lines (PANC-1 and BXPC-3) present different behavior against GEM. In particular, the presence of GEM induces a higher amount of ROS in BXPC-3 and reduces its viability more efficiently than in PANC-1 (Figure S20). In other words, as already mentioned above, PANC-1 presents chemoresistance to GEM, and for this reason, this cell line is often employed to assess novel therapeutics against pancreatic cancer. Thus, PANC-1 and BXPC-3 cancer cells were incubated with the three trimers and GEM for 48 h. In this case, a significant increase of ROS levels in both cell lines was observed when treated with **C-1** or **C-3** in the presence of GEM, compared to the ROS generated when treated only with GEM (Figure 4C,D). On the other hand, the combination **C-2** and GEM enhanced ROS only in PANC-1 cells. The ROS activity of the combined treatment was confirmed by fluorescent microscopy, where an increase in fluorescent signal was observed compared to that of the cells treated only with GEM (Figure 4E). Remarkably, when the iron salt and GEM were used, the ROS production was lower than in the cases where **C-1** and **C-3** were employed. This result highlights the advantage of the CPs evaluated herein compared to the iron salt $Fe(OTs)_2$ (Figure S21). Then, similar experiments were also carried out in HaCaT cells. In this case, only **C-3** was able to increase ROS, and only weakly (Figure S22A). However, in the presence of GEM, a significant increase of ROS was observed in the case of **C-1** and **C-3**, as previously shown in pancreatic cancer cells (Figure S22B).

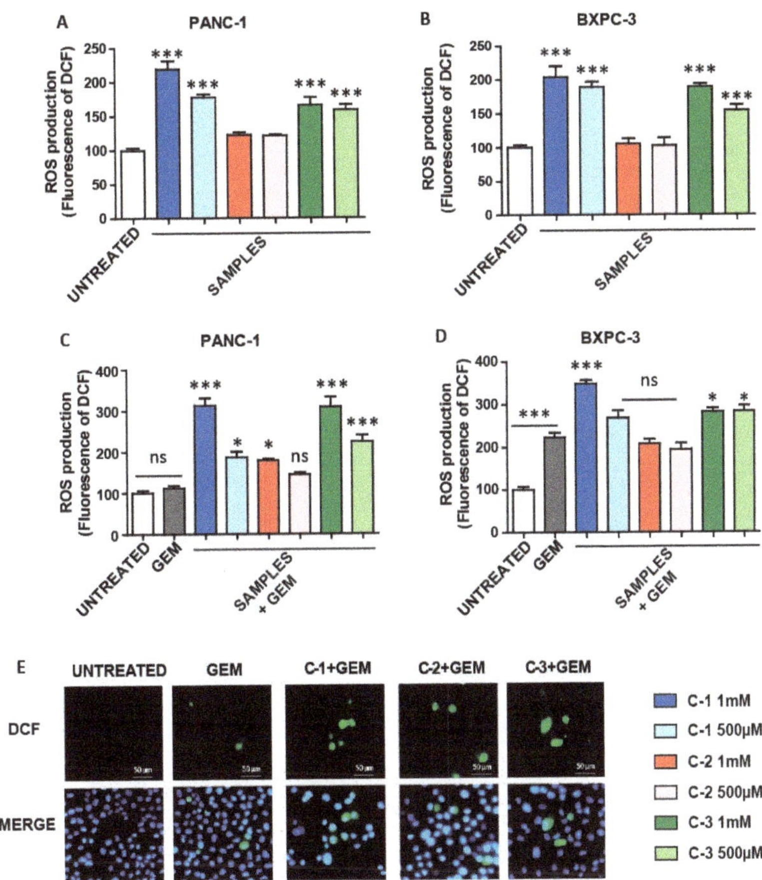

Figure 4. Reactive oxygen species (ROS) generated by coordination complexes. (**A,B**) ROS generation in pancreas cancer cell lines incubated with **C-1, C-2,** and **C-3** for 48 h. The DCF fluorescence (ROS production) of treated cells was normalized to untreated controls and reported as mean ±SEM. (**C,D**) ROS generation after combination therapy (incubation with GEM (4.5 µM) and **C-1, C-2,** and **C-3** for 48 h). The DCF fluorescence of treated cells was normalized to that of untreated controls and reported as mean ± SEM. Statistical analysis was performed using one-way ANOVA (each group vs. control). (*** $p < 0.001$, * $p < 0.05$). (**E**) Fluorescence images of PANC-1 cells untreated and treated with the combined therapy. DCF, corresponding to ROS production levels are shown in green, and the nuclei are labeled in blue by Hoechst staining.

3.5. Combined Treatment Attenuates the mTOR Pathway in an ROS-Dependent Manner

Finally, the effect of ROS generated by the therapeutic mixtures in the regulation of cell proliferation was assessed by the addition of the standard inhibitor of oxidative stress, *N*-acetyl cysteine (NAC), which does not affect cell viability (Figure S23). In this case, the effect on cell viability by the therapeutic mixtures was attenuated after NAC administration in PANC-1 cancer cells (Figure 5A). Intriguingly, when the mixture contained **C-2** and GEM, the removal of ROS had no significant effect on cell viability, suggesting that its synergistic effect was independent of oxidative stress. Accordingly, the increasing of ROS after the combined therapy was reduced after NAC treatment (Figure S24A). Furthermore, the phosphorylation at Ser371 of p70S6K, was also dramatically recovered by the addition of NAC (Figure 5B). These results highlight the role of ROS in the limitation of cancer cell growth mediated by the compounds described herein.

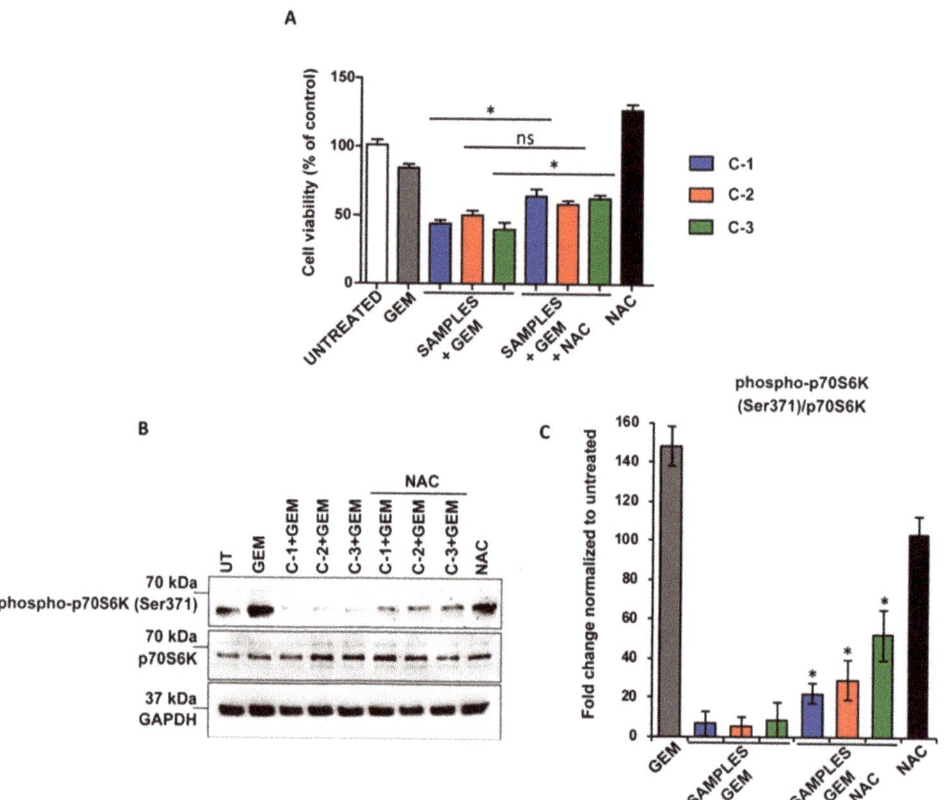

Figure 5. Combined treatment attenuates the mTOR pathway in an ROS-dependent manner. (**A**) PANC-1 cancer cell lines treated with 4.5 μM GEM, 500 μM NAC, and 1 mM **C-1** (blue), **C-2** (red), and **C-3** (green) for 48 h. Their viability was assessed with the alamarBlue test. The values of treated cells were normalized to that of untreated controls and reported as mean ± SEM. (**B**) Western blot analysis of phospho-p70SK6 and p70S6K of PANC-1 cells after treatment with 4.5 μM GEM, 500 μM NAC, and 1 mM of **C-1**, **C-2**, and **C-3** for 48 h. GAPDH protein level in the same extract was used as a control loading. (**C**) Densitometry of bands was performed using NIH Image J software and reported as fold change with respect to the untreated condition. The values reported are the mean of three experiments and were reported as mean ± SEM). Statistical analysis was performed using one-way ANOVA (C-1+GEM, C-2+GEM, C-3+GEM vs C-1+GEM+NAC, C-2+GEM+NAC, C-3+GEM+NAC)., * $p < 0.05$).

4. Discussion

In this work, we used a particularly interesting family of iron CPs, which is the chain-like polymeric Fe(II) system, with general formula $\{[Fe(Rtrz)3]X2\}n$, (Rtrz = substituted triazole; X = monovalent anion), which has been largely studied in a solid state. These Fe(II) systems present the physical phenomenon of spin-crossover, which is accompanied by a strong thermochromism and a magnetic transition between diamagnetic and paramagnetic behavior in the solid state, either as a powder or as nanoparticles. These CPs have not been studied very much in the aqueous solution, mainly due to their poor solubility. To overcome this limitation, we used in this report shorter polymeric chains (more precisely, trinuclear oligomers), achieving unprecedented water-soluble derivatives (**C-1**, **C-2**, **C-3**).

Once these oligomers were synthesized and characterized by a multitude of techniques, their anticancer activities were studied for the first time. Remarkably, these derivatives were able to increase the activity of GEM in a pancreatic cancer cell line (PANC-1). Intriguingly, in BXPC-3, another pancreatic cancer cell line, the combined therapy did not enhance the chemotherapy effect, and the reduction observed was only due to GEM toxicity without any synergistic or additive effects. Furthermore, the specificity of our therapeutic systems was evaluated by assessing the activity of such derivates in HaCaT keratinocyte cells. In this case, none of the three compounds exhibited toxicity in this non-tumoral cell line and did not enhance the cytotoxic activity of GEM. These observations suggest that these derivates are safe in non-tumor models and have therapeutic potential in the selective treatment of pancreatic cancer cells resistant to GEM. The drastic difference between the pancreas cancer cell lines versus the non-tumoral model may be explained by, among other causes, acquired selective mutations, which lead to the regulation of different signaling pathways after GEM treatment. For example, chemoresistance of pancreatic cancer has been attributable to hyperactivation of MAPK [39], STAT-3, and NF-κB signaling pathways [40]. Moreover, the effects of GEM are mediated by transporters to cross extracellular or intracellular membranes, and oncogenic mutations occurring in solute carrier (SLC) and the ATP-binding cassette (ABC) superfamilies also have been linked to drug resistance in pancreatic cancer [41].

We investigated the molecular mechanisms underlying the activity of our CPs in pancreatic cancer cells. Interestingly, we observed that the CPs were able to attenuate chemoresistance in PANC-1 by inducing apoptosis and inhibiting mTOR signaling, which we detected through the phosphorylation of p70S6K, a classic downstream readout of this pathway. Curiously, although all three compounds were able to increase Annexin-v fluorescence when combined to GEM, the reduction of Bcl-2 protein, a classical inhibitor of apoptosis, was observed only when the therapeutic mixture contained C-1. This partial discrepancy could be explained in part because Bcl-2's role in apoptosis is ambiguous. Indeed, Bcl-2 can also act as an apoptotic inducer blocking Bax/Bak oligomerization [42] and plays a role in other non-canonical functions including mitochondrial metabolism and biogenesis and autophagy regulation [43]. Another explanation could be that even if the translocation of phosphatidylserine is produced after the treatment, a process that is usually associated with the early stage of apoptosis, this does not result in the complete activation of the intrinsic apoptosis pathway and the formation of apoptosomes, which precedes caspase's activation [44].

ROS play a key role in various cellular processes, including proliferation, growth, apoptosis, and migration [45]. Importantly, ROS levels are frequently increased in cancer cells because of their high metabolic activity, mitochondrial dysfunction, and hyper-activation of oncogenes [46]. However, while ROS facilitate carcinogenesis and cancer progression with mild-to-moderate elevated levels, their excessive production may damage cancer cells dramatically through the modification of lipids, proteins, or DNA and lead to cell death [28]. Hence, ROS signaling and the antioxidant program play key roles in the response to chemotherapy and represent important targets to overcome drug-resistance [47].

In this study, the effect of our derivates on ROS metabolism in pancreatic cancer cells was investigated. Once we established that such structures were able to increase ROS production, both alone and in combination with GEM, we studied whether the cytotoxic effect of the combined therapy was due, at least in part, to the CPs' capacity to enhance ROS levels. To unveil this, experiments using the ROS scavenger NAC were performed, which showed that NAC could recover cell viability and the mTOR pathway affected by trimers and GEM.

Overall, these data allow us to speculate that both apoptosis induction and mTOR inhibition, resulting after combined therapy, might contribute to the synergistic reduction of cell viability observed in pancreatic cancer cells, and also that the cytotoxic effect is due to the increase in oxidative stress.

5. Conclusions

In summary, we have demonstrated that CPs can be used as adjuvants to enhance synergistically the activity of GEM in pancreatic cancer cell lines, which present intrinsic resistance to this chemotherapeutic through modulation of ROS levels. Interestingly, in contrast to that observed in PANC-1, the combined therapy did not enhance the chemotherapy effect either in BXPC-3, a pancreatic cancer cell line not resistant to GEM, nor in HaCaT keratinocyte cells, and the reduction of cell viability was only due to GEM toxicity without any synergistic or additive effect.

Furthermore, we studied the signaling pathways and molecular mechanisms underlying the chemosensitivity to GEM. We observed that the combined treatment increased apoptosis and reduced the phosphorylation of p70S6k, a well-established functional readout of the mTOR pathway involved in cell growth.

Finally, to unveil whether the increased ROS detected after combined therapy could have a role in the CPs' therapeutic effect, we performed functional studies using NAC, a standard inhibitor of oxidative stress. Notably, the addition of NAC significantly reverted the reduction of cell viability and the mTOR pathway observed after combined therapy, suggesting the CPs' effect was, at least in part, due to exacerbated ROS production in PANC-1 GEM-resistant cell lines.

Supplementary Materials: The following are available online at https://www.mdpi.com/2076-3921/10/1/66/s1, Figure S1: Scheme of the reaction of the ligands TE-Tria and PE-Tria, Figure S2: ^{1}H NMR (400 MHz, DMSO-d6, 298 K) of TE-Tria, Figure S3: Infrared spectrum of TE-Tria, Figure S4: ^{1}H NMR (400 MHz, DMSO-d6, 298 K) of the PE-Tria, Figure S5: Infrared spectrum of PE-Tria, Figure S6: UV-vis spectra of (a) **C-1** in water, (b) **C-1** in PBS, (c) **C-2** in water, (d) **C-2** in PBS, (e) **C-3** in water, and (f) **C-3** in PBS, Figure S7: Solution of **C-1**, **C-2**, and **C-3** (10^{-2} M) in water initially and aged for 24 and 48 h (a) with ascorbic acid and (b) without ascorbic acid, Figure S8: Infrared spectrum of **C-1**, Figure S9: Infrared spectrum of **C-2**, Figure S10: Infrared spectrum of **C-3**, Figure S11: Powder X-ray Diffraction of **C-1**, **C-2**, and **C-3**, Figure S12: Cell viability studies in pancreas cells, Figure S13: Dose response curve of GEM, Figure S14: Dose response studies in pancreas cancer cells, Figure S15: Cell viability studies in human keratinocytes, Figure S16 CP administered alone does not affect Bcl-2 protein level, Figure S17: DCF fluorescence generated by coordination complexes, Figure S18: DCF fluorescence generated by salt and hydrogen peroxide, Figure S19: ROS generation in pancreas cancer cell lines, Figure S20: Different responses to gemcitabine in pancreatic cancer cells, Figure S21: ROS generation in pancreatic cancer cell lines after combined treatment, Figure S22: ROS generation in human keratinocytes, Figure S23: Cell viability studies in pancreas cancer cells, Figure S24: Combined treatment attenuates the mTOR pathway in an ROS-dependent manner.

Author Contributions: Conceptualization, M.C., J.S.C. and Á.S.; Data curation, M.C., P.M.-R. and E.R.-U.; Funding acquisition, J.S.C. and Á.S.; Investigation, M.C., P.M.-R. and E.R.-U.; Methodology, M.C., P.M.-R. and E.R.-U.; Resources, L.S. and A.B.; Validation, M.C., P.M.-R. and E.R.-U.; Writing—original draft, M.C., J.S.C. and Á.S.; Writing—review and editing, M.C., E.R.-U., A.G., J.S.C. and Á.S. All authors have read and agreed to the published version of the manuscript.

Funding: This work was supported by the Spanish Ministry of Economy and Competitiveness (SAF2017-87305-R, CTQ2016-80635-P, RYC-2014-16866), Comunidad de Madrid (IND2017/IND-7809, PEJD-2017-PRE/IND-4037, P2018/NMT-4321), Asociación Española Contra el Cáncer, and IMDEA Nanociencia. IMDEA Nanociencia acknowledges support from the 'Severo Ochoa' Programme for Centres of Excellence in R&D (MINECO, Grant SEV-2016-0686).

Institutional Review Board Statement: Not applicable.

Informed Consent Statement: Not applicable.

Data Availability Statement: Data are contained within the article. The raw data of blots are available from the corresponding author upon reasonable request.

Conflicts of Interest: The authors declare no conflict of interest.

References

1. Rawla, P.; Sunkara, T.; Gaduputi, V. Epidemiology of Pancreatic Cancer: Global Trends, Etiology and Risk Factors. *World J. Oncol.* **2019**, *10*, 10–27. [CrossRef]
2. Willett, C.G.; Czito, B.G.; Bendell, J.C.; Ryan, D.P. Locally advanced pancreatic cancer. *J. Clin. Oncol.* **2005**, *23*, 4538–4544. [CrossRef]
3. Burris, H.A.; Moore, M.J.; Andersen, J.; Green, M.R.; Rothenberg, M.L.; Modiano, M.R.; Cripps, M.C.; Portenoy, R.K.; Storniolo, A.M.; Tarassoff, P.; et al. Improvements in survival and clinical benefit with gemcitabine as first-line therapy for patients with advanced pancreas cancer: A randomized trial. *J. Clin. Oncol.* **1997**, *15*, 2403–2413. [CrossRef]
4. Kleger, A.; Perkhofer, L.; Seufferlein, T. Smarter drugs emerging in pancreatic cancer therapy. *Ann. Oncol.* **2014**, *25*, 1260–1270. [CrossRef]
5. Chen, X.-M. Assembly Chemistry of Coordination Polymers. In *Modern Inorganic Synthetic Chemistry*; Elsevier: Amsterdam, The Netherlands, 2011; pp. 207–225. ISBN 9780444535993.
6. Liu, F.; He, X.; Chen, H.; Zhang, J.; Zhang, H.; Wang, Z. Gram-scale synthesis of coordination polymer nanodots with renal clearance properties for cancer theranostic applications. *Nat. Commun.* **2015**, *6*, 8003. [CrossRef]
7. Hu, Y.; Lv, T.; Ma, Y.; Xu, J.; Zhang, Y.; Hou, Y.; Huang, Z.; Ding, Y. Nanoscale Coordination Polymers for Synergistic NO and Chemodynamic Therapy of Liver Cancer. *Nano Lett.* **2019**, *19*, 2731–2738. [CrossRef]
8. Rezaei, M.; Abbasi, A.; Dinarvand, R.; Jeddi-Tehrani, M.; Janczak, J. Design and Synthesis of a Biocompatible 1D Coordination Polymer as Anti-Breast Cancer Drug Carrier, 5-Fu: In Vitro and in Vivo Studies. *ACS Appl. Mater. Interfaces* **2018**, *10*, 17594–17604. [CrossRef]
9. He, C.; Poon, C.; Chan, C.; Yamada, S.D.; Lin, W. Nanoscale coordination polymers codeliver chemotherapeutics and sirnas to eradicate tumors of cisplatin-resistant ovarian cancer. *J. Am. Chem. Soc.* **2016**, *138*, 6010–6019. [CrossRef]
10. Imaz, I.; Rubio-Martínez, M.; García-Fernández, L.; García, F.; Ruiz-Molina, D.; Hernando, J.; Puntes, V.; Maspoch, D. Coordination polymer particles as potential drug delivery systems. *Chem. Commun.* **2010**, *46*, 4737–4739. [CrossRef]
11. Duan, X.; Chan, C.; Han, W.; Guo, N.; Weichselbaum, R.R.; Lin, W. Immunostimulatory nanomedicines synergize with checkpoint blockade immunotherapy to eradicate colorectal tumors. *Nat. Commun.* **2019**, *10*, 1899. [CrossRef]
12. Liu, D.; Poon, C.; Lu, K.; He, C.; Lin, W. Self-assembled nanoscale coordination polymers with trigger release properties for effective anticancer therapy. *Nat. Commun.* **2014**, *5*, 4182. [CrossRef] [PubMed]
13. Liu, J.; Wang, H.; Yi, X.; Chao, Y.; Geng, Y.; Xu, L.; Yang, K.; Liu, Z. pH-Sensitive Dissociable Nanoscale Coordination Polymers with Drug Loading for Synergistically Enhanced Chemoradiotherapy. *Adv. Funct. Mater.* **2017**, *27*, 1703832. [CrossRef]
14. Liu, J.; Tian, L.; Zhang, R.; Dong, Z.; Wang, H.; Liu, Z. Collagenase-Encapsulated pH-Responsive Nanoscale Coordination Polymers for Tumor Microenvironment Modulation and Enhanced Photodynamic Nanomedicine. *ACS Appl. Mater. Interfaces* **2018**, *10*, 43493–43502. [CrossRef] [PubMed]
15. Xu, S.; Liu, J.; Li, D.; Wang, L.; Guo, J.; Wang, C.; Chen, C. Fe–salphen complexes from intracellular pH-triggered degradation of Fe3O4@Salphen-InIII CPPs for selectively killing cancer cells. *Biomaterials* **2014**, *35*, 1676–1685. [CrossRef] [PubMed]
16. Suárez-García, S.; Arias-Ramos, N.; Frias, C.; Candiota, A.P.; Arús, C.; Lorenzo, J.; Ruiz-Molina, D.; Novio, F. Dual T1/T2 Nanoscale Coordination Polymers as Novel Contrast Agents for MRI: A Preclinical Study for Brain Tumor. *ACS Appl. Mater. Interfaces* **2018**, *10*, 38819–38832. [CrossRef] [PubMed]
17. Chen, Y.; Ai, K.; Liu, J.; Ren, X.; Jiang, C.; Lu, L. Polydopamine-based coordination nanocomplex for T1/T2 dual mode magnetic resonance imaging-guided chemo-photothermal synergistic therapy. *Biomaterials* **2016**, *77*, 198–206. [CrossRef]
18. Liu, D.; Huxford, R.C.; Lin, W. Phosphorescent Nanoscale Coordination Polymers as Contrast Agents for Optical Imaging. *Angew. Chemie Int. Ed.* **2011**, *50*, 3696–3700. [CrossRef]
19. Roubeau, O. Triazole-based one-dimensional spin-crossover coordination polymers. *Chem. A Eur. J.* **2012**, *18*, 15230–15244. [CrossRef]
20. Costa, J.S. Macroscopic methods: Magnetic, optical, and calorimetric techniques. *Comptes Rendus Chim.* **2018**, *21*, 1121–1132. [CrossRef]
21. Gütlich, P.; Gaspar, A.B.; Garcia, Y. Spin state switching in iron coordination compounds. *Beilstein J. Org. Chem.* **2013**, *9*, 342–391. [CrossRef]
22. Salmon, L.; Catala, L. Spin-crossover nanoparticles and nanocomposite materials. *Comptes Rendus Chim.* **2018**, *21*, 1230–1269. [CrossRef]
23. Bräunlich, I.; Sánchez-Ferrer, A.; Bauer, M.; Schepper, R.; Knüsel, P.; Dshemuchadse, J.; Mezzenga, R.; Caseri, W. Polynuclear iron(II)-aminotriazole spincrossover complexes (polymers) in solution. *Inorg. Chem.* **2014**, *53*, 3546–3557. [CrossRef] [PubMed]
24. Barrett, S.A.; Kilner, C.A.; Halcrow, M.A. Spin-crossover in [Fe(3-bpp)2][BF4]2 in different solvents—A dramatic stabilisation of the low-spin state in water. *Dalt. Trans.* **2011**, *40*, 12005–12016. [CrossRef] [PubMed]
25. Giannoni, E.; Buricchi, F.; Raugei, G.; Ramponi, G.; Chiarugi, P. Intracellular Reactive Oxygen Species Activate Src Tyrosine Kinase during Cell Adhesion and Anchorage-Dependent Cell Growth. *Mol. Cell. Biol.* **2005**, *25*, 6391–6403. [CrossRef] [PubMed]
26. Cooke, M.S.; Evans, M.D.; Dizdaroglu, M.; Lunec, J. Oxidative DNA damage: Mechanisms, mutation, and disease. *FASEB J.* **2003**, *17*, 1195–1214. [CrossRef] [PubMed]
27. Klaunig, J.E.; Kamendulis, L.M.; Hocevar, B.A. Oxidative stress and oxidative damage in carcinogenesis. *Toxicol. Pathol.* **2010**, *38*, 96–109. [CrossRef]

28. Zhang, L.; Li, J.; Zong, L.; Chen, X.; Chen, K.; Jiang, Z.; Nan, L.; Li, X.; Li, W.; Shan, T.; et al. Reactive Oxygen Species and Targeted Therapy for Pancreatic Cancer. *Oxid. Med. Cell. Longev.* **2015**, *2016*, 1616781. [CrossRef]

29. González-Bártulos, M.; Aceves-Luquero, C.; Qualai, J.; Cussó, O.; Martínez, M.A.; De Mattos, S.F.; Menéndez, J.A.; Villalonga, P.; Costas, M.; Ribas, X.; et al. Pro-oxidant activity of amine-pyridine-based iron complexes efficiently kills cancer and cancer stem-like cells. *PLoS ONE* **2015**, *10*, e0137800. [CrossRef]

30. Ye, J.; Ma, J.; Liu, C.; Huang, J.; Wang, L.; Zhong, X. A novel iron(II) phenanthroline complex exhibits anticancer activity against TFR1-overexpressing esophageal squamous cell carcinoma cells through ROS accumulation and DNA damage. *Biochem. Pharmacol.* **2019**, *166*, 93–107. [CrossRef]

31. Di Veroli, G.Y.; Fornari, C.; Wang, D.; Mollard, S.; Bramhall, J.L.; Richards, F.M.; Jodrell, D.I. Combenefit: An interactive platform for the analysis and visualization of drug combinations. *Bioinformatics* **2016**, *32*, 2866–2868. [CrossRef]

32. Tokarev, A.; Salmon, L.; Guari, Y.; Nicolazzi, W.; Molnár, G.; Bousseksou, A. Cooperative spin crossover phenomena in [Fe(NH2trz)3](tosylate)2 nanoparticles. *Chem. Commun.* **2010**, *46*, 8011. [CrossRef] [PubMed]

33. Gómez, V.; De Pipaón, C.S.; Maldonado-Illescas, P.; Waerenborgh, J.C.; Martin, E.; Benet-Buchholz, J.; Galán-Mascarós, J.R. Easy Excited-State Trapping and Record High TTIESST in a Spin-Crossover Polyanionic FeII Trimer. *J. Am. Chem. Soc.* **2015**, *137*, 11924–11927. [CrossRef] [PubMed]

34. Chen, W.-B.; Leng, J.-D.; Wang, Z.-Z.; Chen, Y.-C.; Miao, Y.; Tong, M.-L.; Dong, W. Reversible crystal-to-crystal transformation from a trinuclear cluster to a 1D chain and the corresponding spin crossover (SCO) behaviour change. *Chem. Commun.* **2017**, *53*, 7820–7823. [CrossRef]

35. Fiorini, C.; Cordani, M.; Padroni, C.; Blandino, G.; Di Agostino, S.; Donadelli, M. Mutant p53 stimulates chemoresistance of pancreatic adenocarcinoma cells to gemcitabine. *Biochim. Biophys. Acta Mol. Cell Res.* **2015**, *1853*, 89–100. [CrossRef]

36. Yamaguchi, R.; Lartigue, L.; Perkins, G. Targeting Mcl-1 and other Bcl-2 family member proteins in cancer therapy. *Pharmacol. Ther.* **2019**, *195*, 13–20. [CrossRef] [PubMed]

37. Janku, F.; Yap, T.A.; Meric-Bernstam, F. Targeting the PI3K pathway in cancer: Are we making headway? *Nat. Rev. Clin. Oncol.* **2018**, *15*, 273–291. [CrossRef]

38. Dixon, S.J.; Stockwell, B.R. The role of iron and reactive oxygen species in cell death. *Nat. Chem. Biol.* **2014**, *10*, 9–17. [CrossRef]

39. Lee, S.; Rauch, J.; Kolch, W. Targeting MAPK signaling in cancer: Mechanisms of drug resistance and sensitivity. *Int. J. Mol. Sci.* **2020**, *21*, 1102. [CrossRef]

40. Zhang, X.; Ren, D.; Wu, X.; Lin, X.; Ye, L.; Lin, C.; Wu, S.; Zhu, J.; Peng, X.; Song, L. miR-1266 Contributes to Pancreatic Cancer Progression and Chemoresistance by the STAT3 and NF-κB Signaling Pathways. *Mol. Ther. Nucleic Acids* **2018**, *11*, 142–158. [CrossRef]

41. Luo, W.; Yang, G.; Qiu, J.; Luan, J.; Zhang, Y.; You, L.; Feng, M.; Zhao, F.; Liu, Y.; Cao, Z.; et al. Novel discoveries targeting gemcitabine-based chemoresistance and new therapies in pancreatic cancer: How far are we from the destination? *Cancer Med.* **2019**, *8*, 6403–6413. [CrossRef]

42. Singh, R.; Letai, A.; Sarosiek, K. Regulation of apoptosis in health and disease: The balancing act of BCL-2 family proteins. *Nat. Rev. Mol. Cell Biol.* **2019**, *20*, 175–193. [CrossRef] [PubMed]

43. Akl, H.; Vervloessem, T.; Kiviluoto, S.; Bittremieux, M.; Parys, J.B.; De Smedt, H.; Bultynck, G. A dual role for the anti-apoptotic Bcl-2 protein in cancer: Mitochondria versus endoplasmic reticulum. *Biochim. Biophys. Acta Mol. Cell Res.* **2014**, *1843*, 2240–2252. [CrossRef]

44. Shlomovitz, I.; Speir, M.; Gerlic, M. Flipping the dogma—Phosphatidylserine in non-apoptotic cell death. *Cell Commun. Signal.* **2019**, *17*, 139. [CrossRef] [PubMed]

45. Denicola, G.M.; Karreth, F.A.; Humpton, T.J.; Gopinathan, A.; Wei, C.; Frese, K.; Mangal, D.; Yu, K.H.; Yeo, C.J.; Calhoun, E.S.; et al. Oncogene-induced Nrf2 transcription promotes ROS detoxification and tumorigenesis. *Nature* **2011**, *475*, 106–109. [CrossRef] [PubMed]

46. Schieber, M.; Chandel, N.S. ROS function in redox signaling and oxidative stress. *Curr. Biol.* **2014**, *24*, R453–R462. [CrossRef] [PubMed]

47. Liu, Y.; Li, Q.; Zhou, L.; Xie, N.; Nice, E.C.; Zhang, H.; Huang, C.; Lei, Y. Cancer drug resistance: Redox resetting renders a way. *Oncotarget* **2016**, *7*, 42740–42761. [CrossRef] [PubMed]

MDPI

St. Alban-Anlage 66

4052 Basel

Switzerland

Tel. +41 61 683 77 34

Fax +41 61 302 89 18

www.mdpi.com

Antioxidants Editorial Office

E-mail: antioxidants@mdpi.com

www.mdpi.com/journal/antioxidants